Use R!

This series of inexpensive and focused books on R is aimed at practitioners. Books can discuss the use of R in a particular subject area (e.g., epidemiology, econometrics, psychometrics) or as it relates to statistical topics (e.g., missing data, longitudinal data). In most cases, books combine LaTeX and R so that the code for figures and tables can be put on a website. Authors should assume a background as supplied by Dalgaard's Introductory Statistics with R or other introductory books so that each book does not repeat basic material.

How to Submit Your Proposal

Book proposals and manuscripts should be submitted to one of the publishing editors in your region per email – for the list of statistics editors by their location please see https://www.springer.com/gp/statistics/contact-us. All submissions should include a completed Book Proposal Form.

***

For general and technical questions regarding the series and the submission process please contact Faith Su (faith.su@springer.com) or Veronika Rosteck (veronika.rosteck@springer.com).

Duccio Rocchini
Editor

R Coding for Ecology

Springer

Editor
Duccio Rocchini
BIOME Lab, Department of Biological, Geological and Environmental Sciences
Alma Mater Studiorum University of Bologna
Bologna, Italy

ISSN 2197-5736 ISSN 2197-5744 (electronic)
Use R!
ISBN 978-3-031-99664-1 ISBN 978-3-031-99665-8 (eBook)
https://doi.org/10.1007/978-3-031-99665-8

This Springer imprint is published by the registered company Springer Nature Switzerland AG
The registered company address is: Gewerbestrasse 11, 6330 Cham, Switzerland

Contents

Part I

Introduction

Introduction to the Book

Duccio Rocchini

1 Aim of the Book

This book is devoted to ecological data analysis with R, with examples created by the same original developers of the described packages. R software is now an important reality in ecological informatics. Specific packages have been developed to manage, analyze, and visualize ecological data. This book is a comprehensive catalogue of ecological R packages with running examples. The frustration of seeing outputs without functions is now out, with this step-by-step procedure book, which is intended for students and researchers, with previous experience in R, who have passion with coding their own ecological analyses. The code is provided at: https://github.com/RCodingForEcology/.

2 Why This Book

This is the first book in ecology with running examples made by the authors of the presented packages. The main branches of Ecology are included in a single book, e.g., community ecology, spatial ecology, and ecological remote sensing. This book will lead young and experienced researchers to build their own code starting from the presented example under a free and open-source software philosophy.

D. Rocchini (✉)
BIOME Lab, Department of Biological, Geological and Environmental Sciences, Alma Mater Studiorum University of Bologna, Bologna, Italy

Faculty of Environmental Sciences, Department of Applied Geoinformatics and Spatial Planning, Czech University of Life Sciences Prague, Praha, Suchdol, Czech Republic
e-mail: duccio.rocchini@unibo.it

D. Rocchini (ed.), *R Coding for Ecology*, Use R!,
https://doi.org/10.1007/978-3-031-99665-8_1

3 Structure of the Book

In order to cover several aspects of ecological modeling considering several spatial scales, the book is divided into two main sections: (i) species distribution modeling, population, and community ecology and (ii) ecological remote sensing and spatial ecology. The goal is to show the characteristics of different R packages related to such themes:

- Species distribution modeling, population, and community ecology:
  - The `sdm` R package for species distribution modeling, a powerful tool specifically developed for species distribution modeling. This chapter explores the package's functionalities, including how it facilitates the analysis and prediction of species habitats by integrating various environmental and spatial data, making it an essential resource for ecologists and conservationists working in biodiversity assessment and environmental management.
  - The `ecospat` package for spatial ecology, which offers a comprehensive suite of pre-, core-, and post-modeling tools designed to investigate species niches and distributions. This chapter details how the package supports ecological niche analysis by providing methods for niche quantification, overlap assessment, and spatial distribution modeling, making it a valuable resource for researchers studying species-environment relationships and biogeographical patterns.
  - The `ecodist` package for dissimilarity-based analysis of ecological data, which focuses on dissimilarity-based analysis of ecological data. The chapter and the relative package offer a range of functions including ordination techniques and Mantel tests, specifically designed for analyzing spatial and community ecological datasets. This chapter highlights how `ecodist` facilitates the exploration of ecological patterns through dissimilarity measures, providing crucial support for ecologists analyzing spatial interactions and community dynamics.
  - The `USE` package for a uniform sampling of the environmental space to improve species distribution models by generating ecologically meaningful pseudo-absence data. This chapter highlights the importance of carefully selecting pseudo-absences to reduce sampling bias and avoid class overlaps, thereby enhancing the ecological realism of model predictions. By providing tools for more reliable pseudo-absence generation, `USE` supports ecologists and modelers in producing robust and interpretable estimates of species distributions and ecological niches.
  - The `Rarefy` package for rarefaction methods tailored for ecological community data, allowing researchers to standardize species richness estimates across samples with varying sizes. By applying rarefaction techniques, the chapter helps ecologists to compare biodiversity metrics more accurately, minimizing biases caused by differences in sampling effort. This makes `Rarefy` an

important tool for studies focused on community ecology, species diversity assessment, and conservation planning.

- Ecological remote sensing and spatial ecology:
  - The `biodivMapR` package for spectral diversity mapping, through remote-sensing data. It enables computing and mapping a variety of diversity metrics based on remotely sensed vegetation traits, including α- and β-diversity, as well as functional diversity maps measuring richness, evenness, and divergence. By providing flexible methods and examples, `biodivMapR` supports ecologists and conservationists in mapping and understanding biodiversity patterns at multiple spatial scales.
  - The `rsmove` package to bridge remote sensing and movement ecology, which serves as a bridge between remote-sensing data and movement ecology. This chapter enables researchers to integrate spatially and temporally explicit remote-sensing information with animal movement data, facilitating the study of how environmental factors influence movement patterns. By combining these data sources, `rsmove` supports advanced analyses of animal behavior, habitat use, and ecological dynamics, helping ecologists to better understand the complex interactions between organisms and their changing environments.
  - The `rasterdiv` package for measuring diversity from space, and more in general from spatial data. This chapter provides methods to calculate various diversity indices on numerical matrices using principles from information theory. By applying these techniques to raster data, `rasterdiv` enables ecologists to quantify spatial patterns of diversity from remote sensing and other geospatial datasets, offering valuable insights into landscape-level ecological processes and biodiversity distribution.
  - The `landscapemetrics` and `motif` packages for measuring landscape patterns and processes, two powerful tools for measuring and analyzing landscape patterns and ecological processes. The `landscapemetrics` package provides a tidy, reproducible workflow for calculating landscape metrics on categorical landscape patterns, serving as an efficient R-based alternative to FRAGSTATS. It also supports advanced calculations of landscape complexity through five information-theoretic metrics: marginal entropy, conditional entropy, joint entropy, mutual information, and relative mutual information. Complementing this, the `motif` package focuses on local pattern analysis of categorical raster data within any defined regular or irregular spatial areas. It quantitatively describes spatial patterns using built-in signatures based on co-occurrence matrices, while also allowing users to apply custom functions. Additionally, `motif` enables advanced spatial analyses such as pattern search, change detection, and clustering, offering deeper insights into the dynamics of landscape structures. Together, these packages equip researchers and land managers with comprehensive and reproducible tools to quantify, analyze, and interpret complex landscape patterns and processes across various spatial scales.

- The CAST package for model training and prediction using machine learning, which extends the popular machine learning package caret to better handle spatial and spatial-temporal modeling tasks. Designed to support model training and prediction in spatial contexts, CAST includes advanced features such as the novel 'Nearest Neighbor Distance Matching' cross-validation method, which improves the estimation of spatial prediction model performance. The package also offers spatial variable selection tools that identify the most relevant predictors for enhancing spatial model accuracy. Additionally, CAST provides functionality to estimate the spatial area of applicability for prediction models, helping users understand where models make reliable predictions. The methods and tools described in the chapter are a vital resource for ecologists and data scientists working with spatially explicit predictive modeling.
- The imageRy package for teaching remote sensing in a straightforward way, through a user-friendly approach. The chapter describes function to easily manipulate, visualize, and share raster images directly within R, making remote-sensing concepts more accessible to students and researchers alike. imageRy streamlines the handling of spatial raster data and supports interactive workflows, helping educators deliver straightforward, hands-on remote-sensing education.
- The cblinplot package for building meaningful ecological maps by colorblind people, designed to foster visually inclusive science for people with color vision deficiencies. The tool allows users to input any image or plot within a coding workflow, specify their type of color blindness, and instantly generate a colorblind-friendly version of the visual. The chapter begins by explaining the nature of color vision deficiencies and the challenges they pose in data visualization, followed by a detailed, step-by-step example demonstrating how to use the package effectively. This tool significantly improves accessibility and inclusivity in scientific communication and data presentation.
- The cartogram package for mapping ecological patterns, creating cartograms within R. The chapter enables users to construct both continuous and noncontiguous area cartograms, allowing spatial data to be transformed so that the size of geographic regions reflects a particular variable of interest, such as species abundance or environmental factors. By offering flexible cartogram creation methods, the cartogram package helps ecologists and spatial analysts produce intuitive, informative maps that reveal spatial trends and patterns in ecological data.

4 To Whom This Book Is Dedicated

This book is mainly dedicated to students and scientists passionate about ecological informatics. The main research areas on which the book will focus are population ecology, community ecology, spatial ecology, and ecological remote sensing.

These will also be the sections of the book. The main zoological, botanical, and geographical societies would certainly be interested in the type of book we are proposing.

5 The Teaching Concept Behind the Book

The teaching concept behind our book refers to the use of open-source code. We want to provide examples in R for analyzing ecological data and understanding ecological processes, under a free and open-source coding perspective. Nowadays, students and researchers are familiar with software to analyze data, but they need to be guided when aiming at analyzing data and see what the software is exactly doing. This is the general aim of the book: teaching in a step-by-step manner how to analyze ecological data in R, by the same maintainers of the presented R packages.

6 Typical Courses in Line with This Book

This book might be adopted because it represents a friendly and robust approach in ecology. Moreover, the examples provided will make use of simple-to-manage data provided as a virtual appendix to the book. Typical courses adopting the book would be:

- Ecological informatics
- Ecology
- Computational ecology
- Spatial ecology
- Ecological modeling

The level could spread from final-year bachelor students as well as master or PhD students, with the possibility to extend the audience to researchers.

Part II

Species Distribution Modelling, Population and Community Ecology

The sdm R Package for Species Distribution Modeling

Babak Naimi, Elham Ebrahimi, and Miguel B. Araújo

1 Introduction

Global biodiversity change is one of the most pressing environmental issues of our time (Pecl et al. 2017; Dornelas et al. 2023). The planet is facing the sixth mass extinction (Kolbert 2014), with the potential loss of up to half of all species (Hughes 2023), which is ranked among the top three global risks to humanity over the next decade (World Economic Forum, 2022). In response to this crisis, the Kunming-Montreal Global Biodiversity Framework (KM-GBF) was adopted at the 15th Conference of the Parties (COP15) with the ambitious goals to halt and reverse biodiversity loss by 2030 and 2050, and to restore ecosystems, conservation of species, and ensure the sustainable use of natural resources.

Quantitative analyses and sophisticated modeling tools play a significant role to inform biodiversity conservation strategies and policy decisions (e.g., Whittaker

B. Naimi (✉)
Quantitative Biodiversity Dynamics (QBD), Department of Biology, Utrecht University, Utrecht, The Netherlands

E. Ebrahimi
Wildlife Ecology and Conservation Organization Department of Environmental Sciences, Wageningen University & Research, Wageningen, The Netherlands

Wildlife Ecology & Nature Restoration, Department of Biology, Utrecht University, Utrecht, The Netherlands

M. B. Araújo
Department of Biogeography and Global Change, National Museum of Natural Sciences, CSIC, Madrid, Spain

Rui Nabeiro Biodiversity Chair, MED Institute, University of Évora, Évora, Portugal

Theoretical Sciences Visiting Program, Okinawa Institute of Science and Technology Graduate University, Onna, Japan

D. Rocchini (ed.), *R Coding for Ecology*, Use R!,
https://doi.org/10.1007/978-3-031-99665-8_2

et al. 2005). Species Distribution Models (SDMs), the most common and widely used approach for biodiversity modeling (Araújo et al. 2019), have emerged as pivotal tools in biodiversity research and assessments, offering robust methods to predict the current geographic distributions of species across landscapes and project them into different times and spaces. These models, also known as bioclimatic envelope models, ecological niche models, and habitat suitability models (for a review of terminologies see Araújo et al. 2012), explore the relationships between species observations (usually in the form of presence-only, presence-absence, or abundance) and environmental variables to provide insights into the potential distribution of species across geographic space and time (Ceballos et al. 2017; Dirzo et al. 2014; Guisan et al. 2005, 2000). The application of SDMs spans a wide range of tasks, such as forecasting or hindcasting the effects of climate change on biodiversity (Ebrahimi et al. 2022; Thorup et al. 2021), inferring risks from species invasion (Naimi et al. 2022), selecting sites for species conservation (Araújo et al. 2000), habitat restoration (Zellmer et al. 2019), rewilding (Araújo et al. 2024), and species translocation (Chauvenet et al. 2013), as well as quantifying combined effects of anthropogenic factors on biodiversity (e.g., Hof et al. 2011; Taheri et al. 2021), among many other applications (for a review see Peterson et al. 2012).

The implementation of SDMs involves a workflow that integrates biological and environmental data, statistical and/or machine learning methods, and computational tools to predict species distributions. The `sdm` R package (Naimi and Araújo 2016) provides a comprehensive and extensible platform for developing SDMs using three commonly used types of species data: presence-absence, presence-only, and abundance. The package facilitates implementing the entire species distribution modeling workflow including data integration, modeling and evaluating their performance using multiple accuracy metrics, and using the models for certain applications (Fig. 1). This package supports a variety of modeling algorithms,

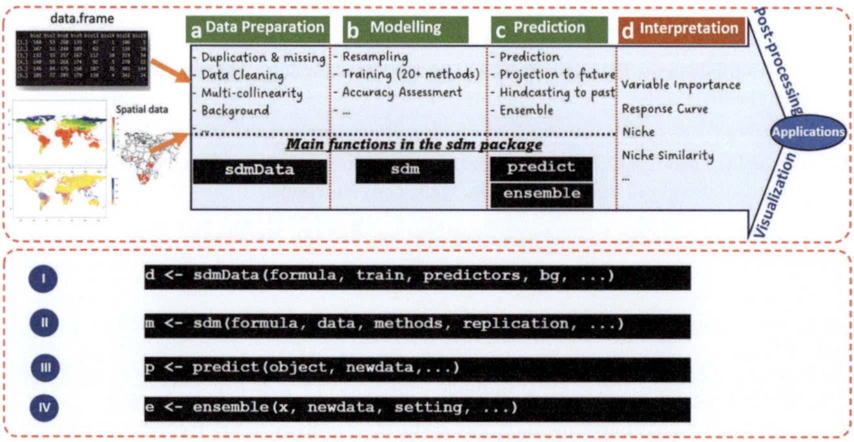

Fig. 1 A schematic representation of the species distribution modeling workflow with the main functions, followed by their basic usages, in the "*sdm*" R package

including machine learning techniques such as Random Forests (RF, Breiman 2001), Boosted Regression Trees (BRT, Friedman 2001), Flexible Discriminant Analysis (FDA, Hastie et al. 1994), Mixture Discriminant Analysis (MDA, Hastie et al. 1996), Radial Basic Function Neural Network (RBF, Hudak 1992), Multi-layer Perceptron Neural Networks (MLP, Rosenblatt 1958), Maxlike (Royle et al. 2012), MaxEnt (Phillips et al. 2006), and MaxNet (Phillips et al. 2017), as well as classical statistical methods such as Generalized Linear Models (GLM, McCullagh 2019) and Generalized Additive Models (GAM, Wood 2017), and also older family of SDM methods such as bioclim (Busby 1991), and Domain (Carpenter et al. 1993). Furthermore, multiple ensemble forecasting methods (Araujo and New 2007) are supported in the package to combine multiple predictions from different algorithms and replications to generate a consensus spatial distribution of species. By offering a flexible and user-friendly interface, the `sdm` R package supports multiple common formats of spatial and temporal data, facilitates developing SDMs by simultaneously incorporating multiple algorithms, enables automation of model fitting for multiple species simultaneously and comparison of their performance, generates different types of consensus across ensembles, and provides tools for interpretation of models, all enhancing the robustness and accuracy of predictions.

This book chapter describes the workflow of species distribution modeling and practical steps using the `sdm` R package. We provide a comprehensive guide, from data preparation and selection of environmental variables to model building, evaluation, and prediction. Emphasizing practical application, this chapter walks the reader through the utilization of various modeling algorithms supported by the sdm package, including techniques for optimizing model performance and interpreting results. By illustrating the usage of the package with a case study, we highlight the versatility and effectiveness of SDMs in biodiversity research and conservation planning.

1.1 An Overview of the SDM Workflow in the sdm R Package

The `sdm` R package is a comprehensive framework that provides a set of user-friendly functions for implementing the entire species distribution modeling workflow to develop data-driven SDMs and begins with providing data for both species (response or dependent variable) and environment (predictor or explanatory variables; also called covariates). The workflow consists of three major steps to build SDMs including data preparation, modeling, and prediction. The `sdm` package offers a user-friendly function for each step in the workflow with arguments controlling the tasks given a user's inputs.

(a) *Data preparation* involves controlling and integrating species and environmental data, identifying biases and cleaning issues with data (Rocchini et al. 2023) such as mismatch in coordinate reference systems, duplication of records, missing values, and testing collinearity among predictor variables (Naimi et al. 2014). Different types of species data are supported in the sdm package including **presence-only**, **presence-absence**, and **abundance** data. Since presence-only is the most widely used data type, a case study in this book chapter is demonstrated based on using this data type. In case of presence-only, pseudo-absence or background records can be generated at this stage. The main function for this stage in the sdm package is `sdmData` which supports both tabular (e.g., `data.frame`) and spatial data formats and can do all the abovementioned tasks for cleaning and preparing data.

(b) *Modeling* utilizes various (20+) statistical and machine learning algorithms to construct SDMs based on the prepared clean data in the previous step (a). Once the models are built (also called training of the models), they undergo rigorous evaluation to assess their performance and predictive accuracy. This step often includes resampling procedure for dividing data into training and test partitions using multiple methods such as subsampling, cross-validation, and bootstrapping (Naimi and Araújo 2016), which employ widely used evaluation metrics, like the Area Under the Curve (AUC) and True Skill Statistic (TSS), to ensure robust model validation (Allouche et al. 2006; Fielding and Bell 1997). The single function of `sdm` in the package integrates all the modeling and evaluation tasks defined in this phase that can be simultaneously applied for multiple species, multiple modeling methods, and multiple replications of data.

(c) *Prediction*: After a successful evaluation, the models are used for predicting (or projecting or hindcasting) species distributions across geographic areas and time periods under current and future environmental scenarios. The predict function in the `sdm` package can generate the predicted values (e.g., probability of occurrence) based on each trained model given the predictor variables as inputs. When multiple modeling methods are employed, the predictions of individual models can be combined into a single consensus prediction using ensemble prediction approach (Araujo and New 2007). The ensemble function in the sdm package offers several methods to combine individual models such as weighted and unweighted mean, median, mean of predicted presence-absence, two-step mean, etc. (check the help page of the function in the sdm package for more information).

(d) *Interpretation*: Finally, the workflow may include post-processing steps for model interpretation and visualization. This can involve mapping predicted distributions, analyzing variable importance, and generating graphical representations of model outputs which facilitate understanding and communication of results to address specific applied questions. The `sdm` R package streamlines these processes, providing researchers and practitioners with powerful tools to develop, evaluate, and apply SDMs effectively to inform various applications, such as identifying priority areas for conservation, assessing the impacts of climate change, and guiding habitat restoration efforts (Fig. 1).

1.2 Case Study: Species Distribution Modeling of *Snow Leopard*

To demonstrate the SDM workflow, this section is dedicated to assess and explore **impacts of climate change on geographical distribution of Snow Leopard** (scientific name: *Panthera uncia*) (Fig. 2). We use the sdm R package, along with some other packages, to conduct the study, and here, a step-by-step guideline is provided to discuss the practical solution for implementing the workflow and generating results in the forms of maps, tables, and graphs.

The solution to conduct the study has been provided following the main steps in the workflow defined in Fig. 1. The solution is provided through the following sections:

(i) *Downloading data*: Here, we use R and the geodata package to download species data from GBIF and climate data (for both the current and future times) from WorldClim.

(ii) *Preparing data*: Data are checked for some issues and get ready as spatial datasets for conducting SDMs.

(iii) *Developing models using the sdm R package*: The main workflow in the sdm package (Naimi and Araújo 2016) to develop and evaluate SDMs is demonstrated.

(iv) *Predict/project the map of habitat suitability in current and future times*: We predict the current distribution of the species (habitat suitability) and project its distribution into a future climatic scenario (for the year 2100).

(v) *Assess range shift in response to climate change*: Given the habitat suitability maps in both the current and future times, we can assess the magnitude and distribution of changes, and quantitatively measure range shifts due to climate change.

Fig. 2 Snow Leopard (*Panthera uncia*)

(i) Downloading Data

In this section, we will use R to download occurrence data for the Snow Leopard (*Panthera uncia*) from the Global Biodiversity Information Facility (GBIF) and climate data from the WorldClim database. These datasets will serve as the foundation for developing species distribution models and assessing the potential impacts of climate change on the Snow Leopard's habitat.

Species Occurrence Data We will use the sp_occurrence function in the geodata package to download Snow Leopard occurrences from GBIF. Occurrences are the locations where the species has been observed (mostly cover presence-only records), providing spatial points essential for building SDMs. Some descriptions are provided as comments within the code:

```
library(geodata)
```

```
## Loading required package: terra

## terra 1.7.78
```

```
# let's first check whether any records available on GBIF without
↪   downloading:

sp_occurrence("Panthera","uncia", download = F)
```

```
## Loading required namespace: jsonlite

## [1] 197
```

```
# we have 196 records available, let's download them as a data.frame:

sp <-  sp_occurrence("Panthera","uncia")
```

```
## 197 records found

## 0-197
## 197 records downloaded
```

```
#-------
# the sp data.frame has over 100 columns, corresponding to different
↪   information
# Some of the columns can be used to filter and exclude some records.
# For example, the column "basisOfRecord" is a Darwin Core term that
↪   refers to
# the specific nature of the record and can refer to one of 6 classes

# Let's check our records to see their frequency for different classes:

table(sp$basisOfRecord)
```

```
##
## HUMAN_OBSERVATION  MATERIAL_CITATION    MATERIAL_SAMPLE
↪  PRESERVED_SPECIMEN
##               116                  8                  4
↪  69
```

```
# You may find the definition from the GBIF website:
# https://docs.gbif.org/course-data-use/en/basis-of-record.html

#------
# You can also check the years of records in our dataset:

table(sp$year)
```

```
##
## 1899 1979 1998 2000 2004 2007 2008 2009 2010 2011 2012 2013 2014 2015
↪  2016 2017
##    1    1    1    1    2    2    2    1    1    9    6   17    9   14
↪  15   18
## 2018 2019 2020 2021 2022 2023 2024
##   16   25    6    5    6   24   11
```

```
# let's also check occurrenceStatuse column:

table(sp$occurrenceStatus) # we are sure now all records are presence!
```

```
##
## PRESENT
##     197
```

```
# if it was not all "PRESENT", we could keep only "PRESENT" to use data
↪  as
# a presence-only dataset
#---------------------------
###### Filtering some records:

# Given the information avaialble in "basisOfRecords" and "year" columns,
↪  we
# want to filter data and keep appropriate records. We may use other
↪  columns
# like coordinateUncertaintyInMeters; or we may use some specific
↪  packages
# designed to clean GBIF data (e.g., CoordinateCleaner). But here, we do
↪  our
# filtering based on the "basisOfRecords" and "year" columns.

# Based on the information we checked above, we keep records with
# basisOfRecord == "HUMAN_OBSERVATIO" & year > 1970, we use the which
↪  function:
```

```
w <- which(sp$basisOfRecord == "HUMAN_OBSERVATION" & sp$year > 1970)

# w keeps the record number (row number in sp data.frame) that we need to
↪  keep

sp <- sp[w,] # only keep records specified in w

#------
## Get rid of all columns except coordinate columns (lon & lat)
# from all columns, we only keep 2 columns (lon, lat)

sp <- sp[,c('lon','lat')]

# we add another column (its name can simply be "species") to which we
↪  assign
# 1 to all records in the column as they are presence-only:

sp$species <- 1

# let's check the head of the table (data.frame) to make sure it is fine:

head(sp)
```

```
##        lon      lat species
## 1 74.60183 42.50146       1
## 2 98.36646 35.58493       1
## 3 77.90942 32.34409       1
## 4 98.46910 35.55448       1
## 5 77.62451 33.91132       1
## 6 76.86968 34.31075       1
```

```
# now, we convert the sp data.frame to Spatial Points using the vect
↪  function in
# the terra package (generates SpatVector object):

sp <- vect(sp, c('lon','lat'))

sp # sp, now, is a SpatVector object containing spatial points of species
↪  records
```

```
##  class       : SpatVector
##  geometry    : points
##  dimensions  : 116, 1  (geometries, attributes)
##  extent      : 71.76065, 105.8244, 27.41172, 49.8562  (xmin, xmax,
↪  ymin, ymax)
##  coord. ref. :
##  names       : species
##  type        :   <num>
##  values      :       1
##                      1
##                      1
```

```
# let's generate a map to see the locations of the species records:

# We had the world map but if you don't, you can find it on Internet!
wld <- vect('world.shp')


plot(wld)

points(sp,col='red')
```

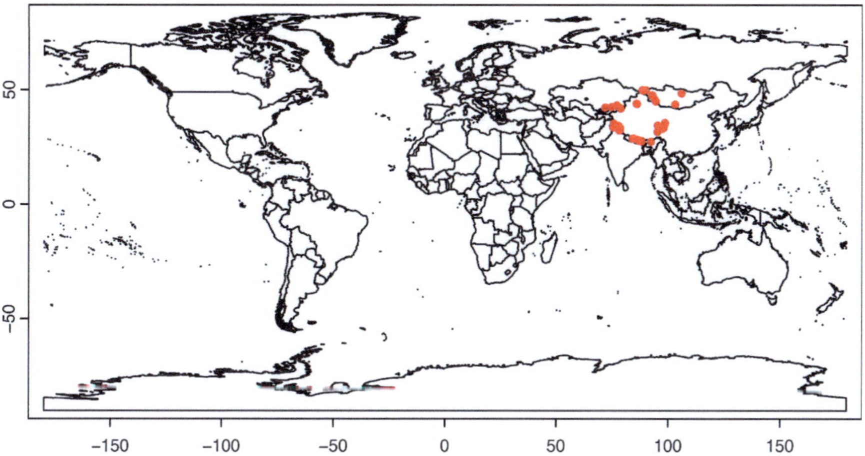

Downloading Climate Data

We are going to use climate data with 19 bioclimatic variables (represent different aspects of climate condition including mean annual temperature and precipitation, their seasonality, and extreme conditions) available in the WorldClim dataset. The bioclim variables are extracted from monthly weather data (monthly minimum and maximum temperature and mean precipitation). The variables are described here: https://www.worldclim.org/data/bioclim.html

Given that the purpose of the instructions provided here is to demonstrate how the SDM workflow can be implemented, we will use a coarse resolution (10 minutes; approximately 20 km) dataset to make the computation fast. However, in practice, the resolution should be selected based on the scale of the study to ensure accurate results. The WorldClim dataset is available on the *worldclim.org* website.

We can download the dataset for both current and future time periods. The future climate conditions are estimated using General Circulation Models (GCMs), which simulate the Earth's climate based on various physical processes and interactions within the atmosphere, ocean, and land surface. These models consider different scenarios known as Shared Socioeconomic Pathways (SSPs), which reflect various potential future conditions based on socioeconomic trends and policies. SSPs provide a comprehensive framework to analyze the impacts of different socioeconomic developments on climate change and to explore mitigation and adaptation strategies.

The climate data for the current time (baseline) represent mean climate condition generated from weather data recorded between 1980–2000. To download the climate data for future times, we need to select the appropriate GCMs (we may select several or even all GCMs for a real study) and SSP scenarios that align with the objectives of our study. For example, we might choose SSP2-4.5, which represents a moderate scenario where socioeconomic trends lead to stabilization in global emissions by mid-century. Alternatively, we could select SSP5-8.5, which represents a high-emission scenario. Under SSP5-8.5, greenhouse gas emissions continue to rise throughout the century, leading to significant increases in temperature and more extreme climate conditions (also called business as usual or worst-case scenario). This scenario is useful for understanding the potential impacts of unmitigated climate change and helps in planning for worst-case outcomes.

For this exercise, let's download climate data for both the current time (baseline) and future time (year 2100; it is generated based on 20 years of data between 2081-2100) for one of GCMs and SSP5-8.5. We can use the *worldclim_global* function to download climate data for the current time, and cmip6_world to download for the future. Check the help page of the functions for more details about the usage of these functions:

```
# the following function, download bioclim layers with the resolution of
↪ 10 min.

#########################################################
##-----> Downloadin climate for the Baseline (current time):
#########################################################

bio <- worldclim_global(var="bio",res=10, path = getwd())

bio
```

```
## class       : SpatRaster
## dimensions  : 1080, 2160, 19  (nrow, ncol, nlyr)
## resolution  : 0.1666667, 0.1666667  (x, y)
## extent      : -180, 180, -90, 90  (xmin, xmax, ymin, ymax)
## coord. ref. : lon/lat WGS 84 (EPSG:4326)
## sources     : wc2.1_10m_bio_1.tif
##               wc2.1_10m_bio_2.tif
##               wc2.1_10m_bio_3.tif
##               ... and 16 more source(s)
## names       : wc2.1~bio_1, wc2.1~bio_2, wc2.1~bio_3, wc2.1~bio_4,
↪ wc2.1~bio_5, wc2.1~bio_6, ...
## min values  :   -54.72435,     1.00000,    9.131122,       0.000,
↪ -29.68600,   -72.50025, ...
## max values  :    30.98764,    21.14754,  100.000000,    2363.846,
↪ 48.08275,    26.30000, ...
```

```
# let's check the names of layers in the bio object
names(bio)
```

```
##  [1] "wc2.1_10m_bio_1"  "wc2.1_10m_bio_2"  "wc2.1_10m_bio_3"
↪  "wc2.1_10m_bio_4"
##  [5] "wc2.1_10m_bio_5"  "wc2.1_10m_bio_6"  "wc2.1_10m_bio_7"
↪  "wc2.1_10m_bio_8"
##  [9] "wc2.1_10m_bio_9"  "wc2.1_10m_bio_10" "wc2.1_10m_bio_11"
↪  "wc2.1_10m_bio_12"
## [13] "wc2.1_10m_bio_13" "wc2.1_10m_bio_14" "wc2.1_10m_bio_15"
↪  "wc2.1_10m_bio_16"
## [17] "wc2.1_10m_bio_17" "wc2.1_10m_bio_18" "wc2.1_10m_bio_19"
```

```
# As you can see, the bio object contains 19 bioclim variables
↪  representing
# climate conditions for the current time. The names are long, have
↪  information
# about the version of the WorldClim dataset (WC_2.1) and resolution.

# Let's make the names simpler (i.e., bio1, bio2, ..., bio19)

# We can simply assign the new names:

names(bio) <- paste0('bio',1:19) # Do you know why we used the paste0
↪  function?

# now, let's check again the names of our data object:
names(bio)
```

```
##  [1] "bio1"  "bio2"  "bio3"  "bio4"  "bio5"  "bio6"  "bio7"  "bio8"
↪  "bio9"
## [10] "bio10" "bio11" "bio12" "bio13" "bio14" "bio15" "bio16" "bio17"
↪  "bio18"
## [19] "bio19"
```

```
# Let's visualise the first layer (Mean Annual Temperature):


plot(bio[[1]],main='Bio1: Mean Annual Temperature (Current time)')
```

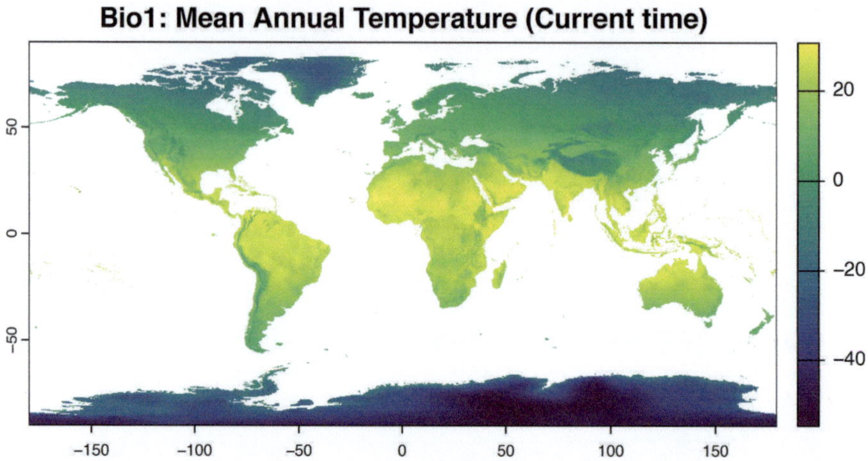


```
# and 12th layer (Mean Annual Precipitation)

# for this one, let's define an appropriate color palette:

cl <- colorRampPalette(c('gray','yellow','green','blue','darkblue'))


plot(bio[[12]],col=cl(200),main='Bio12:Mean Annual Precipitation (Current
↪  time)')
```

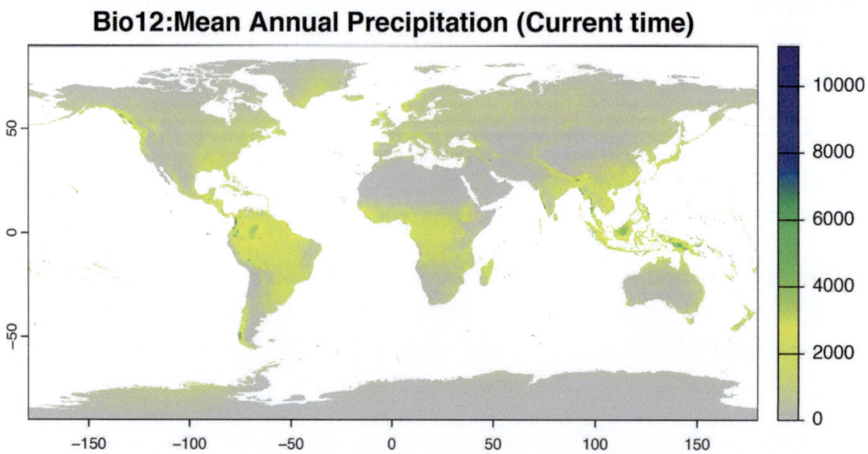


```
###########################################################
#----> Downloading bioclim variables for the Future time:
###########################################################

# To download bioclim for a future time, we need the following
↪  information:
```

```
## GCM: Which Climate Model? (20+ GCMs are available!, here we chose:
↪  "MIROC6")

## SSP: Which Climate Scenario? (4 different scenarios are available!)
#-------> ssp: 126, 245, 370, 585 (from "optimistic"....to....
↪  "pessimistic")

## Year: Which future time (e.g., "2081-2100")

# we can also specify variables (here, we need bioclim; "bioc"),

## res: resolution (10 arc min ~ equivalent to almost 20 km)
#-----------
# we can use cmip6_world function in the geodata package to download
↪  climate
# variables for the future times:

biof <- cmip6_world(model="MIROC6", ssp='585',
                    time = "2081-2100", var = 'bioc', res=10,
                    ↪  path=getwd())

biof
```

```
## class       : SpatRaster
## dimensions  : 1080, 2160, 19  (nrow, ncol, nlyr)
## resolution  : 0.1666667, 0.1666667  (x, y)
## extent      : -180, 180, -90, 90  (xmin, xmax, ymin, ymax)
## coord. ref. : lon/lat WGS 84 (EPSG:4326)
## source      : wc2.1_10m_bioc_MIROC6_ssp585_2081-2100.tif
## names       : bio01, bio02, bio03,  bio04, bio05, bio06, ...
## min values  : -50.3,  -1.9, -10.6,   13.4,   -26, -68.4, ...
## max values  :  36.3,  22.5,  95.4, 2250.5,    55,  27.8, ...
```

```
# let's check the names of the biof object which contains bioclim
↪  variables of
# future (year: 2100) for the worst-case scenario (ssp585):

names(biof)
```

```
##  [1] "bio01" "bio02" "bio03" "bio04" "bio05" "bio06" "bio07" "bio08"
↪  "bio09"
## [10] "bio10" "bio11" "bio12" "bio13" "bio14" "bio15" "bio16" "bio17"
↪  "bio18"
## [19] "bio19"
```

```
# For our modelling practice, the names of bioclim layers in the current
↪  and
# future time should be exactly the same, so since the order of the
↪  layers in
```

```
# bio and biof is the same, we can simply assign the names of biof to be
↪  the
# same as bio:

names(biof) <- names(bio)


# let's visualise precipitation for the future time:

plot(biof[[12]],col=cl(200),
     main="Bio12: Mean Annual Precipitation (Year:2100; SSP-585)")
```

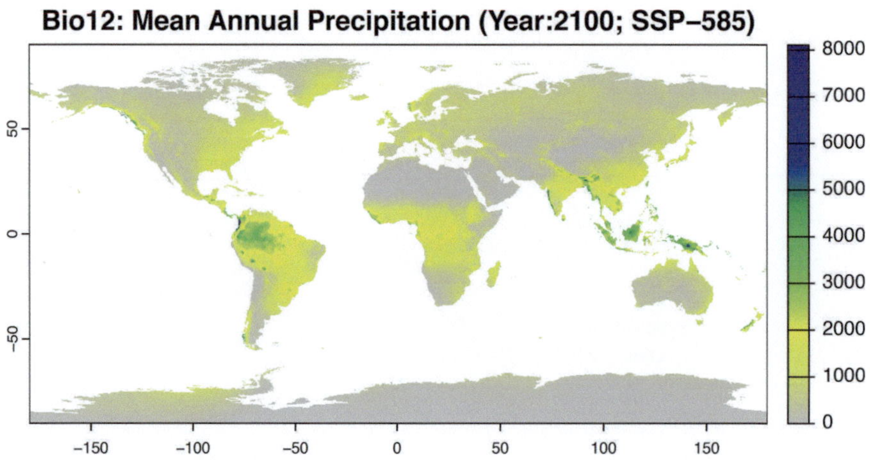

(ii) Data Preparation

After downloading data, we need to take some steps to prepare data for modeling.

While statistical methods can be employed to identify outliers of the species data, a visual inspection of the map with species locations suggests that the points in North America may be outliers. This suspicion can be further validated by comparing these points with the species' range map provided by the IUCN. Therefore, we can exclude the outlier records. However, if the study area to train the model is selected and the outliers are located outside of the study area, the outliers will not be considered in the modeling procedure (i.e., no need to exclude them).

Selecting Study Area

Although the purpose of our case study is to generate **global** potential distribution of species, we may select a realistic study area which can be an area potentially accessible to the species. Since the background records are generated within the study area, it is important to select a reasonable area to train SDMs. We may use a biogeographical map to overlap presence records and select the polygons intersected which species presence locations and consider them as the study area. Alternatively, we can approximately specify the area by choosing its upper left and lower right which give us the minimum and maximum x and y coordinates to make an extent object using the terra package.

```
library(terra)

plot(wld)

bnd <- ext(c(60,120,10,60)) # spatial extent with c(X_min, X_max, Y_min,
↪  Y_max)

points(sp,col='red')

plot(bnd,add=T,border='blue')
```

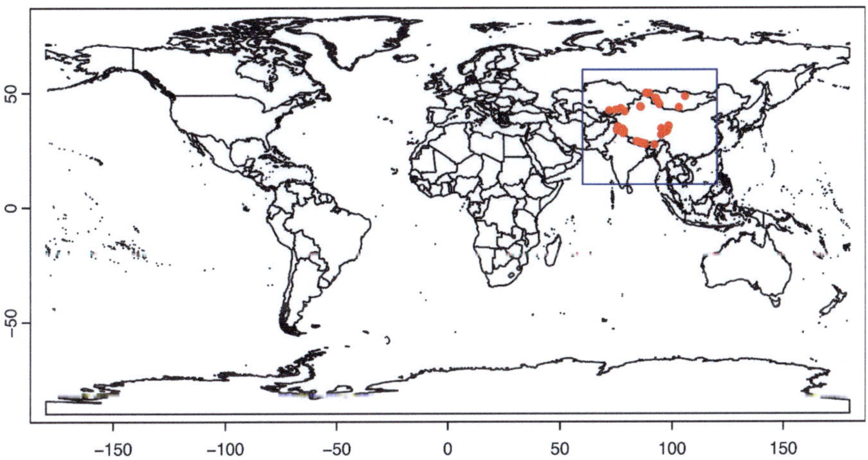

Cropping Climate Data to the Extent Defined as the Study Area

As mentioned earlier, when the species records are presence-only, we need to consider a realistic area (areas potentially accessible for the species) to calibrate (train) SDMs where the background records are drawn. Here, the specified extent (the rectangle) is used as the study area to train SDMs but we will use the models to predict/project across the world.

The following code crops the bioclim variables of the current time within the specified extent (bnd object). We don't crop the future dataset (biof) as the model training uses on the variables in the current time:

```
# the extent specified earlier:
bnd
```

```
## SpatExtent : 60, 120, 10, 60 (xmin, xmax, ymin, ymax)
```

```
bioc <- crop(bio, bnd) # crop bio object within the bnd extent

# let's visualise a layer from "bioc" and add the species occurrences to
↪   the map:

plot(bioc[[1]],main='Bio1 (cropped)')

points(sp,col='red')
```

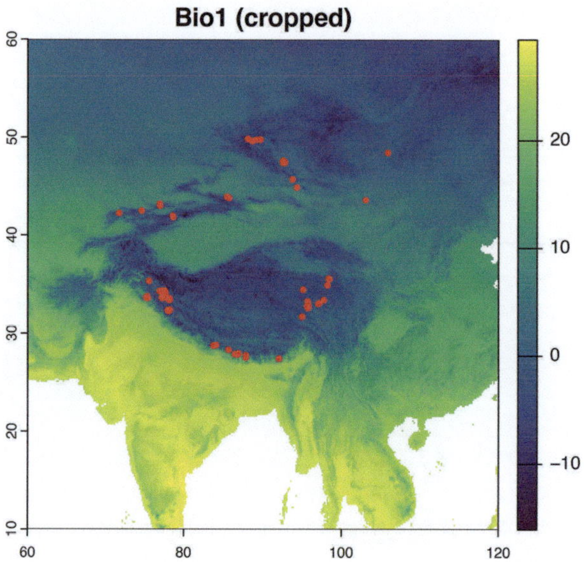

Multi-collinearity

When we have two or more numerical predictor variables that are strongly correlated, the data are subjected to the issue of multicollinearity (also called collinearity) (Dormann et al. 2013). Some modeling methods, such as GLMs, may be sensitive to the collinearity issue as their parameterization can be affected. Even if a modeling method is not sensitive, excluding the collinear variables is a good practice given the parsimony rule to avoid redundant information in the model which makes it unnecessarily complex.

Several approaches are available to deal with collinearity issue. One approach is to transform the environmental variables into a new set of variables using Principle Component Analysis (PCA), an ordination method for reducing data dimensions, and generate new, uncorrelated variables. This can be done using the `pca` function or by specifying the `pca` term in the formula in the `sdmData` function.

Another approach is to identify and exclude variables that show signs of collinearity. To identify collinear variables, a simple pairwise correlation test can be conducted between each pair of variables (a correlation greater than 0.7 is indicative of collinearity). Alternatively, the variance inflation factor (VIF) can be measured for each variable, representing how much of a variable's variance can be explained by other variables (a VIF greater than 10 is a sign of collinearity).

A popular hybrid approach was introduced by Naimi et al. (2014) to measure VIF or a combination of VIF and correlation coefficient through a stepwise procedure (Naimi et al. 2014). Two functions implemented in the usdm package can be used here including `vifstep` and `vifcor`. To deal with collinearity, one of these two methods should be used. `vifstep` checks the VIF metric for all numerical predictor variables and identifies the variable with the maximum VIF for exclusion if it exceeds the specified threshold (default=10). This procedure is repeated step by step until all the remained variables have VIF values below the threshold.

Alternatively, `vifcor` checks the correlation coefficient for all possible pairs of variables and identifies the pair with the maximum correlation. If it exceeds the specified threshold (e.g., 0.7), one of the two variables in the pair is excluded. To make decision that which one should be excluded, VIF is calculated for both variables in the pair, and the one with the greater VIF is excluded. The procedure is repeated step by step until none of the remaining pairs have a correlation greater than the threshold (Naimi et al. 2014).

Here, we test the collinearity issue among the 19 bioclim variables in the current time using the `vifstep` function and exclude collinear variables from all datasets (current and future times):

```
library(usdm) # contains the functions to deal with collinearity

v <- vifstep(bioc, th=10) # checks collinearity with vifstep and
↪  threshold of 10

# here is the output of vifstep, reporting which variables should be
↪  excluded:
v
```

```
## 10 variables from the 19 input variables have collinearity problem:
##
## bio6 bio11 bio10 bio7 bio16 bio1 bio12 bio17 bio4 bio5
##
## After excluding the collinear variables, the linear correlation
↪  coefficients ranges between:
## min correlation ( bio3 ~ bio2 ):  -0.005749341
## max correlation ( bio18 ~ bio13 ):  0.721448
##
## ---------- VIFs of the remained variables --------
##   Variables      VIF
## 1      bio2 2.537522
## 2      bio3 2.681313
## 3      bio8 1.431894
## 4      bio9 2.126017
## 5     bio13 4.116040
## 6     bio14 2.463484
## 7     bio15 2.264739
```

```
## 8     bio18 2.866524
## 9     bio19 1.371750
```

```
# Now, let's exclude the collinear variables from all data objects:
# ... exclude collinear variables from bioc (cropped bio) based on "v"
bioc <- exclude(bioc,v)

# now, you can see that bioc has the selected variables:

bioc
```

```
## class       : SpatRaster
## dimensions  : 300, 360, 9  (nrow, ncol, nlyr)
## resolution  : 0.1666667, 0.1666667  (x, y)
## extent      : 60, 120, 10, 60  (xmin, xmax, ymin, ymax)
## coord. ref. : lon/lat WGS 84 (EPSG:4326)
## source(s)   : memory
## names       :     bio2,     bio3,      bio8,      bio9, bio13, bio14,
↪  ...
## min values  :  3.83750, 14.90886, -15.74217, -32.20467,     3,     0,
↪  ...
## max values  : 18.35242, 76.99110,  36.88192,  34.43004,  2381,    77,
↪  ...
```

```
# let's do it for objects the objects of global scale (bio, biof):
bio <- exclude(bio, v)

biof <- exclude(biof, v)

# If you want to save the new data objects, you can use the writeRaster
↪  function
# Example:....> bio <- writeRaster(bio,
↪  filename="bioclim_selected_current.tif")

#-----------
# as you can see, the selected variables are in bio (and in biof and
↪  bioc):

bio
```

```
## class       : SpatRaster
## dimensions  : 1080, 2160, 9  (nrow, ncol, nlyr)
## resolution  : 0.1666667, 0.1666667  (x, y)
## extent      : -180, 180, -90, 90  (xmin, xmax, ymin, ymax)
## coord. ref. : lon/lat WGS 84 (EPSG:4326)
## sources     : wc2.1_10m_bio_2.tif
##               wc2.1_10m_bio_3.tif
##               wc2.1_10m_bio_8.tif
##               ... and 6 more source(s)
## names       :     bio2,       bio3,      bio8,      bio9, bio13, bio14,
↪  ...
```

```
## min values  :  1.00000,   9.131122, -66.29942, -54.26688,    0,     0,
↪   ...
## max values  : 21.14754, 100.000000,  37.70479,  37.43358, 2381,   484,
↪   ...
```

(iii) Model Development

Now, we have both species (response) and climate (predictors) variables ready to implement the modeling workflow using the sdm package. The package can be installed normally from CRAN or from GitHub using `devtools::install_github('babaknaimi/sdm'))`. **To get all modeling methods available for training SDMs, some additional packages also need to be installed. You may use the `installAll()` function one time after installing sdm to get all required packages installed on your machine.**

In the sdm package, two user-friendly functions should be used to develop the models. First, an sdmdata object should be created using the `sdmData` function, and then, models are trained and evaluated using the `sdm` function.

The usage of the function:

```
sdmData(formula,train,predictors,test,bg,filename,crs,impute,
metadata,...)
```

The `sdmData` function requires both species and predictor variables, and several issues with data (e.g., duplications, missing values) and inconsistencies (e.g., mismatch in coordinate reference systems) are checked and fixed by this function. It has a `formula` interface (check the following box) which provides flexibility for users to control data structure and formats and apply several functions on data.

Box 1 Formula Interface in the sdmData Function

The "formula" interface in R is a powerful and versatile tool that simplifies the specification of statistical models. Central to many modeling functions in R, the formula interface allows users to define the relationship between the dependent and independent variables in a clear and concise manner. This interface uses a symbolic syntax to represent models, making it intuitive and accessible for users to specify complex models without delving into the underlying mathematical details.

At its core, the formula interface uses a simple syntax: The tilde ( ) operator separates the response variable on the left from the predictor variables on the right. For example, in the formula y x1 + x2, y is the dependent variable (species), and x1 and x2 are the independent variables (predictors; environmental variables). In the sdm R package, the formula interface has been extended to support specifying different types of data (e.g., categorical/factor variables, time, and coordinates) and provide flexibility in controlling data processing and modeling by users.

(continued)

Box 1 (continued)

The formula interface is implemented as the first argument in the two main functions in the sdm package, including `sdmData` and `sdm`. The following syntax illustrates how the formula in the sdm package provides a flexibile interface, which contributes to ease of use. The formula in both the `sdmData` and `sdm` functions is used to specify the species in the left-hand side and predictors in the right-hand side. Let us assume the name of species in our data is "species," and we used three predictors named "bio1," "bio2," and "bio3"; then, the formula can be defined as

```
sdmData(species ~ bio1 + bio2 + bio3, ...)
```

We can simply use "." in the right-hand side to avoid typing all the variable names, meaning use all existing variables:

```
species ~ .
```

Categorical Variables If we have categorical variables in our dataset, we can explicitly specify them by using the term of "factor" or simply "f" in the formula (which refers to factor data type that is used in R for handling categorical data). Let us assume our categorical variable is "landuse"; then the formula would be

```
species ~ . + f(landuse)
```

Spatial Coordinates When we use spatial data objects (e.g., spatial points and raster), coordinates of species locations are extracted from the dataset, but when the input dataset is a data.frame and if spatial coordinates are available (e.g., two columns of the data.frame are coordinates with the names of "lon" and "lat"), we can specify them in the formula:

```
species ~ . + f(landuse) + coords(lon+lat)
```

Time (Temporal Data) If data/time records are available (e.g., date of species observations), it can be specified using the term of "time":

```
species ~ . + f(landuse) + coords(lon+lat) +
time(eventDate)
```

Multiple Species The sdm package supports modeling for simultaneously multiple species. When we have multiple species in the dataset, their names can be included in the left-hand side of the formula as

```
species1 + species2 + species3 ~ . + f(landuse)
```

If the left-hand side of the formula is left empty, the package tries to detect the available species data in the input dataset by assuming all variables having binomial records (1 and 0) are species variables (presence-absence):

```
~ . + f(landuse)
```

Selecting Variables Another formula term in the sdm package, "select," can be used in the data preparation procedure to select a subset of variables using a variable selection procedure. For instance, assessing the issue of multicollinearity in data may be resulted to exclude some problematic variables

(continued)

Box 1 (continued)

and keep a subset. The multicollinearity assessment in the sdm package uses two different approaches, "vifcor" and "vifstep" (Naimi et al. 2014), implemented in the usdm R package. By using "select," a user can specify names of (numerical) variables for which the collinearity should be checked: The following example checks collinearity for all numeric variables using the vifstep function given the threshold of 10:

```
species ~ . + select(., method="vifstep",th=10)
```

The following example checks collinearity for all numeric variables except "bio1" using the vifcor function given the threshold of 0.7:

```
species~ . + select(. - bio1, method="vifcor", th=0.7)
```

Scaling the Variables Given that the range (gradient) of predictor variables is usually different and not directly comparable to each other, transforming them to a similar range (scaling) can significantly enhance model convergence and performance. Scaling ensures that each predictor variable contributes equally to the model, preventing variables with larger ranges from disproportionately influencing the results. This step is especially important for algorithms that are sensitive to the scale of input data, as it facilitates optimisation and leads to a faster and more stable parameter convergence.

In the sdm package, it is straightforward and easy to use the scale term in the formula to standardize predictor variables before model fitting. By using the scale function, users can ensure that their models are robust, efficient, and interpretable, as this step is a best practice in the species distribution modeling workflow, reinforcing the reliability and effectiveness of the models developed using the sdm package. Two scaling methods are available in the package, "minmax" (default) transforms the data to a range between 0 and 1, and "center" scales each variable by subtracting its mean, divided by its standard deviation:

```
species ~ . + scale(.) # scaling of all variables using
the default method ("minmax")
species ~ . + scale(bio1 + bio2) # scaling of only bio1
and bio2 variables
species ~ . + scale(., method="center") # scaling all
using the method of "center"
```

Principle Component Analysis (PCA) PCA is a valuable technique for reducing the dimensionality of data and addressing issues of multicollinearity among predictor variables. In the context of species distribution modeling, PCA transforms the original numerical variables into a set of uncorrelated components, which can then be used as predictors in the model. This transformation helps in simplifying the model by retaining only the most significant components, thereby enhancing computational efficiency and

(continued)

Box 1 (continued)
model interpretability. The sdm R package facilitates the incorporation of PCA directly within the formula interface. By specifying a PCA term in the formula, users can automatically transform the numerical variables into principal components, selecting either the first *n* components or those that explain a specified portion of the total variance (e.g., 80%). The option of scaling variables in the PCA analysis ensures that each variable contributes equally to the analysis. This integration of PCA within the sdm package streamlines the workflow, allowing for more effective data preprocessing and robust model development:

The following line of code rescales and transforms all numeric variables using PCA and selects the first two generated components to be used as predictor variables in the SDM workflow:

```
species ~ pca(. , n=2 , scale = TRUE)
```

Alternatively, if n = "auto", the function identifies the best number of components which is the default behavior; that is, the following line of code is equivalent to species ~ pca(.):

```
species ~ pca(. , n= "auto", scale = TRUE)
```

Or you may specify the minimum percentage of variability explained by the selected components:

```
species ~ pca(. , n=0.8 , scale = TRUE)
```

By using the formula interface, users can leverage R's extensive statistical modeling capabilities with ease. This interface not only promotes consistency across different modeling functions but also enhances readability and reproducibility of the code. As such, the formula interface is an essential feature of R that empowers users to build, analyze, and interpret a wide array of statistical models efficiently.

The sdmdata object, generated by the `sdmData` function, is like a database containing all species and environmental records, as well as additional information (spatial coordinates, temporal dimension, grouping factors, coordinate reference system, metadata, etc.).

Let us create the `sdmdata` object given the species data (`SpatVector` object: `sp`) and the climate variables (`SpatRaster` object: `bioc`). In the species data (`sp`), we had the presence-only values (1) assigned to a column named `species`.

Backgrounds (Pseudo-Absences) Since the species dataset is presence-only, we need to generate background (pseudo-absence) records. Several methods are available to generate backgrounds (you may check the help page of the `background` function in the sdm package). For this purpose, we use the method of `gRandom` which generates background randomly in the geographic space. The setting to generate background records can be provided through the `bg` argument to the `sdmData` function using a list. In addition to the method, `n` (the number of background records) and some optional arguments can be provided (e.g., `bias` that is a raster file can be used for a target-based background generation).

```
library(sdm)
```

```
## sdm 1.2-52 (2024-10-23)
```

```
## the first step in the sdm package is to create the data object:
# ---> 500 background records will be generated using "gRandom":
d <- sdmData(species~.,train=sp,predictors=bioc,
bg=list(n=500,method='gRandom'))

# you may get a summary from the created object by executing the object:

d
```

```
## class                                 : sdmdata
## ===========================================================
## number of species                     :  1
## species names                         :  species
## number of features                    :  9
## feature names                         :  bio2, bio3, bio8, ...
## type                                  :  Presence-Background
## has independent test data?             :  FALSE
## number of records                     :  616
## has Coordinates?                      :  TRUE
```

Modeling and Evaluation] After creating the data object (`sdmdata`), the `sdm` function can be used to train and evaluate multiple models in parallel. More than 20 modeling methods are available in the package. The models can be extended by users through using the `add` function. The list of existing methods can be checked using the getmethodNames function:

```
getmethodNames(alt=F)
```

```
##  [1] "bioclim"       "bioclim.dismo" "brt"           "cart"
##  [5] "domain.dismo"  "fda"           "gam"           "glm"
##  [9] "glmnet"        "glmpoly"       "mahal.dismo"   "mars"
## [13] "maxent"        "maxlike"       "maxNet"        "mda"
## [17] "mlp"           "ranger"        "rbf"           "rf"
## [21] "rpart"         "svm"
```

Usage of `sdm`:

```
sdm(formula, data, methods,...)
```

In the sdm function, `formula` (to specify the structure of the model), `data` (the output of `sdmData`), and `methods` (to specify the names of modeling methods) are the minimum inputs to get the function works. Some additional inputs can be provided to specify settings for the `replication` methods which split data into training and test partitions for evaluating the models. The following subsection provides some details about model evaluation:

Replication To evaluate the models, if an independent dataset is available, they can be used to test the performance of models. If that is the case, the independent test data can be provided to the `sdmData` function. However, such independent data are not available in most situations. Therefore, an alternative solution might be used which is to split the data into two partitions using a resampling method. The resampling methods available in the sdm package include (i) "subsampling" (which uses ....), (ii) "bootstrapping" (which uses....), and (iii) "cross-validation" (which devides....) A user can specify one or more of these methods through the `replication` argument, and the procedure can be repeated multiple times if a user adds the `n` argument with the number of replications.

Here is an example of using the sdm function where we chose five modeling methods, also used subsampling with 30% of data for test dataset, and repeated the procedure three times:

```
## second step: to train and evaluate the models:

#----> 5 modelling methods are used:
# =======>> glmp: polynomial generalised linear model;
# =======>> brt: Boosted Regression Trees;
# =======>> rf: Random Foress;
# =======>> maxent: Maximum Entropy;
# =======>> svm: Support Vector Machine;

#----> Replication method: sub-sampling (with 30 % of test data; repeats
↪   3 times)

m <- sdm(species~., d, methods = c('glm','brt','rf','maxent','svm'),
         replication='sub', test.p=30, n = 3)


# to see a summary of the modelling outputs, you may check the object:

m
```

```
## class                                 : sdmModels
## ========================================================
## number of species                     :  1
## number of modelling methods           :  5
## names of modelling methods            :  glm, brt, rf, maxent, svm
## replicate.methods (data partitioning) :  subsampling
## number of replicates (each method)    :  3
```

```
## toral number of replicates per model  :  3 (per species)
## test percentage (in subsampling)      :  30
## ------------------------------------------
## model run success percentage (per species)  :
## ------------------------------------------
## method          species
## ----------------------
## glm        :         100   %
## brt        :         100   %
## rf         :         100   %
## maxent     :         100   %
## svm        :         100   %
##
## ###################################################################
## model Mean performance (per species), using test dataset (generated using
↪  partitioning):
## -------------------------------------------------------------------------------
##
##  ## species    :  species
## =========================
##
## methods    :      AUC     |     COR     |     TSS     |     Deviance
## -------------------------------------------------------------------------
## glm        :      0.87    |     0.58    |     0.62    |     0.65
## brt        :      0.93    |     0.74    |     0.73    |     0.59
## rf         :      0.96    |     0.82    |     0.83    |     0.36
## maxent     :      0.94    |     0.74    |     0.81    |     0.47
## svm        :      0.95    |     0.77    |     0.84    |     0.42
```

As you can see above, in the summary information provided from the `sdmModels` object, three models were trained for each modeling method (15 models in total), and they were evaluated using multiple accuracy metrics (e.g., AUC, TSS). Mean values of these metrics can provide a quick way to compare the performance of different methods.

There is a user-friendly and interactive way to explore part of the results available within the sdmModels object through using the `gui` function. This function opens a graphical user interface with different parts (windows/tabs) which allows a user to explore the results (evaluation, variable importance, etc.). You may simply use the following line of code to use the function:

```
gui(m)
```

Box 2 Model Evaluations for SDMs

Evaluating the performance of models is a crucial step in species distribution modeling (SDM) to ensure the accuracy and reliability of predictions. Several evaluation metrics are commonly used to assess how well a model predicts the occurrence of species based on known data. Here are some of the most widely used methods:

(continued)

Box 2 (continued)

Confusion Matrix and Threshold-Based Metrics

The confusion matrix is a tool used to evaluate the performance of a models when the response variable is categorical (or classes such as presence and absence). The confusion matrix tool allows to compare the observed values (presence/absences) with the predicted presence and absence. Since SDMs are usually generating probability of occurrences (or habitat suitability values) rather than presence/absence values, a threshold is needed to first convert the predicted probabilities to presence/absence, then make the confusion matrix, and calculate the accuracy metrics. Therefore, the metrics generated based on a confusion matrix is also called **threshold-based** metrics. The matrix is composed of four key components (Fig. 3):

Description of the Metrics:

- *True Positives (TPs)*: The number of observed presences that are correctly predicted as presence by the model (i.e., the model correctly identifies the presence of the species)
- *True Negatives (TNs)*: The number of observed absences that are correctly predicted as absence by the model (i.e., the model correctly identifies where the species is absent)
- *False Negatives (FNs)*: The number of observed presences that are incorrectly predicted as absence (i.e., missed presences)
- *False Positives (FPs)*: The number of absences where the model incorrectly predicts presence

Key Metrics Derived from the Confusion Matrix:

- *Accuracy*:

$$\text{Accuracy} = \frac{TP + TN}{TP + TN + FP + FN}$$

Accuracy measures the overall correctness of the model by calculating the proportion of correctly classified instances (both presences and absences) out of all predictions.

- *Sensitivity (True Positive Rate or Recall)*:

$$\text{Sensitivity} = \frac{TP}{TP + FN}$$

(continued)

Box 2 (continued)

Sensitivity evaluates how well the model identifies actual presences (true positives). A higher sensitivity indicates fewer missed presences.

- *Specificity (True Negative Rate)*:

$$\text{Specificity} = \frac{TN}{TN + FP}$$

Specificity assesses how well the model identifies actual absences (true negatives). A higher specificity means fewer false positives (wrongly predicted presences).

- *Precision or ppv (Positive Predictive Value)*:

$$\text{Precision} = \frac{TP}{TP + FP}$$

Precision measures the proportion of predicted presences that are true presences. A higher precision means that when the model predicts presence, it is more likely to be correct.

- *F1 Score*:

$$\text{F1 Score} = 2 \times \frac{\text{Precision} \times \text{Recall}}{\text{Precision} + \text{Recall}}$$

The **F1** score is the harmonic mean of precision and recall, giving a balanced measure of the model's accuracy when there is a trade-off between false positives and false negatives.

- *Matthews Correlation Coefficient (MCC)*:

$$\text{MCC} = \frac{(TP \times TN) - (FP \times FN)}{\sqrt{(TP + FP)(TP + FN)(TN + FP)(TN + FN)}}$$

The **MCC** is a more balanced metric that considers all elements of the confusion matrix (TP, TN, FP, and FN). It is particularly useful for imbalanced datasets, as it captures the relationship between true positives, true negatives, false positives, and false negatives. MCC ranges from -1 to 1, where 1 indicates perfect prediction, 0 means no better than random, and

(continued)

Box 2 (continued)

−1 means the total disagreement between predictions and observations.

- *True Skill Statistic or TSS*:

$$\text{TSS} = \text{Sensitivity} + \text{Specificity} - 1$$

TSS is a widely used metric which evaluates model performance by combining sensitivity and specificity into a single measure. TSS ranges from -1 to 1, where 1 indicates perfect prediction, 0 indicates no better than random, and negative values indicate performance worse than random.

Threshold Optimization:

As mentioned, predictions from SDMs (those that are developed using presence-absence data) are in the form of probability (which may be interpreted as habitat suitability). Since probability values are not comparable with observed presence/absence, a threshold is needed to convert the probability values to presence-absence. Then, the question is that what the best threshold is! A threshold optimization method may be used to find the best threshold given some criteria available for this purpose. In the sdm R package, 15 different criteria are available (it can be specified in different functions through using the "opt" argument; check the help page of the "evaluates" function for more information). Here is a list of some widely used methods:

For instance, we may find the threshold under which [sensitivity + specificity] is maximized (which is known as max[Se+Sp]).

Threshold optimization is crucial when converting continuous model predictions (e.g., probabilities of species presence) into binary classifications (presence/absence). Several approaches can be used for threshold optimization:

(1) *Se = Sp*: This is one of the commonly used approaches for threshold optimization, and it finds the threshold where the sensitivity and specificity are equal. In the sdm package, this threshold can be used by setting "opt = 1."
(2) *max[Se + Sp]*: This is probably the most widely used approach, and it is the threshold that maximizes [sensitivity + specificity] which is equivalent to maximizing TSS. In the sdm package, this threshold can be used by setting "opt = 2" (default).
(3) *P10, P5, P1, and P0*: These approaches select the threshold based on percentiles of the predicted probabilities for presence records in the evaluation dataset. P10, P5, P1, and P0 correspond to the 10th, 5th, 1st, and 0th percentiles, respectively.

(continued)

Box 2 (continued)

Threshold-Independent Metrics

The metrics mentioned so far are all threshold-based metrics, relying on finding the best threshold to first convert probability values to presence/absence, and then creating a confusion matrix based on which multiple metrics can be calculated. Therefore, depending on which optimization approach is used, the metrics can vary. There is another group of metrics available to measure the performance of SDMs, with the advantage that there is no need to select a threshold. These threshold-independent metrics are often preferred when the goal is to evaluate the overall predictive ability of the model, regardless of specific classification cutoffs to convert probabilities to presence/absence.

- *Area Under the Receiver Operating Characteristic Curve (AUC)*:
  AUC (area under the ROC curve) is probably the most widely used metric that measures the ability of SDMs to discriminate between presence and absence locations. AUC ranges between 0 and 1, with a value of 0.5 indicating a performance equivalent to random guessing, while values closer to 1 indicate better discriminative ability. AUC is calculated based on the ROC which plots the true positive rate (sensitivity) against the false positive rate (1-specificity) across all possible threshold values. AUC summarizes the curve into a single value, providing an overall assessment of model performance across thresholds (Fig. 4).
- *Correlation (COR)*:
  Correlation is another threshold-independent metric that assesses the agreement between predicted and observed values. The most common measure used is Pearson's correlation coefficient, which ranges from -1 to 1, with 1 showing a perfect positive correlation, 0 showing no correlation, and -1 representing perfect negative correlation.
- *Deviance*:
  This metric represents the average deviation of model predictions from observations; therefore, a smaller value of it represents a better model (which can be interpreted as smaller error).
- *Continuous Boyce Index (CBI)*:
  The Continuous Boyce Index (CBI) is an evaluation metric specifically designed for **presence-only** test dataset. It assesses how much better the predicted model is compared to a random model. CBI ranges from -1 to 1, where values close to 1 indicate a model with high predictive ability, 0 indicates no better than random, and negative values suggest the model is predicting worse than random.
- *Calibration*:
  Calibration measures reliability of predicted probabilities related to observed species prevalence or proportion of presences across the range of

(continued)

Box 2 (continued)

probabilities predicted by a model. If predictions are perfectly calibrated, we expect that the estimated probabilities represent the actual prevalence of data. For instance, across sites with a predicted probability of 0.7, we expect 70% of observations to be presences and 30% of them to be absences. The calibration metric is a novel method implemented in the sdm package to quantify the level of calibration, which ranges between 0 and 1. A value of 1 refers to a perfect calibration, and a value of 0.5 refers to the perfect discrimination (all observations greater than a threshold are presence and below a threshold are absence). A negative value may also be generated for rare or not-likely situations such as generating opposite predictions (i.e., high probabilities for absences and low probabilities for presences). The following figure shows a realistic calibration plot (on the left) with some extreme examples to show the range of calibration metrics (small plots on the right) (Fig. 5).

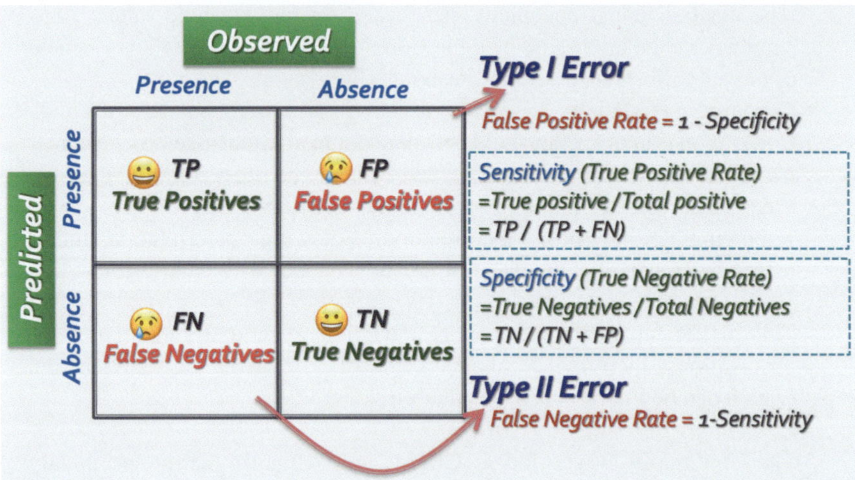

Fig. 3 Confusion matrix compares observed presence and absence records with predicted presence and absence records to measure multiple metrics representing accuracy of a model (e.g., sensitivity and specificity)

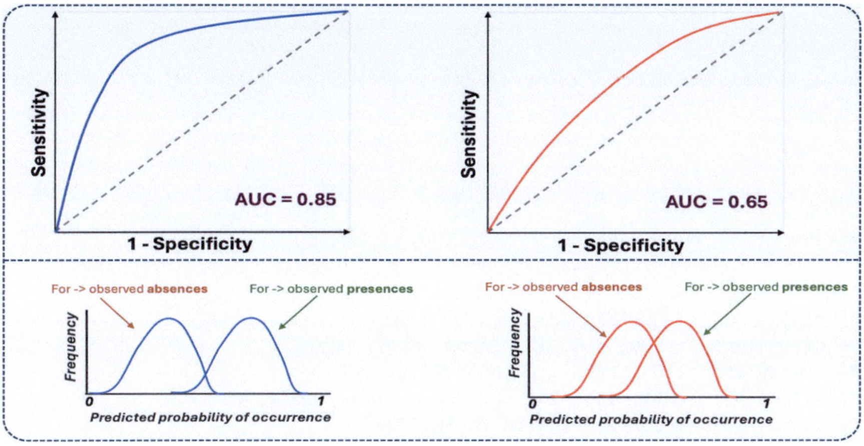

Fig. 4 Two examples of ROC curve with corresponding AUC (area under the ROC curve) values. AUC is a metric assessing discriminative capacity of a model. The lower panel shows the density plots corresponding to the ROC curves in the upper panel, representing the distribution of predicted values across presence and absence locations in an evaluation (test) dataset. When AUC is greater, the model has greater power in discriminating presence locations from absence locations

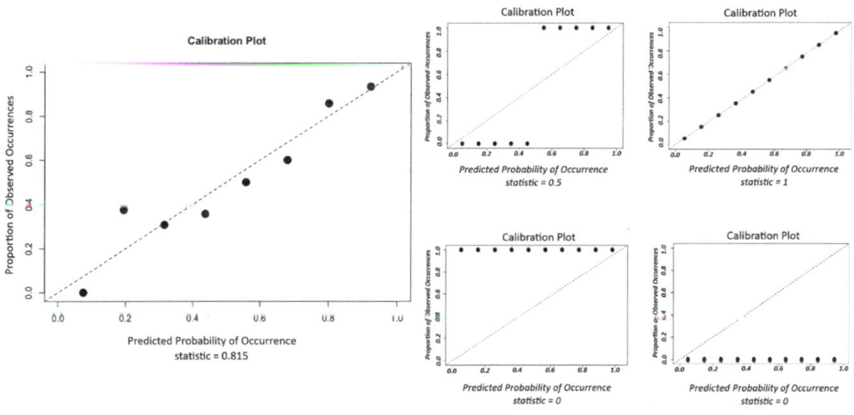

Fig. 5 The calibration plot with a novel metric developed in the sdm package representing the degree of calibration. The practical range is between 0.5 and 1

(iv) Predict/Project Maps of Habitat Suitability in the Current and Future Times

After the models were trained and evaluated successfully, we can now use them to predict the probability of occurrence (it may be interpreted as the degree of habitat suitability) for each pixel across the study area. Here, the aim is to assess the geographic distribution of the species globally; therefore, we use the non-cropped predictor variables (`bio`) to predict the values in the current time:

```
## Predict habitat suitability of Snow leopard in the current time:

p <- predict(m, bioc)

# since we had 15 models in the `m` object, `p` is a SpatRaster with 15
↪  layers
# corresponding to the 15 models:

p
```

```
## class       : SpatRaster
## dimensions  : 300, 360, 15  (nrow, ncol, nlyr)
## resolution  : 0.1666667, 0.1666667  (x, y)
## extent      : 60, 120, 10, 60  (xmin, xmax, ymin, ymax)
## coord. ref. : lon/lat WGS 84 (EPSG:4326)
## source(s)   : memory
## names       :  id_1_~_subs,  id_2_~_subs,  id_3_~_subs, id_4_~_subs,
↪  id_5_~_subs, id_6_~_subs, ...
## min values  : 2.220446e-16, 2.220446e-16, 1.079028e-13,  0.08533683,
↪  0.08229015,  0.08046598, ...
## max values  : 1.000000e+00, 1.000000e+00, 1.000000e+00,  0.72582535,
↪  0.72132055,  0.71911020, ...
```

```
# We have 5 modelling methods, and 3 models for each method (15 in
↪  total),
# the first 3 layers are for the first method ("glmp"), the secord 3 (4,
↪  5, and 6)
# are related to the second method ("brt"), and so on.

# let's visualise one of the outputs for each method:

# appropriate color schemes:
cl <- colorRampPalette(c('gray','orange','yellow','green','blue'))

## rasterVis for visualisation:
# So far, we used the simple plot function to visualise rasters,
# but with rasterVis, you may generate nicer plots:

library(rasterVis)
```

```
## Loading required package: lattice
```

```
levelplot(p[[c(1,4,7,10,13)]],col.regions=cl(200),margin=F)
```

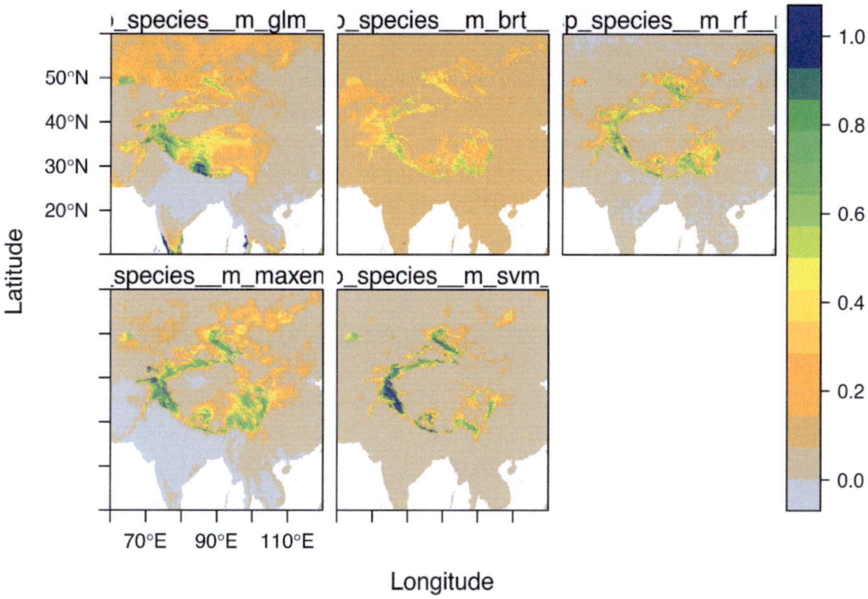

Ensemble Forecasting Employing multiple modeling methods for species distribution modeling which is usually applied to several replications of data (e.g., in the example above, we produced 15 models) often results in varying predictions. While these models may all perform well according to evaluation metrics (e.g., AUC), their predictions can still differ significantly. These variations in outcomes of models are referred to model-based uncertainty, which is one of the key sources of uncertainty affecting SDMs.

To address this uncertainty, an effective approach is ensemble forecasting, where the predictions from multiple models are combined into a single, consensus prediction (Araujo and New 2007). Ensemble forecasting helps to smooth out the variability in individual models, providing a more robust and reliable prediction of species distribution. According to Araujo and New (2007), ensemble methods can improve the accuracy and reliability of SDMs by leveraging the strengths of different models while minimizing their individual weaknesses.

Ensemble Forecasting Methods In the sdm package, the `ensemble` function can be used for generating the consensus prediction that supports multiple methods for combining the predictions. Several approaches are available to combine predictions from multiple models in the `ensemble` function (the method is specified within a list passed through the `setting` argument:

- **Mean** (simple averaging): This is the simplest method, where the predictions from all models are averaged to produce the final consensus prediction. Each model contributes equally to the final result. To use this method, `method =`

'mean' or method = 'unweighted' can be used in the setting argument of the ensemble function.

- **Weighted Mean**: In this method, models with better performance (as determined by evaluation metrics such as AUC or TSS) are given more weight in the final prediction. For example, models that perform better based on AUC would contribute more significantly to the final ensemble prediction than models with lower AUC scores. Examples of settings in the ensemble function include setting = list(method = 'weighted', stat = 'auc'), which uses AUC values as weight, OR setting = list(method = 'weighted', stat = 'tss', opt=2), which uses TSS values as weight. Since TSS is a threshold-based metric, the opt argument is also added which specifies which threshold optimization method should be used to select the best threshold for calculating TSS. Fifteen optimization methods are available in the sdm package (check the help page of the evaluates function), and opt = 2 refers to the second method that selects threshold under which [sensitivity + specificity] is maximized (i.e., max[se+sp]).
- **Median**: This method takes the median of predictions from all models. To use this method (method = 'median').
- **Thresholding and stacking**: This method involves thresholding predictions to determine areas of presence or absence given the best threshold for each model specified using a threshold optimization method (the opt argument). The predicted presence-absence values from multiple models are stacked and averaged (method = 'pa'). Example: setting = list(method = 'pa', opt=2).
- **Two-step averaging**: When multiple predictions are generated for each modeling method (from multiple data replications), the averaging of predictions can be done in two steps, across replications for each modeling method and then across modeling methods. The available two-step methods include mean-weighted, mean-unweighted, median-weighted, and median-unweighted. For instance, using the method = 'mean-weighted' in the setting argument first takes the mean of different replications of each modeling methods; then they are combined through using a weighted-mean function.
- **Assess variability in predictions**: In addition to generating a consensus prediction, several methods are available to explore the variability of predictions for each location. These methods include coefficient of variation (method = 'cv'), standard deviation (method = 'stdev'), and confidence interval (method = 'ci'; this generates the range of confidence interval which is "upper range–lower range") and uncertainty (method = 'uncertainty'). The uncertainty method can characterize model-based uncertainty or inconsistency among predictions of different models. An entropy metric is used to characterize uncertainty, ranging between 0 and 1. The value of 1 at a certain location refers to maximum inconsistency which is the case when half of the models predict "presence" for the location and the other half predict "absence."

The abovementioned methods can be employed together at the same time.

The following are examples of how to generate an ensemble prediction using the `ensemble()` function in the sdm package. Please **note** that the `ensemble` function *first calls the "predict" function to generate predictions* and then use an ensemble method to combine them into a consensus map. However, if the predictions are generated beforehand, the output of the predict function can be introduced as the second argument to the ensemble function (in our case, `p`); otherwise, the predictor object (in our case, `bioc`) should be used as the second argument:

```
## Ensemble:
#----> first argument: the model (output of the sdm function)
#---->  second argument: either the output of the predict function (if
↪  available)
#---------------------- OR the predictors object (bioc; )
#----> third argument (setting): a list with settings for ensemble
↪  procedure

### We first try 3 methods and compare them in a plot:

## (i) Weighted-mean using values of AUC as weights
en1 <- ensemble(m, p, setting = list(method='weighted', stat= 'auc'))

#----
## (ii) Weighted-mean using values of TSS as weights
# opt=2 refers to the threshold obtained using max[Sensitivity +
↪  Specificity]

en2 <- ensemble(m, p, setting = list(method='weighted', stat= 'tss',
↪  opt=2))

## (iii) PA: thresholding and stacking:
en3 <- ensemble(m, p, setting = list(method='pa', opt=2))

#------
## Visualising the outputs of 3 different ensembles:


en <- c(en1,en2,en3) # 3 objects stacked in a single object
names(en) <- c('ens_AUC','ens_TSS','ens_PA')

levelplot(en,col.regions=cl(200),margin=F,
          main=c('Weighted-Mean (AUC)','Weighted-Mean
          ↪  (TSS)','Thresholding (PA)'))
```

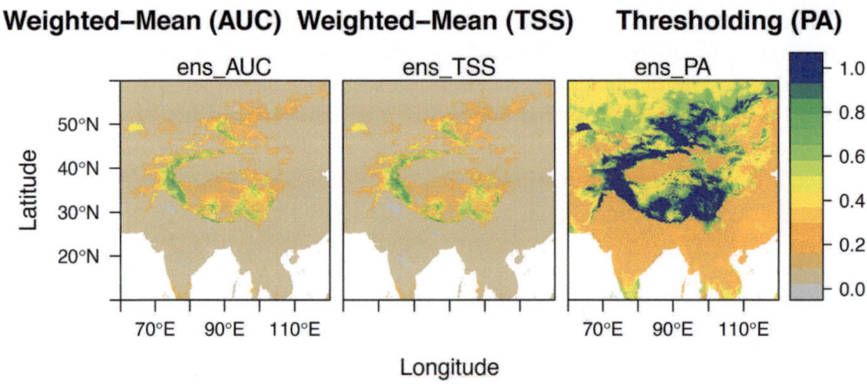


```
#-------------------------------
## Assessing variability among predictions:

en4 <- ensemble(m, p, setting = list(method=c('mean','ci'), opt=2))

levelplot(en4,col.regions=cl(200),margin=F,
          main=c('Mean','Range of Confidence-interval'))
```

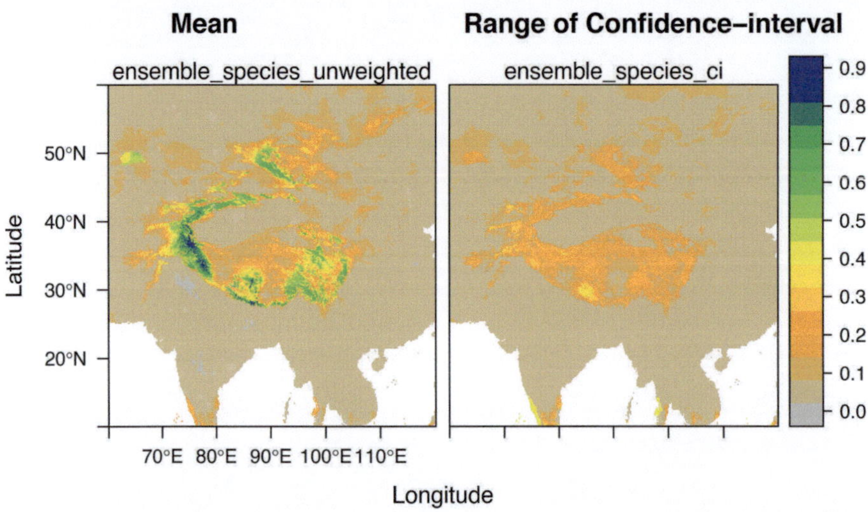


```
## to assess the upper and lower limits of confidence intervals:
ci.lower <- en4[[1]] - (en4[[2]] / 2)
ci.upper <- en4[[1]] + (en4[[2]] / 2)
ci <- c(ci.lower,ci.upper)
names(ci) <- c('ci_lower','ci_upper')

levelplot(ci,margin=F,
          main=c('CI.lower (alpha = 0.05)','CI.upper (alpha = 0.05)'))
```

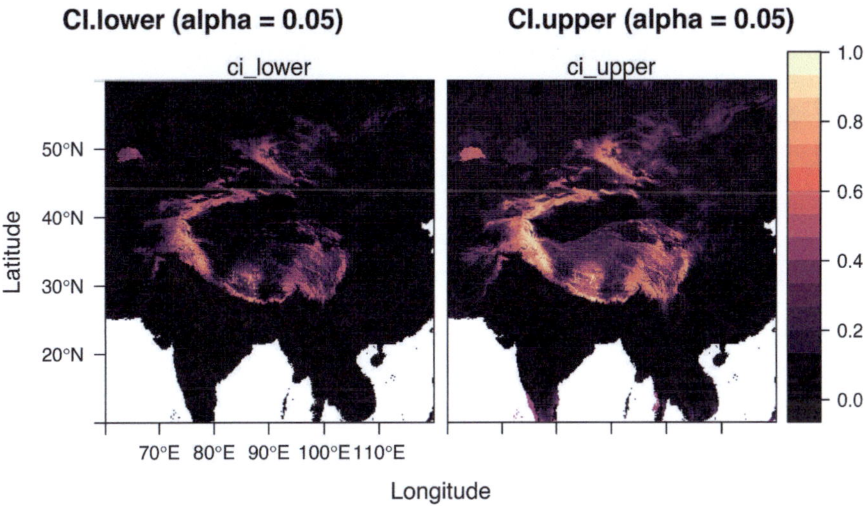


```
#----------
## Uncertainty
en5 <- ensemble(m, p, setting = list(method='uncertainty', opt=2))

levelplot(en5, margin=F, main='Model-based Uncertainty')
```

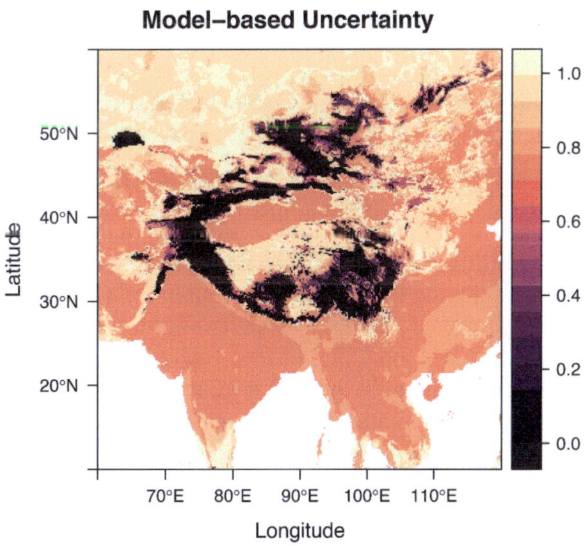

As you can see above, a single map is generated by combining the 15 predicted rasters (output of the predict function) using the ensemble method specified in the setting. There are other optional arguments available to control the ensemble procedure including:

- `id`: By default, predictions from all models are contributing to generate the consensus map unless a user specifies the `id` of certain predictions (based on their `modelIDs`). For example, adding `id = 1:10` to the `setting` list only uses the first ten predictions in the above example.
- `expr`: An expression can be specified by a user to control or limit which predictions should contribute into generating the consensus map. For example, using `expr = auc > 0.7` uses the specified condition to only include well-performing models (those with AUC > 0.7) contributing to the ensemble procedure.
- `power`: This argument is used to change weights by calculating `weights = weights ^ power`. A value of greater than 1 gives more emphasize on better performing models.
- Check the help page of the `ensemble` function in the `sdm` package for the other arguments and more details about the function.

(v) Range Shift Assessments in Response to Climate Change
In this section, the models are used to generate the spatial distribution of species globally across time (i.e., for both the current and future periods). SDMs can project potential changes in species' geographic ranges by incorporating climate change scenarios (e.g., different future climate scenarios or SSPs). These projections allow researchers to estimate how a species' suitable habitat might shift under changing climate conditions.

Key Considerations for Assessing Range Shifts:

(a) **Baseline Distribution vs. Future Projections**: The first step in assessing range shifts involves comparing the baseline distribution of a species (i.e., its current range) with its projected distribution under future climate scenarios. These projections are typically made for various future time periods, such as 2050 or 2100 (here in this chapter, we only do it for the year 2100), based on different climate models (GCMs) and emission scenario SSPs (Shared Socioeconomic Pathways). The difference between the two distributions highlights potential range expansions, contractions, or shifts.

(b) **Types of Range Shifts**:

- **Latitudinal Shifts**: Species may shift poleward in response to rising temperatures. For example, species in the northern hemisphere may move northward, while those in the southern hemisphere might shift southward.
- **Altitudinal Shifts**: As temperatures increase, species living in mountainous regions may move to higher elevations where cooler conditions prevail.
- **Range Contractions**: Some species might experience range contractions due to unsuitable environmental conditions, leading to local **extinctions**.
- **Range Expansions**: Conversely, some species may benefit from climate change, expanding (colonizing) into new areas that were previously too cold or otherwise unsuitable.

(c) **Metrics for Quantifying Range Shifts**:

- **Magnitude of Change**: The change in the degree of suitability between the future and current times can be simply calculated as a map by subtracting the current probability from the future one. Each pixel shows the magnitude of gains (positive values) and losses (negative values) in the future. Visualizing the map highlights areas where habitat suitability increases (gain), decreases (loss), or remains unchanged. This can help to visualize the areas most affected by climate change.

$$\begin{aligned}\text{Magnitude of Change} = &\ \text{Habitat Suitability in Future} \\ &- \text{Habitat Suitability in Current}\end{aligned}$$

- **Assessing Expansion/Contraction**: By converting the habitat suitability values in both times to presence/absence, given a threshold, and comparing the values at each pixel, a pixel in future will be expanded (colonized) when species will be presence in the future, while it is absence in the current time. Oppositely, the contraction (extinction) of a pixel in the future will be the case when species is presence in the current time, while it will become absence in the future.
- **Centroid Shift**: The geographic centroid of the species' range is calculated for both the current and future projections. The distance and direction of the shift in the centroid can provide insight into whether the species is moving poleward, altitudinally, or in another direction. To calculate distance, we can use a Euclidean distance function (the coordinates should be metric):

$$\text{Centroid shift} = \sqrt{(X_f - X_c)^2 + (Y_f - Y_c)^2}$$

where X_f, Y_f represent the coordinates of the future centroid, and X_c, Y_c represent the current centroid coordinates.

- **Range Overlap**: The degree of overlap between the current and future predicted range can be used to assess how much of the species' original habitat remains suitable under climate change. Low overlap indicates substantial changes in habitat suitability.

$$\text{Range overlap (\%)} = \frac{\text{Area of overlap}}{\text{Current range area}} \times 100$$

- **Range Area Change**: The change in the total area of suitable habitat is another key metric. This can be calculated by comparing the size of the species' predicted range in the future with its current range.

$$\text{Range area change (\%)} = \frac{\text{Future range area} - \text{Current range area}}{\text{Current range area}} \times 100$$

Assessing Range Shifts for our case study:

For our case study, we have already fitted SDMs across a defined study area, but now we want to use these models to predict the potential distribution of the species **globally** for both the current and future times. We have already downloaded bioclimatic data for both the current period and future projections (year 2100 under the SSP585 scenario, which represents a high greenhouse gas emission scenario). This will allow us to assess how the species' range might shift under a potential future climate.

Area of Applicability Given that the SDMs were initially fitted for a study area, which was a subset of the entire globe, predicting across the global scale can introduce new challenges. When applying these models globally, it is likely that some areas will exhibit environmental conditions that were not represented within the original study area. This phenomenon is referred to as environmental extrapolation. It can affect the reliability of predictions and should be carefully considered when interpreting the results. To handle this, we can identify and restrict our predictions to areas where the environmental conditions are within the area of applicability (AOA) of the model, meaning that the conditions are similar to those found within the training dataset. To calculate AOA, the `aoa` function in the sdm package can be used which generates a raster map with pixel values ranging between 0 and 1. A value of 1 means that the pixel is within the range used to fit the model. Deviating from 1 toward 0 refers to a more dissimilarity of the environmental condition of a pixel compared to the condition used to fit SDMs.

Let us follow the range shift assessment globally step by step:

```
## Predict spatial distribution for the current time (globally):

# the model object: m,
# the global bioclim dataset for the current time: bio
# the global bioclim dataset for the current time: biof

#---> we directly use the ensemble (no need to first run predict
↪  function):
#------> so, we use bio (predictors) as the second argument

hc <- ensemble(m, bio, setting=list(method="weighted",stat="auc"))
names(hc) <- 'Current'

## Project spatial distribution to the future
hf <- ensemble(m, biof, setting=list(method="weighted",stat="auc"))
names(hf) <- 'Future'

cl <- colorRampPalette(c('gray','orange','yellow','green','blue'))


levelplot(c(hc,hf),col.regions=cl(1000),margin=F)
```

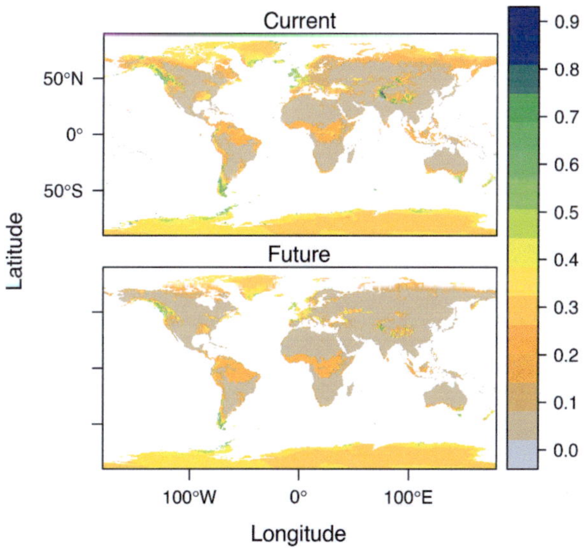


```
## Area of applicability:

# to identify the unreliable pixels (check the Area of Applicability
↪  section):
# The aoa function needs the following arguments:
# ----> x: the raster predictors
#-----> d: the output of sdmData (or sdmModels)
```

```
#-----> vi: optional variable importance of the predictor variables

#-- let's first extract variable importance from the variables:

# in the id argument, we may specify modelIDs or "ensemble":
vi <- getVarImp(m,id='ensemble')

plot(vi)
```

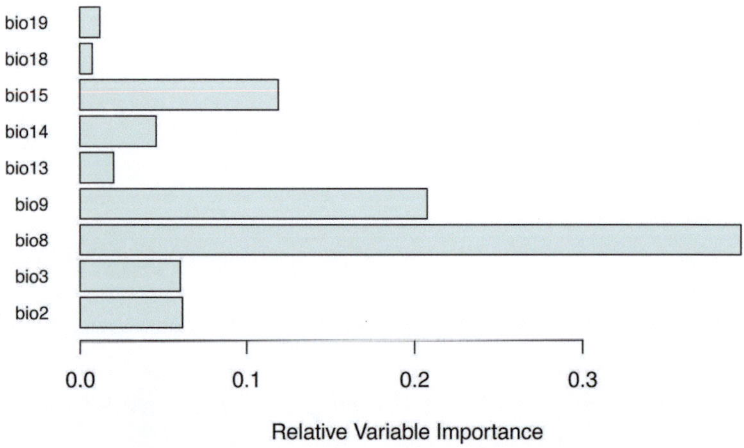


```
vi@varImportance$corTest
```

```
## [1] 0.0619 0.0606 0.3951 0.2075 0.0200 0.0461 0.1193 0.0073 0.0119
```

```
vimp <- vi@varImportance$corTest # get the vector of importances

ac <- aoa(bio, m, vi=vimp) # the output is the raster layer (current
↪  time)



af <- aoa(biof, m, vi=vimp) # for the future

# let's combine them into a single object for visualisation:
a <- c(ac,af)
names(a) <- c('AOA_current','AOA_future')

levelplot(a,margin=F,par.settings = RdBuTheme)
```

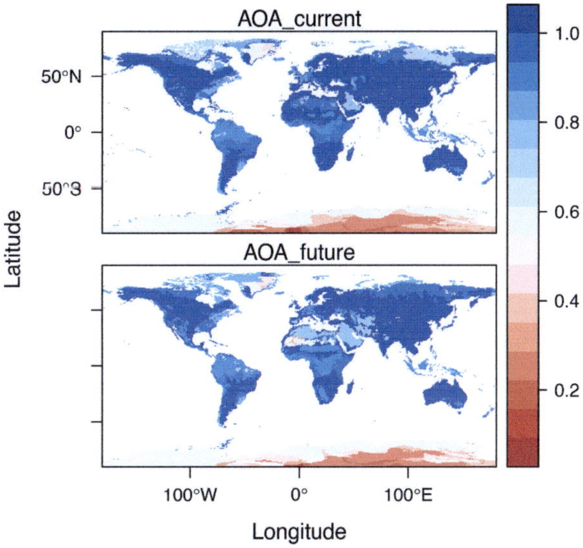


```
# the pixels less than 0.8 (or less than 1 to be conservative) are not
↪ reliable:

# so, let's make a mask:

acm <- ifel(ac >= 0.8,1, NA) # a mask to only keep reliable pixels
↪ (current time)
afm <- ifel(af >= 0.8,1, NA) # a mask to only keep reliable pixels
↪ (future time)

plot(acm,main='Mask of AoA for the current time')
```

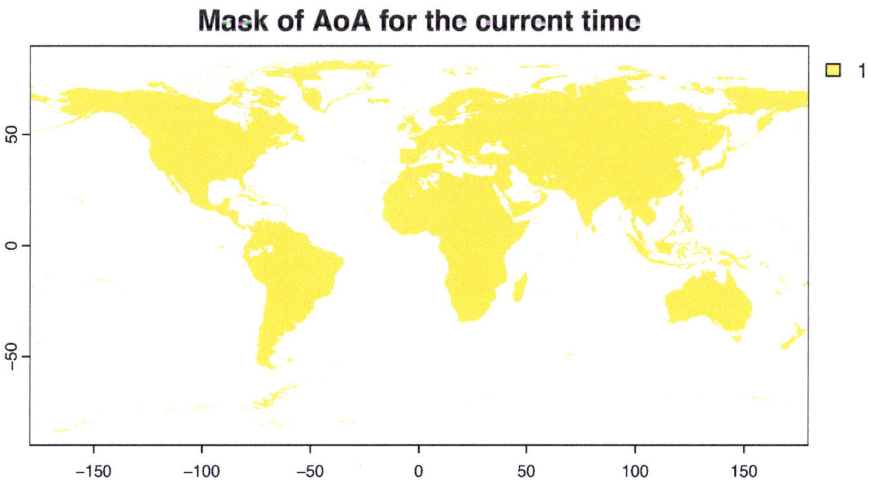

```
# let's adjust the predictions/projections (hc, and hf):

hc <- mask(hc, acm) # exclude unreliable pixels for the current time
hf <- mask(hf, afm) # exclude unreliable pixels for the current time

levelplot(c(hc,hf),col.regions=cl(1000),margin=F)
```

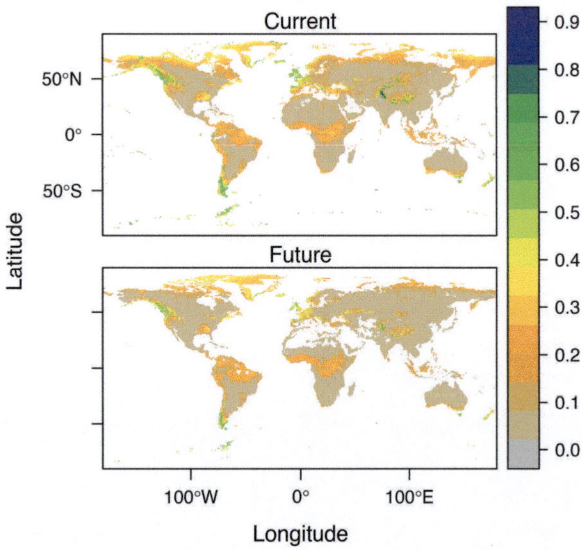


```
##---------------------------------------------------

## Assess Magnitude of changes:

# we can simply subtract current suitability from the future suitability:

ch <- hf - hc

cl2 <- colorRampPalette(c('black','darkred','red','orange',
                          'yellow','gray','blue','darkblue'))

levelplot(ch,col.regions=cl2(1000),margin=F,main='Magnitude of
↪  Gains/Losses')
```

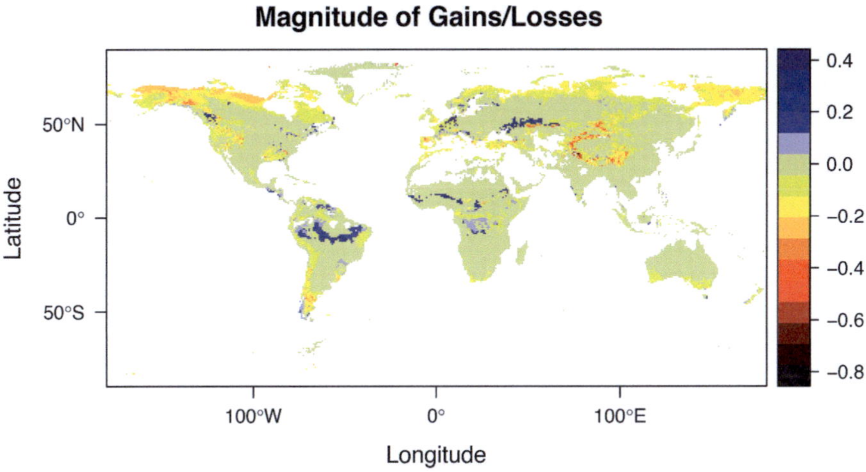


```
## Species range mapping and assess range shifts:

# to map ranges, we can convert the habitat suitabilities to
↪  presence/absences (P/A)
# the `pa` function can be used for this purpose:
# pa uses a threshold to convert suitability values to P/As

# to extract the threshold, we use the function of `threshold`.
#--- Arguments of getThreshold:
#------> x: sdmModels (m: output of the sdm function)
#------> id: From which model the threshold is extracted (it can be
↪  'ensemble')
#------> opt: which threshold optimisation method should be considered
#------> when id = "ensemble", we provide setting used for the ensemble
↪  function


th <- getThreshold(m,id='ensemble',opt=2,
setting=list(method='weighted',stat='auc'))

th # the threshold to convert probability values to P/As using the `pa`
↪  function:
```

```
## [1] 0.30561
```

```
#--- Arguments of pa:
#------> x: raster of habitat suitability
#------> y: threshold value

pac <- pa(hc, th) # P/As for the current time (baseline)

paf <- pa(hf, th) # P/As for the future time

#------------
```

```
# let's visualise them:

# defining a colorKey for the legend in map:
colKey <- list(at=c(0,0.5,1), ## where the colors change in legend
                labels=list(
                  at=c(0,1) ## where to print labels
                ))

levelplot(c(pac,paf),col.regions=c('gray','green'),
margin=F,colorkey=colKey)
```

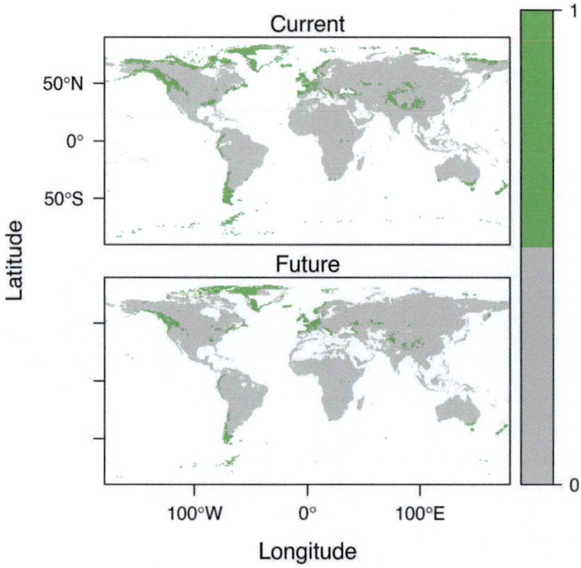


```
#----------------------

## Range shift metrics:

# by subtracting the current P/As from the future one, we can get the
↪  following
# values in the map:

#-------------------------
# Future - Current = Change
#-------------------------
#   1    -   0     = 1; Expansion (P in future; A in current)
#   1    -   1     = 0; No Change; persistence (P in future; P in
↪  current)
#   0    -   0     = 0; No Change; Unsuitable (A in; A in current)
#   0    -   1     = -1; Contraction or Extinction (A in future; P in
↪  current)

#--------------------------
```

```
# As you can see above, No change in suitable and unsuitable sites cannot
↪  be
# separated, a solution can be to change the 1/0 values in the future
↪  layer to
# a different value, e.g., 100/0 (so, presence sites in the future layer
↪  has a
# value of 100)! To do so, we can simply multiply paf to 100:

paf2 <- paf * 100

# Now:
#--------------------------
# Future - Current = Change
#--------------------------
#   100  -  0   = 100; Expansion (P in future; A in current)
#   100  -  1   = 99; No Change; Refugia (P in future; P in current)
#   0    -  0   = 0; No Change; Unsuitable (A in; A in current)
#   0    -  1   = -1; Contraction or Extinction (A in future; P in
↪  current)

#---------------------------
# so, we have 4 possible values representing range shifts

pach <- paf2 - pac


cls <- data.frame(id=c(-1,0,99,100),
                  range=c('Extinction', 'Unsuitable_NoChange',
                          'Refugia','Expansion'))

levels(pach) <- cls


levelplot(pach,att='range',
col.regions=c('red','gray','green','blue'),margin=F)
```

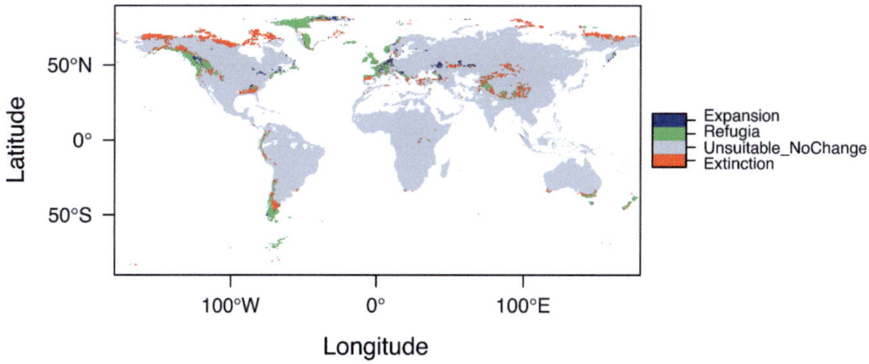

```
# Refugia is important for biodiversity conservation as it indicates
↪  potential
# locations for species to persist amid climate change, These locations
↪  enhance
# the likelihood of survival for species during climate shifts.

##-----------------------------------------

## Metrics of Range Shift assessment:
##-------------------------------------

# Here, we quantitatively assess range shifts:

## Area of Contraction (Extinction) and Expansion (Colonization):
# to calculate the area of the pixels representing either of the
↪  contraction
# or expansion classes, we first need to identify the pixels of these
↪  classes

# By multiplying the number of pixels of each class to the area/size of a
↪  pixel,
# we may roughly estimate the area, however, since our dataset uses a
↪  lat/lon
# coordinate reference system, the area of pixels are varying across
↪  latitudes,
# therefore, we may first calculate the area of each pixel, and then use
↪  the areas
# to calculate the size of contraction or expansion:

# we can use the cellSize function to measure the size of pixels:
area <- cellSize(pach, unit='km') # a raster with areas assigned to
↪  pixels

# which cells belong to the class of contraction?
# in pach, the value of -1 represents contraction (extinction):

w1 <- which(pach[] == -1) # which cells in pach has the value of -1
↪  (extinction)?

head(w1) # w1 is a vector with cell numbers of pixels for the class of
↪  extinction
```

```
## [1] 82972 82973 82974 82975 82976 82977
```

```
sum(area[w1]) # Area of Extinction (km2) in the year 2100 based on SSP5
↪  scenario!
```

```
## [1] 5398468
```

```
w2 <- which(pach[] == 100) # which cells in pach has the value of 100
↪   (expansion)?

sum(area[w2]) # Area of Expansion (km2) in the year 2100 based on SSP5
↪   scenario!
```

```
## [1] 947459
```

```
##############################
## Calculate centroid shift:

# to calculate centroid shift, we need to first calculate Xf,Xc, Yf, and
↪   Yc which
# are the mean X and Y (longitudes and latitudes) across pixels of
↪   species range
# in the Current (c) and Future times (f)

# Let's first identify cells of species range for the current and future
↪   time

# which pixels in pac (presence/absence of current time) is equal to 1?
Wc <- which(pac[] == 1) # pixel numbers of species range in the current
↪   time
Wf <- which(paf[] == 1) # pixel number of species range in the future
↪   time

XYc <- xyFromCell(pac, Wc)# X and Y coordinates of the current time
↪   across Wc cells
XYf <- xyFromCell(paf, Wf)# X and Y coordinates of the future time across
↪   Wf cells


head(XYc) # let's see how they look like
```

```
##              x        y
## [1,] -38.91667 83.58333
## [2,] -38.75000 83.58333
## [3,] -38.58333 83.58333
## [4,] -38.41667 83.58333
## [5,] -37.58333 83.58333
## [6,] -37.41667 83.58333
```

```
Xc <- mean(XYc[,1],na.rm=T) # mean of the X column
Yc <- mean(XYc[,2],na.rm=T) # mean of the Y column

Xf <- mean(XYf[,1],na.rm=T)
Yf <- mean(XYf[,2],na.rm=T)

Xc
```

```
## [1] -27.3959
```

```
Yc
```

```
## [1] 41.37305
```

```
Xf
```

```
## [1] -37.26772
```

```
Yf
```

```
## [1] 44.43572
```

```
# We can calculate Euclidean distance to measure the shift distance
↪   between the
# future and  current times. However, since the coordinate reference
↪   system (CRS)
# of our layers is geographic, the coordinates of the centroids are also
# in the unit of decimal degrees (based on latlon CRS). To calculate
↪   distance in
# meter, we can use the distGeo function from the package of geosphere
# to calculate metric distance between two points when they are in
↪   geographic
# coordinates.

# To do so, first let's load the geosphere package (install it if you
↪   don't have it)
library(geosphere)

# distance between two points (centroids) based on their coordinates:
dis <- distGeo(c(Xc,Yc),c(Xf,Yf)) / 1000 # divided by 1000 to get
↪   distance in KM


dis
```

```
## [1] 874.3126
```

```
## Direction of change:

# we can use the bearing (from the geosphere package) function to get the
↪
# azimut (direction) from the centroid of current time to the future:


azimuth_deg <- bearing(c(Xc,Yc),c(Xf,Yf))

# we can convert it to a range between 0 to 360 to get
# a clockwise direction--> 0: north; 90: west, 180: south; 270: west;
↪   360: north
```

```
if (azimuth_deg < 0) azimuth_deg <- azimuth_deg + 360


print(paste("Centroid shift distance:", round(dis, 2), "km"))
```

```
## [1] "Centroid shift distance: 871.31 km"
```

```
print(paste("Direction of shift:", round(azimuth_deg, 2), "degrees"))
```

```
## [1] "Direction of shift: 296.19 degrees"
```

```
## Latitudinal Shift and
##----------------------
# we can simply measure the changes between the mean latitude of future
↪   (Yf)
# and the mean latitude of the current time (Yc) to get an idea about the
↪
# latitudinal shift:

latitudinal_shift <- Yf - Yc

if (latitudinal_shift > 0) {
    direction <- "northward"
} else if (latitudinal_shift < 0) {
    direction <- "southward"
} else {
    direction <- "no change"
}

cat("The latitudinal shift is", abs(latitudinal_shift), "degrees,",
↪   direction, ".\n")
```

```
## The latitudinal shift is 3.062674 degrees, northward .
```

```
## Range Area Change:
##--------------------

# Above we extracted the pixel number of species range in both times (Wc
↪   and Wf)
# so, we can calculate the area of species range in both times which can
↪   be used
# to calculate another metric of range shift assessment: "range area
↪   change"

areaC <- sum(area[Wc]) # area of the species range in current time
areaC
```

```
## [1] 12640913
```

```
areaF <- sum(area[Wf]) # area of the species range in Future
areaF
```

```
## [1] 6910517
```

```
rangeCh <- ((areaF - areaC) / areaC) * 100
rangeCh # percentage of range change
```

```
## [1] -45.33214
```

```
# as you can see, almost 26% contraction will be happening to species
↪   range in
# response to climate change (given the worst case scenario: SSP585)

## Area of Overlap:
##-----------------
# Overlap refers to regions that are suitable in both times (refugia),
# We already mapped these areas in pach (value of 99), so let's get their
↪   pixel
# number:

wr <- which(pach[] == 99)

areaR = sum(area[wr]) # Area (km2) of Refugia
areaR
```

```
## [1] 5267669
```

```
ov <- (areaR / areaC) * 100

ov # Area of overlap (percentage)
```

```
## [1] 41.67158
```

2 Conclusion

In this chapter, we have explored the comprehensive workflow for Species Distribution Modeling (SDM), focusing on the practical application of these models to assess the impacts of climate change on the spatial distribution of species. Starting from downloading and preparing environmental and species occurrence data to developing robust predictive models and ultimately applying these models to analyze range shifts, we have highlighted the key methods, tools, and considerations needed to understand and predict how species respond to environmental changes.

One of the core applications of SDM discussed here is the ability to predict current and future distributions of species under different future climate change scenarios. By comparing these distributions, we can assess **potential range shifts**—a crucial step in understanding species vulnerability, adaptation capacity, and potential future habitat availability. These range shifts are often characterized by latitudinal and altitudinal movements, range contractions, and expansions. Through the case study, we demonstrated how to utilize SDM techniques to analyze these shifts, providing valuable insights into the direction and extent of movement, as well as the loss or gain of suitable habitats.

Moreover, the chapter underscored the importance of identifying the **Area of Applicability (AOA)** when projecting species distributions globally. This step is critical to ensure that predictions are accurate and minimize the risks of making extrapolations into regions with unfamiliar environmental conditions.

Throughout this chapter, we used the `sdm` R package (Naimi and Araújo 2016), a versatile tool that supports the modeling of SDMs and offers a wide range of functions for species distribution analysis. While we focused on specific aspects of the package, it has much more to explore, including advanced modeling options, evaluation metrics, and ensemble forecasting techniques. We encourage readers to refer to the standards and best practices for developing SDMs as outlined in Araújo et al. (2019), which provides comprehensive guidelines to ensure robust and reliable models (Araújo et al. 2019). To do so, other relevant packages are available in R to be used along with the `sdm` package to enhance the quality of the models. For instance, the `usdm` package (Naimi 2015) offers functions to understand the effects of positional uncertainty on SDMs (Naimi et al. 2011, 2014), `elsa` provides an entropy-based novel approach (Naimi et al. 2019), along with other widely used methods, to measure spatial autocorrelation in data, and `rasterdiv` offers tools to characterize ecosystem heterogeneity (Rocchini et al. 2021). Additionally, we recommend the `climetrics` R package (Taheri et al. 2024), which introduces functions to assess multiple dimensions of climate change. This package offers more possibilities to evaluate the impact of climate change on biodiversity (Taheri et al. 2016), enabling researchers to capture a broader perspective on how changing climatic patterns may affect species distributions (Lemes et al. 2022; González-Trujillo 2024; Ebrahimi et al. 2023).

Climate change represents one of the most significant drivers of biodiversity loss, and the application of SDMs offers a powerful approach to predict and mitigate its effects. Understanding potential range shifts enables conservationists, ecologists, and policymakers to make informed decisions about species management, habitat protection, and the development of wildlife corridors that enhance connectivity between fragmented habitats. By preparing for these shifts, we can better manage species' resilience and adaptability, reducing the likelihood of local extinctions.

Looking forward, there are several emerging techniques and future directions that can further enhance the utility of SDMs in climate change studies or for many other applications. These include the integration of biotic interactions (e.g., competition, predation, and mutualism), land-use change variables, and incorporating genomic data to assess adaptive potential. Additionally, advances in machine learning and artificial intelligence offer new ways to develop more sophisticated models that can handle complex and dynamic ecological processes.

In conclusion, species distribution modeling is a crucial tool in the fight against climate change and biodiversity loss. By accurately predicting the effects of environmental changes on species distributions, we can proactively implement strategies that conserve not only individual species but also entire ecosystems. The knowledge and techniques presented in this chapter provide a solid foundation for conducting effective SDM analyses and underscore the need for continued research, innovation, and collaboration across disciplines to address the multifaceted challenges that lie ahead.

The codes for this chapter is available in the following GitHub repositories: https://github.com/RCodingForEcology https://github.com/babaknaimi/sdm_BookChapter

References

Allouche O, Tsoar A, Kadmon R (2006) Assessing the accuracy of species distribution models: prevalence, kappa and the true skill statistic (TSS). J Appl Ecol 43(6):1223–1232

Araujo M, New M (2007) Ensemble forecasting of species distributions. Trends Ecol Evol 22(1):42–47. https://doi.org/10.1016/j.tree.2006.09.010

Araújo MB, Williams PH (2000) Selecting areas for species persistence using occurrence data. Biol Conservat 96(3):331–345

Araújo MB, Peterson AT (2012) Uses and misuses of bioclimatic envelope modeling. Ecology 93(7):1527–1539

Araújo MB, Anderson RP, Márcia Barbosa A, Beale CM, Dormann CF, Early R, Garcia RA et al (2019) Standards for distribution models in biodiversity assessments. Sci Adv 5(1):eaat4858. https://doi.org/10.1126/sciadv.aat4858

Araújo MB, Alagador D (2024) Expanding European protected areas through rewilding. Current Biol 34(17):3931–3940

Breiman L (2001) Random forests. Mach Learn 45:5–32

Busby JR (1991) BIOCLIM-A Bioclimate Analysis and Prediction System.In: Margules CR, Austin MP (eds) Nature Conservation: Cost Effective Biological Surveys and Data Analysis, CSIRO, Canberra, pp 64–68

Carpenter G, Gillison AN, Winter J (1993) DOMAIN: a flexible modelling procedure for mapping potential distributions of plants and animals. Biodivers Conserv 2:667–680

Ceballos G, Ehrlich PR, Dirzo R (2017) Biological annihilation via the ongoing sixth mass extinction signaled by vertebrate population losses and declines. Proc Natl Acad Sci 114(30):E6089–E6096. https://doi.org/10.1073/pnas.1704949114

Chauvenet ALM, Ewen JG, Armstrong DP, Blackburn TM, Pettorelli N (2013) Maximizing the success of assisted colonizations. Animal Conservat 16(2):161–169

Dirzo R, Young HS, Galetti M, Ceballos G, Isaac NJB, Collen B (2014) Defaunation in the Anthropocene. Science 345(6195):401–406. https://doi.org/10.1126/science.1251817

Dornelas M, Chase JM, Gotelli NJ, Magurran AE, McGill BJ, Antão LH, Blowes SA, Daskalova GN, Leung B et al (2023) Looking back on biodiversity change: lessons for the road ahead. Philosoph Trans R Soc B Biol Sci. https://doi.org/10.1098/rstb.2022.0199

Dormann CF, Elith J, Bacher S, Buchmann C, Carl G, Carré G, García Marquéz JR et al (2013) Collinearity: a review of methods to deal with it and a simulation study evaluating their performance. Ecography 36(1):27–46

Ebrahimi E, Ranjbaran Y, Sayahnia R, Ahmadzadeh F (2022) Assessing the climate change effects on the distribution pattern of the azerbaijan mountain newt (*Neurergus crocatus*). Ecol Compl 50:100997

Ebrahimi E, Araújo MB, Naimi B (2023) Flood susceptibility mapping to improve models of species distributions. Ecol Indicat 157:111250

Fielding AH, Bell JF (1997) A review of methods for the assessment of prediction errors in conservation presence/absence models. Environ Conservat 24(1):38–49

Friedman JH (2001) Greedy function approximation: a gradient boosting machine. Ann Stat 29:1189–1232

González-Trujillo DJ, Naimi B, Assis J, Araújo MB (2024) Reshuffling of Azorean Coastal Marine biodiversity amid climate change. J Biogeography 51:2546–2555

Guisan A, Zimmermann NE (2000) Predictive habitat distribution models in ecology. Ecol Model 135(2–3):147–186. https://doi.org/10.1016/S0304-3800(00)00354-9

Guisan A, Thuiller W (2005) Predicting species distribution: offering more than simple habitat models. Ecol Lett 8(9):993–1009. https://doi.org/10.1111/j.1461-0248.2005.00792.x

Hastie T, Tibshirani R, Buja A (1994) Flexible discriminant analysis by optimal scoring. J Am Stat Assoc 89(428):1255–1270

Hastie T, Tibshirani R (1996) Discriminant analysis by gaussian mixtures. J R Stat Soc Ser B Stat Methodol 58(1):155–176

Hof C, Araújo MB, Jetz W, Rahbek C (2011) Additive threats from pathogens, climate and land-use change for global amphibian diversity. Nature 480(7378):516–519

Hudak MJ (1992) RCE classifiers: theory and practice. Cybern Syst 23(5):483–515

Hughes AC (2023) The post-2020 global biodiversity framework: how did we get here, and where do we go next? Integrat Conservat 2:1–9. https://doi.org/10.1002/inc3.16

Kolbert E (2014) The Sixth Extinction: An Unnatural History. Henry Holt, New York. https://doi.org/10.1038/ngeo1895

Lemes P, Barbosa FG, Naimi B, Araújo MB (2022) Dispersal abilities favor commensalism in animal-plant interactions under climate change. Sci Total Environ 835:155157

McCullagh P (2019) Generalized Linear Models. Routledge, Milton Park

Naimi B (2015) USDM: Uncertainty Analysis for Species Distribution Models. R Package Version 1.1–15. R Documentation

Naimi B, Araújo MB (2016) Sdm: a reproducible and extensible R platform for species distribution modelling. Ecography 39(4):368–375. https://doi.org/10.1111/ecog.01881

Naimi B, Skidmore AK, Groen TA, Hamm NAS (2011) Spatial autocorrelation in predictors reduces the impact of positional uncertainty in occurrence data on species distribution modelling: spatial autocorrelation and positional uncertainty. J Biogeography 38(8):1497–1509. https://doi.org/10.1111/j.1365-2699.2011.02523.x

Naimi B, Hamm NAS, GroenTA, Skidmore AK, Toxopeus AG (2014) Where is positional uncertainty a problem for species distribution modelling? Ecography 37(2):191–203. https://doi.org/10.1111/j.1600-0587.2013.00205.x

Naimi B, Hamm NAS, Groen TA, Skidmore AK, Toxopeus AG, Alibakhshi S (2019) ELSA: entropy-based local indicator of spatial association. Spat Stat 29(March):66–88. https://doi.org/10.1016/j.spasta.2018.10.001

Naimi B, Capinha C, Ribeiro J, Rahbek C, Strubbe D, Reino L, Araújo MB (2022) Potential for invasion of traded birds under climate and land-cover change. Global Change Biol 28(19):5654–566. https://doi.org/10.1111/gcb.16310

Pecl GT, Araújo MB, Bell JD, Blanchard J, Bonebrake TC, Chen I-C, Clark TD et al (2017) Biodiversity redistribution under climate change: impacts on ecosystems and human well-being. Science 355(6332):eaai9214

Phillips SJ, Anderson RP, Schapire RE (2006) Maximum entropy modeling of species geographic distributions. Ecol Model 190(3–4):231–259. https://doi.org/10.1016/j.ecolmodel.2005.03.026

Phillips SJ, Anderson RP, Dudík M, Schapire RE, Blair ME (2017) Opening the black box: an open-source release of maxent. Ecography 40(7):887–893

Rocchini D, Thouverai E, Marcantonio M, Iannacito M, Da Re D, Torresani M, Bacaro G et al (2021) Rasterdiv—an information theory tailored R package for measuring ecosystem heterogeneity from space: to the origin and back. Methods Ecol Evol 12(6):1093–1102

Rocchini D, Tordoni E, Marchetto E, Marcantonio M, Barbosa AM, Bazzichetto M, Beierkuhnlein C et al (2023) A quixotic view of spatial bias in modelling the distribution of species and their diversity. NPJ Biodivers 2(1):10

Rosenblatt F (1958) The perceptron: a probabilistic model for information storage and organization in the brain. Psychol Rev 65(6):386

Royle JA, Chandler RB, Yackulic C, Nichols JD (2012) Likelihood analysis of species occurrence probability from presence-only data for modelling species distributions. Methods Ecol Evol 3(3):545–554

Taheri S, Naimi B, Araújo MB (2016) Did British breeding birds move north in the late 20th century?" Climate Change Respon 3:1–5

Taheri S, Naimi B, Rahbek C, Araújo MB (2021) Improvements in reports of species redistribution under climate change are required. Sci Adv 7(15):eabe1110

Taheri S, Naimi B, Araújo MB (2024) Climetrics: an R package to quantify multiple dimensions of climate change. Ecography 2024(8):e07176

Thorup K, Pedersen L, Da Fonseca RR, Naimi B, Nogués-Bravo D, Krapp M, Manica A et al (2021) Response of an Afro-Palearctic bird migrant to glaciation cycles. Proc Natl Acad Sci 118(52):e2023836118

Townsend PA, Soberón J, Pearson RG, Anderson RP, Martínez-Meyer E, Nakamura M, Araújo MB (2012) Ecological Niches and Geographic Distributions. Princeton University Press, Princeton. https://doi.org/doi:10.1515/9781400840670

Whittaker RJ, Araújo MB, Jepson P, Ladle RJ, Watson JEM, Willis KJ (2005) Conservation biogeography: assessment and prospect. Divers Distribut 11 (1):3–23

Wood SN (2017) Generalized Additive Models: An Introduction with R. Chapman Hall/CRC, Boca Raton

Zellmer AJ, Claisse JT, Williams CM, Schwab S, Pondella DJ (2019) Predicting optimal sites for ecosystem restoration using stacked-species distribution modeling. Front Marine Sci 6:3

The ecospat R Package: A Collection of Pre-, Core-, and Post-Modeling Tools to Investigate Species Niches and Distributions

Olivier Broennimann, Flavien Collart, and Antoine Guisan

Abstract

The R package *ecospat* is a collection of R functions and datasets designed to support spatial ecology analyses, with a focus on pre-, core-, and post-modeling analyses of species distributions (SDMs). Pre-modeling analyses include data extractions, designing sampling strategies, sub-sampling data, assessing spatial autocorrelation, reducing collinearity among predictors, measuring environmental similarity and analog/non-analog situations, assessing niche stability within a species or niche difference among species, and measuring species co-occurrences to investigate community assembly. Core-modeling analyses cover modeling fitting, spatial predictions and projections, and model evaluation. It includes in particular the fitting of SDMs with the "ensemble of small models" (ESMs) approach, which applies especially to species with small sample sizes (e.g., rare and endangered species). The post-modelling analyses comprise variance partitioning, determination of occupied patches and range size using the IUCN criteria, SDM prediction map binarization, and predictions of community properties from stacking individual species, including also functional and phylogenetic diversity. In this book section, we first illustrate two study cases where the ecospat package can be used: (a) modeling rare species distribution and quantifying species range size and (b) anticipating biological invasions via niche comparison. We then discuss other applications of the functions in the

O. Broennimann (✉) · A. Guisan
Department of Ecology and Evolution, University of Lausanne, Lausanne, Switzerland

Institute of Earth Surface Dynamics, University of Lausanne, Lausanne, Switzerland
e-mail: olivier.broennimann@unil.ch; antoine.guisan@unil.ch

F. Collart
Department of Biology, Ecology and Evolution, University of Liége, Liége, Belgium
e-mail: fcollart@uliege.be

D. Rocchini (ed.), *R Coding for Ecology*, Use R!,
https://doi.org/10.1007/978-3-031-99665-8_3

ecospat package and provide some perspectives about the future development of the package.

Keywords

Species distribution model · Niche theory · Ensemble of small models · Rare species · Model evaluation · Response curves · Variable contribution · Range size · Area of occupancy · Extent of occurrence · Niche dynamics · Niche quantification · Map binarization

1 Introduction

Species distribution modeling (SDM; Guisan et al. 2017; Peterson et al. 2011; Franklin 2010) is a field of ecology and biogeography that has grown significantly in the last three decades (Araújo et al. 2019), with important potential conservation applications (Guisan et al. 2013). SDMs work by relating statistically (or sometime in an expert-based way; Di Febbraro et al. 2018, Fourcade et al. 2013) species observations (simple occurrences, presence-absences, or abundances) to a set of environmental attributes (e.g., climate, topography, substrate, landscape, human influences) characterizing the sampling sites. By doing this through a set of locations in the studied region, SDMs implicitly quantify the suitability of the environment for the species (i.e., habitat or environmental suitability), and if done across the whole species' environmental range, it then quantifies the species' realized environmental niche (Araújo and Guisan 2006; Austin et al. 1990; Pulliam 2000; Soberón 2007), which can be used to make geographic projections, e.g., in the future, provided some assumptions are met (e.g., equilibrium, niche stability, no niche truncation, etc.; see Guisan et al. 2017; Zurell et al. 2020). Individual SDMs can be applied across a wide range of uses, including climate change assessments (e.g., Broennimann et al. 2006, Patiño et al. 2023), predicting biological invasions (e.g., Broennimann et al. 2007; Barbet-Massin et al. 2018), identifying suitable areas for translocation (e.g., Ferrarini et al. 2016), and managing rare species (Guisan et al. 2006; Jeliazkov et al. 2022). Individual species predictions can also be stacked across a whole taxonomic group (Dubuis et al. 2011; Kass et al. 2022) to predict community patterns, like richness, composition, and functional or phylogenetic diversity (D'Amen et al. 2015; Ferrier and Guisan 2006; Guisan and Rahbek 2011; Shen et al. 2023), and these multiple species predictions can be then be used to support spatial conservation planning (Moilanen et al. 2022). All these analyses and applications can now be performed, thanks to numerous software packages, most notably in the R framework (Sillero et al. 2023).

One R package to perform SDMs is *ecospat* (Di Cola et al. 2017; Broennimann et al. 2023, currently version 4.0.0), a collection of R functions and datasets designed to support spatial ecology analyses, with a focus on pre-, core-, and post-modeling analyses of species distributions (SDMs) (Table 1). Functions and tools within ecospat can be classified according to the three main groups of analyses that are

Table 1 List of the functions in the ecospat package

Category	Subcategory	Function	Description
Pre-modeling	Terrain sampling preparation	ecospat.rcls.grd	Reclassifies grid files to get a combined stratification from more than one grid
		ecospat.recstrat_prop	Performs a random ecologically stratified sampling proportional
		ecospat.recstrat_regl	Performs a random ecologically stratified sampling equal
	Spatial autocorrelation	ecospat.mantel.correlogram	Investigates spatial autocorrelation of environmental covariables for occurrences
	Variable selection	ecospat.npred	Calculates the maximum number of predictors to include in the model with a desired correlation between predictors
		ecospat.cor.plot	Plots correlations among predictors
	Extrapolation detection	ecospat.mess	Calculates the MESS (i.e., extrapolation) as in Maxent (Elith et al. 2010)
		ecospat.plot.mess	Plots the MESS extrapolation index onto the geographical space
		ecospat.climan	Assesses the climate analogy in a univariate and multivariate environmental space
	Niche quantification	ecospat.grid.clim dyn	Creates a grid z of Rx R pixels (or a vector of R pixels when using scores of dimension 1 or SDM predictions) with occurrence densities, using the scores of two axes from an ordination
		ecospat.niche.overlap	Calculates the overlap metrics D and I based on two species occurrence density grids z1 and z2 created by ecospat.grid.clim.dyn()
		ecospat.niche.equivalency.test	Runs a niche equivalency test (Warren et al. 2008) based on two species occurrence density grids
		ecospat.niche.similarity.test	Runs a niche similarity test (Warren et al. 2008) based on two species occurrence density grids
		ecospat.plot.niche	Plots a niche z created by ecospat.grid.clim.dyn
		ecospat.plot.niche.dyn	Plots niche categories and species density
		ecospat.plot.contrib	Plots the contribution of the initial variables to the analysis (i.e., correlation circle). Typically these are the eigenvectors and eigenvalues in ordinations

(continued)

Table 1 continued

Category	Subcategory	Function	Description
		ecospat.plot.overlap.test	Plots a histogram of observed and randomly simulated overlaps, with p-values of equivalency or similarity tests
		ecospat.niche.dyn.index	Calculates niche expansion, stability, and unfilling
		ecospat.shift.centroids	Draws arrows linking the centroid of the native and exotic (non-native) distribution (continuous line) and between native and invaded extent (dashed line)
		ecospat.Epred	Calculates the weighted mean of several predictions in dataframe format (ensemble models)
		ecospat.maxentvarimport	Computes variable importance for MAXENT models
	Community spatial predictions	ecospat.SESAM.prr	Implements the SESAM framework to predict community composition using a probability ranking rule
		ecospat.CCV.modeling	Creates probabilistic prediction for all species based on SDMs or ESMs
	Model evaluation	ecospat.cv.glm	Performs a K-fold and leave-one-out cross-validation for GLM
		ecospat.permut.glm	Performs permutations to get p-values on GLM coefficients and deviance
		ecospat.cv.gbm	Performs a K-fold and leave-one-out cross-validation for GBM
		ecospat.cv.me	Performs a K-fold and leave-one-out cross-validation for Maxent
		ecospat.cv.rf	Performs a K-fold and leave-one-out cross-validation for randomForest
		ecospat.boyce	Calculates the Boyce index (Hizel et al. 2006)
		ecospat.CommunityEval	Calculates several indices of accuracy of community predictions
		ecospat.adj.D2.glm	Computes the D2 metric in GLM
		ecospat.CCV.communityEvaluation.bin	Calculates a range of community evaluation metrics
		ecospat.CCV.communityEvaluation.prob	Generates a number of community evaluation metrics directly based on the probability
		ecospat.cohen.kappa	Calculates Kappa based on a given threshold
		ecospat.ESM.EnsembleEvaluation	Evaluates ensemble of small models via the pooling procedure (Collart and Guisan 2023)

		ecospat.max.kappa	Computes the maximum value of kappa
		ecospat.max.tss	Computes the maximum value of the true skill statistic
		ecospat.meva.table	Calculates evaluation metrics based on a given threshold
		ecospat.plot.kappa	Plots kappa value in function of the threshold value
		ecospat.plot.tss	Plot TSS values in function of the threshold value
		ecospat.poolingEvaluation	Uses the pooling evaluation to evaluate any models
Post-modeling	Variance partition	ecospat.varpart	Performs variance partitioning for binomial GLM based on the deviance of two groups or predicting variables
	Conservation	ecospat.rangesize	Quantifies the range size of a species using standard IUCN criteria
		ecospat.occupied.patch	Determines the occupied patch of a species using standard IUCN criteria (AOO, EOO) or predictive binary maps from species distribution models
	Other	ecospat.binary.model	Binarizes raster based on threshold
		ecospat.mpa	Calculates the minimum predicted area
Other functions	Co-occurrence analysis	ecospat.co-occurrence	Computes an index of co-occurrences ranging from 0 (never co-occurring) to 1 (always co-occurring)
		ecospat.cons_Cscore	Tests for nonrandom patterns of species co-occurrence and calculates the C-score index for the whole community and for each species pair
		ecospat.Cscore	Computes and tests if C-scores higher than null model
	Phylogenetic diversity measures	ecospat.calculate.pd	Calculates phylogenetic diversity measures (Schweiger et al. 2008)

Adapted from Di Cola et al. (2017)

typically perfomed when building SDMs (Di Cola et al. 2017): (i) pre-modeling analyses, consisting of data preparation and exploratory analyses; (ii) core-modeling analyses, corresponding to model fitting and evaluation; and (iii) post-modeling analyses, which consist of using the models to reach study goals such as combining single-species predictions to reconstruct communities (stacked SDMs; Guisan and Rahbek 2011; D'Amen et al. 2018b; Zurell et al. 2020).

In ecospat, pre-modeling analyses include data extractions, designing sampling strategies, sub-sampling data, assessing spatial autocorrelation, reducing collinearity among predictors, measuring environmental similarity and analog/non-analog situations (MESS analysis), assessing niche stability within a species or niche difference among species (using a large set of functions, a key development in ecospat), and measuring species co-occurrences to investigate community assembly. Core-modeling analyses typically cover modeling fitting, spatial predictions and projections, and model evaluation. In ecospat, it includes fitting SDMs with the "ensemble of small models" (ESMs) approach, which applies especially to species with small sample sizes (e.g., rare and endangered species) but also performs well for other species, and applying the different steps of the SESAM framework (Guisan and Rahbek 2011) for modeling the spatial distribution of species assemblages and communities. It also includes evaluations of individual species models through the computation of cross-validations and evaluation metrics (such as the Boyce index). The post-modelling analyses comprise variance partitioning, determination of occupied patches and range size using the IUCN criteria, SDM prediction map binarization, and predictions of community properties from stacking individual species (e.g., species richness; Dubuis et al. 2011), including also functional and phylogenetic diversity (D'Amen et al. 2018a).

In this section, we first illustrate two study cases where the ecospat package can be used: (a) modeling rare species distribution and quantifying species range size and (b) anticipating biological invasions via niche comparison. We then discuss other applications of the functions in the ecospat package and provide some perspectives about the future development of the package.

2 Case Studies

2.1 Modeling Rare Species

In this section, we will use the ecospat package to employ the method called "ensemble of small models (ESM)," which was developed by Lomba et al. (2010), Breiner et al. (2015), and Breiner et al. (2018) and is particularly suitable for species with small sample size, like rare and endangered species (yet also works particularly well for species with larger sample sizes but at higher computational cost than more traditional approaches; Breiner et al. (2015)). One of the major problems to model rare species is that the number of occurrences is usually scarce. Although some studies reported that species could be accurately modeled with very low sample size (e.g., three occurrences in van Proosdij et al. (2015)), sample size is problematic for the modeling procedure, with the risk of overfitting models when the number of occurrences is low compared to the number of predictors (van de Pol et al. 2016) and for evaluation in the case of test data with small sample sizes (Collart and Guisan 2023). In general, authors are limiting the number of predictors that are put in a model using the rule of thumb that not more than one predictor term should be used per ten occurrences. To avoid this limitation, ESMs compute bivariate models and then combine all possible bivariate models into a weighted ensemble. By averaging simple small models to an ensemble, with their respective cross-validated performance as weights, ESMs avoid overfitting without losing explanatory power through reducing the number of predictor variables, and were shown to perform significantly better than standard SDMs with species having a low number of occurrences (Breiner et al. 2015). For this section, we will be focusing on modeling the ecological niche of the alpine plant *Veronica alpina* in the Western Swiss Alps.

2.1.1 Pre-Modeling

The ESM functions of the *ecospat* package rely on the model fitting functions in the *biomod2* package (Thuiller et al. 2023). We thus need to first format our data by using the *BIOMOD_FormatingData* function, where species occurrences and associated coordinates, the environmental conditions, and the name of the species of interest are given. In this example, we use presence-absence data for 300 plots, where *Veronica alpina* is present in 12 locations, and maps of 5 environmental predictors. The latter variables are: growing degree days above 0 °C, moisture index over the growing season (average values for June to August in mm day-1), annual sum of radiation (in kJ m-2 year-1), slope (in degrees), and topographic position.

2.1.2 Core-Modeling

The function *ecospat.ESM.Modeling* is used to model the ecological niche of the species by generating bivariate models. The argument *data* is for the formatted dataset object generated by *BIOMOD_FormatingData*. The desired algorithms can be provided in the argument *models*. Model parameters can be adapted via the argument *models.options* by giving the object from the function *BIOMOD_ModellingOptions()* of the biomod2 package. As in the package biomod2, ESM can fit 12 different algorithms: Generalized Linear Model ("GLM"), Gradient Boosted Machine ("GBM"), eXtreme Gradient Boosting Training (XGBOOST), Generalized Additive Models (" GAM"), "CTA," Artificial Neural Network ("ANN"), "SRE," "FDA," "MARS," "RF," Maximum entropy ("MAXENT," using the java software or "MAXNET" from the maxnet package). Tuning to obtain the optimal parameters for the model can be realized with the argument *tune*. *Prevalence* can be set to build a "weighted response." If NULL, each observation (presence or absence) will have the same weight. You can also give a specific weight to observations via the argument *Yweights*. To evaluate the models, the function performs a repeated split-sampling cross-validation using the arguments *DataSplit* and *NbRunEval*. *DataSplit* corresponds to the percentage of observations used to calibrate the models. *NbRunEval* indicates the number of times

```
# Load the packages
library(ecospat)
library(biomod2)
library(terra)
library(viridis)

set.seed(123)
data("ecospat.testData")

# coordinates of the plots
xy <- ecospat.testData[,2:3]
# species presences and absences
sp_occ <- ecospat.testData$Veronica_alpina
sum(sp_occ) ## Number of occurrences

[1] 12

# environmental data
env <- rast(system.file("extdata/ecospat.testEnv.tif", package="ecospat"))

# Formatting the data with the BIOMOD_FormatingData() function from the
package biomod2
library(biomod2)
myBiomodData <- BIOMOD_FormatingData(resp.var = as.numeric(sp_occ),
                                     expl.var = env,
                                     resp.xy = xy,
                                     resp.name = "Veronica.Alpina",
                                     filter.raster = TRUE)
```

the split-sampling procedure is replicated. The function also allows user-defined cross-validations by giving a logical matrix in the argument *DataSplitTable*, where each row corresponds to an observation and each column corresponds to a run. A value TRUE means that an observation will be used for model calibration while a FALSE is for model evaluation. *weighting.score* corresponds to the evaluation metric that will be used to weight single bivariate models in the final ensemble model. The available evaluation metrics are "AUC," "SomersD" (2xAUC-1), "Kappa," "TSS," or "Boyce." *which.biva* allows to split the bivariate model procedure in several parts. For example, if which.biva is 1:3, only the three first variable combinations will be modeled. This allows to run different bivariate splits on different computers. However, it is better not to use this option if all models are run on a single computer. If you do so, make sure to give each of your modeling subset a unique *modeling.id.* and avoid space characters. Parallel computing can be enabled with the argument *parallel*.

The following step is to combine all the bivariate models into an ensemble. To do so, we can use the function *ecospat.ESM.EnsembleModeling* which will need the object returned by *ecospat.ESM.Modeling*, the evaluation metric used to weight the bivariate models (*weighting.score*) and a *threshold* to remove poor-performing models. The argument *models* allow to select one or several algorithms to realize the ensemble.

ESM performances resulting from the cross-validations can be observed in the object returned by *ecospat.ESM.EnsembleModeling* (Table 2).

```
my.ESM <-ecospat.ESM.Modeling(data = myBiomodData,
                              NbRunEval = 3,
                              DataSplit = 70,
                              DataSplitTable = NULL,
                              Prevalence = 0.5,
                              weighting.score = "SomersD",
                              models = "GLM",
                              tune = FALSE,
                              modeling.id = as.character(format(Sys.time(), "%s")),
                              models.options = NULL,
                              which.biva = NULL,
                              parallel = FALSE,
                              cleanup = FALSE,
                              Yweights = NULL)
```

```
my.ESM.EF <- ecospat.ESM.EnsembleModeling(ESM.modeling.output = my.ESM,
                                          weighting.score = "SomersD",
                                          threshold = 0,
                                          models = NULL)
```

Table 2 ESM performances based on a mean or standard deviations across bivariate model performances of a same run

	RUN1	RUN2	RUN3
threshold	175.000	525.000	570.000
sensitivity	1.000	1.000	1.000
specificity	0.671	0.847	0.776
Kappa	0.155	0.332	0.238
AUC	0.771	0.926	0.859
sensitivity.sd	0.000	0.000	0.000
specificity.sd	0.051	0.039	0.045
Kappa.sd	0.071	0.127	0.100
AUC.sd	0.059	0.036	0.056
TSS	0.671	0.847	0.776
SomersD	0.541	0.853	0.718
MPA	0.234	0.616	0.612
Boyce	−0.138	0.474	−0.429

However, because a minimum sample size is needed to evaluate models (see Jiménez-Valverde 2020), it is recommended to evaluate ESMs using the recently proposed pooling evaluation approach (Collart and Guisan 2023). The function *ecospat.ESM.EnsembleEvaluation* uses this approach, which consists of pooling the suitability values predicted with the hold-out data (evaluation dataset) across replicates to measure an overall evaluation score. As the same observation (presence or absence or background point) is presumably sampled in several replicates, the suitability values for each data point are consequently averaged across replicates where they were sampled. This procedure generates a series of independent suitability values with a size approximately equal to that of the number of observations (the number of suitability values might be slightly lower than the number of original observations as some data points may not be sampled by chance in any of the n replicates), which agreement to the actual observations can then be measured using

Table 3 ESM performances based on the pooling evaluation

	AUC	SomersD	Boyce	MaxTSS
GLM	0.847	0.694	0.73	0.612

sufficient sample size. This function can compute several metrics, which can be selected with the argument *metrics*. If needed, *EachSmallModels* allows evaluating each bivariate model via the pooling evaluation (Table 3).

```
my.ESM.EF.eval <- ecospat.ESM.EnsembleEvaluation(
                    ESM.modeling.output = my.ESM,
                    ESM.EnsembleModeling.output =
                    my.ESM.EF,metrics = c("SomersD","AUC","MaxTSS","Boyce"),
                    EachSmallModels = FALSE)

## Evaluation dataset obtained by the pooling evaluation
pred.test <- as.data.frame(my.ESM.EF.eval$ESM.fit)

## Evaluation scores of the ESM based on the pooling evaluation
eval<-my.ESM.EF.eval$ESM.evaluations
```

The *ecospat* package also has numerous functions to compute model performances based on model predictions and species observation. For example, the Boyce index that only requires species presences for evaluating predictions (Hirzel et al. 2006) can be calculated with the function *ecospat.boyce*. The argument *obs* should contain the model prediction for the presences, while *fit* should contain the predictions of background points. Correlation measurement can be changed via the argument *method*.

```
boyce.index <- ecospat.boyce(fit = pred.test$Fit_GLM,
                             obs = pred.test$Fit_GLM[pred.test$resp==1],
                             PEplot = FALSE,
                             method = "spearman")
boyce.index$cor
```

```
[1] 0.73
```

MaxTSS and MaxKappa can be calculated via the functions *ecospat.max.tss* and *ecospat.max.kappa*, and the variations of TSS and Kappa metric on a threshold can be done with *ecospat.plot.tss* (Fig. 1) and *ecospat.plot.kappa* (Fig. 2):

Model adequacy can also be checked by observing species response curves to each environmental predictor and assess whether they make ecological sense. To do so, the function *ecospat.ESM.responsePlot* can be used. This function is an adaptation of the Evaluation Strip method proposed by Elith et al. (2005) and needs the objects returned by *ecospat.ESM.Modeling* and *ecospat.ESM.EnsembleModeling*. The statistic used generates the response curve for a single predictor while keeping all others constant can be changed via the argument *fixed.var.metric*. When looking

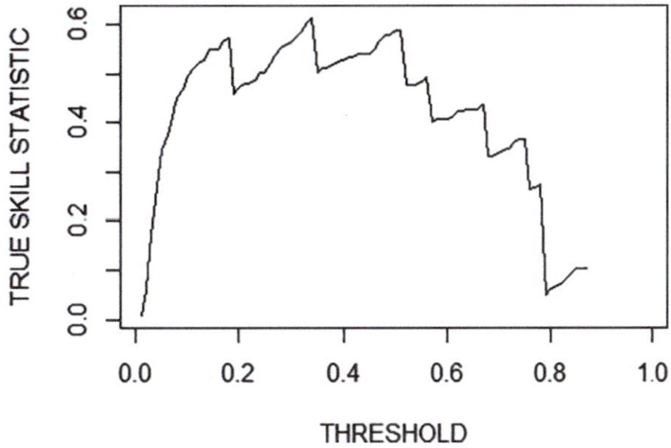

Fig. 1 TSS value as a function of the probability threshold. The maxTSS of 0.61 corresponds to a probability threshold of 0.34

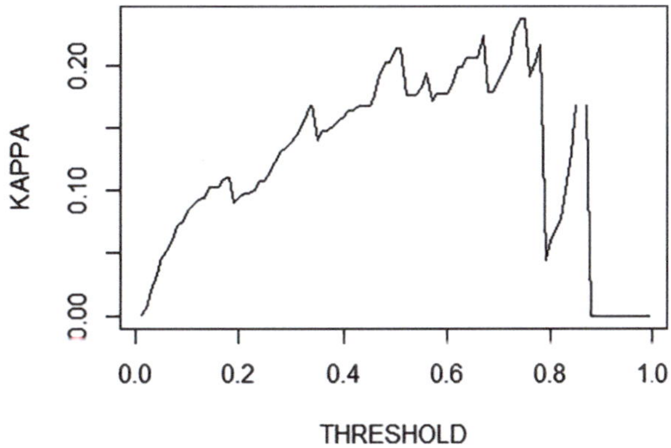

Fig. 2 TSS value as a function of the probability threshold. The maxKappa of 0.24 corresponds to a probability threshold of 0.76

```
MaxTSS <- ecospat.max.tss(Pred = pred.test$Fit_GLM, Sp.occ = pred.test$resp)
MaxTSS$max.TSS
```

[1] 0.611715

```
ecospat.plot.tss(Pred = pred.test$Fit_GLM, Sp.occ = pred.test$resp)
```

```
MaxKappa <- ecospat.max.kappa(Pred = pred.test$Fit_GLM, Sp.occ = pred.test$resp)

MaxKappa$max.Kappa
```

[1] 0.2496419

```
ecospat.plot.kappa(Pred = pred.test$Fit_GLM, Sp.occ = pred.test$resp)
```

at the curves produced with this approach, we can see that two variables seem important for our studied species (Fig. 3). To go deeper into this, we can also compute their variable importance:

To measure the contribution of each variable, the function *ecospat.ESM.VarContrib* can be employed. This function computes the ratio between the sum of weights of bivariate models where a focal variable was used

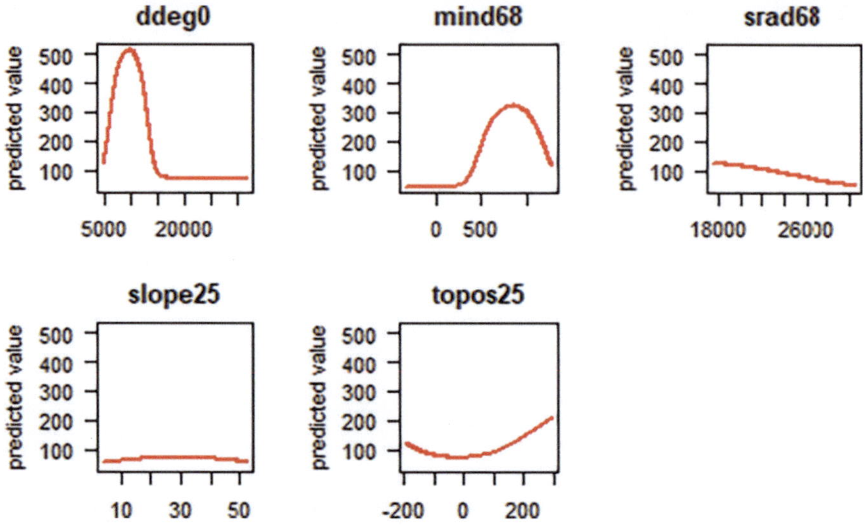

Fig. 3 Response curves to each environmental predictor

```
response.plots <- ecospat.ESM.responsePlot(
                          ESM.EnsembleModeling.output = my.ESM.EF,
                          ESM.modeling.output = my.ESM,
                          fixed.var.metric = 'median')
```

Table 4 Variable contributions to ESMs

	GLM
ddeg0	1.596
mind68	1.326
srad68	0.873
slope25	0.563
topos25	0.898

and the sum of weights of bivariate models where the focal variable was not used. The ratio is corrected for the number of models with or without the focal variable. This ratio gives an indication on the proportional contribution of the variable in the final ensemble model (Breiner et al. 2015). A value of higher than 1 indicates that the focal variable has a higher contribution than average. For the ensemble model, a weighted mean is applied among model algorithms. In this example, we can see that only two variables seem useful to quantify the ecological niche of *Veronica alpina* (Table 4):

```
var.contrib <- ecospat.ESM.VarContrib(ESM.modeling.output = my.ESM,
                                       ESM_EF.output = my.ESM.EF)
```

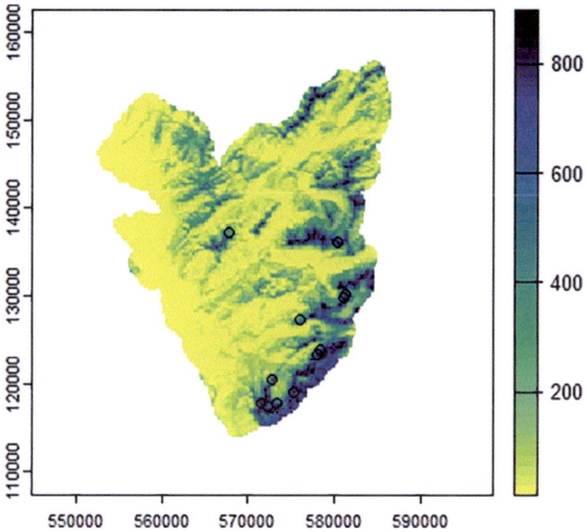

Fig. 4 Projected ESM of Veronica alpina at present time

After looking at the model performances and response curves, models can be projected using two functions: *ecospat.ESM.Projection* that projects each bivariate model and *ecospat.ESM.EnsembleProjection* that generates the ensemble of these bivariate models. The *new.env* argument allows the use of a *SpatRaster* object to project models onto a new area or time period. The argument *name.env* gives a name to the projection. Parallel computing can be enabled with the argument *parallel*. The generated maps correspond to the rounded values of habitat suitability times 1000 (Fig. 4):

```
### Projection of simple bivariate models into new space
my.ESM.proj.current <- ecospat.ESM.Projection(ESM.modeling.output = my.ESM,
                                              new.env = env,
                                              name.env = "currentTime",
                                              parallel = FALSE,
                                              cleanup = FALSE)

### Projection of calibrated ESMs into new space
my.ESM.EF.proj.current <- ecospat.ESM.EnsembleProjection(
                          ESM.prediction.output = my.ESM.proj.current,
                          ESM.EnsembleModeling.output = my.ESM.EF,
                          chosen.models = "all")

plot(my.ESM.EF.proj.current, col = rev(viridis::viridis(50)))
points(xy[sp_occ==1,],col = "black")
```

2.1.3 Post-Modeling

ESM projections can be binarized afterwards (i.e., transforming the continuous suitability index into presence-absence). To binarize these maps, diverse thresholds

Table 5 Various threshold and fit performances of ESM

	Full_GLM_ESM
sensitivity	0.917
specificity	0.845
Kappa	0.280
AUC	0.922
sensitivity.sd	0.083
specificity.sd	0.022
Kappa.sd	0.070
AUC.sd	0.023
SomersD	0.843
Boyce	NA
TSS	0.761
TSS.th	0.645
MPA1.0	0.386
MPA0.95	0.737
MPA0.90	0.776
Boyce.th.min	0.340
Boyce.th.max	0.722

can be computed via the function *ecospat.ESM.threshold* (Table 5). This function also provides evaluation scores for the full model (thus, evaluating the fit of the model but not its predictive power and transferability).

```
Thr <- ecospat.ESM.threshold(ESM.EnsembleModeling.output = my.ESM.EF,
PEplot = FALSE)
```

Model projections can be afterwards binarized with the function *ecospat.binary.model* which need in the arguments *Pred*, a spatial grid (Fig. 5) and *Threshold*, the value of the threshold.

After binarizing the maps, we could quantify the species range size or the occupied patches from ESM maps and IUCN criteria. In the ecospat package, the function *ecospat.rangesize* and *ecospat.occupied.patch* are made for these purposes.

More precisely, *ecospat.rangesize* allows quantifying the area of occupancy (AOO) and the extent of occurrence (EOO) criteria (IUCN 2012; Fig. 6). Numerous parameters are available and described in the help file.

ecospat.occupied.patch quantified the number of patches where species occupied based on species distribution predictions, species occurrences, and a buffer value (in meter) around species occurrences (Fig. 7).

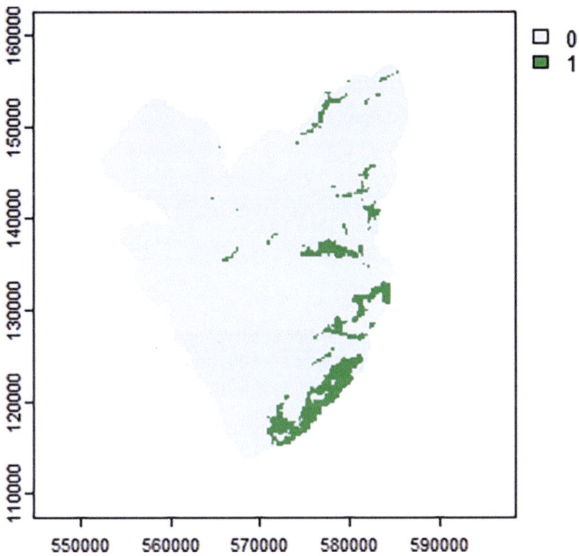

Fig. 5 Binarization of the projected ESM based on the threshold maximizing the TSS

```
my.ESM.EF.proj.current.bin <- ecospat.binary.model(Pred =
my.ESM.EF.proj.current,
                                                    Threshold =
(Thr$TSS.th)*1000)
plot(my.ESM.EF.proj.current.bin)
```

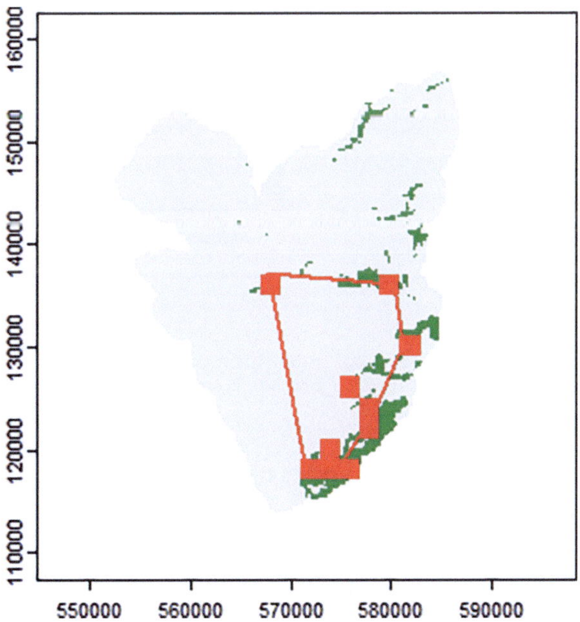

Fig. 6 IUCN criteria. AOO corresponds to areas in green. EOO corresponds to the convex polygon in red

```
rangesize <- ecospat.rangesize(my.ESM.EF.proj.current.bin,
                               xy = xy[sp_occ==1,],
                               AOO.circles = TRUE,
                               lonlat = FALSE)
A00<-rangesize$RangeSize$AOO
A00/1000^2 #area in km2

## [1] 40

EOO<-rangesize$RangeSize$EOO
EOO/1000^2 #area in km2

## [1] 175.4488

plot(my.ESM.EF.proj.current.bin,legend = FALSE)
plot(rangesize$RangeObjects$AOO,add = TRUE, col = "red",legend = FALSE)
plot(rangesize$RangeObjects$EOO@polygons,add = TRUE, border = "red", lwd = 2)
```

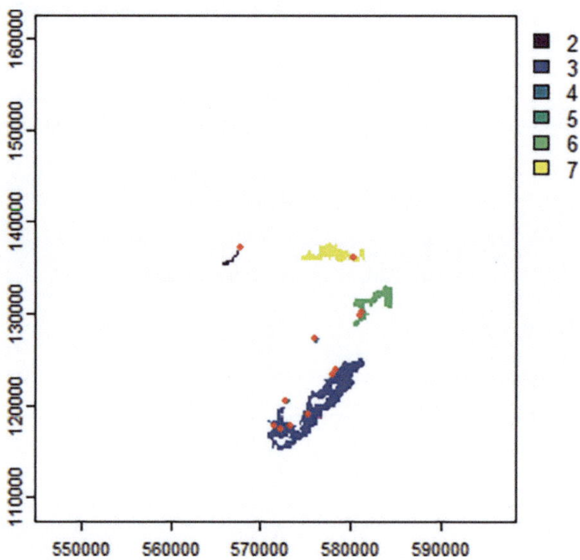

Fig. 7 Occupied patches. Each patch corresponds to a different color

```
ocp  <- ecospat.occupied.patch(my.ESM.EF.proj.current.bin,
                               xy[sp_occ==1,],
                               buffer = 500)
plot(ocp,col = viridis::viridis(50))
points(xy[sp_occ==1,],col = "red",cex = 0.5, pch = 19)
```

2.2 Niche Dynamics of Invasive Species

In this section, we use the *ecospat* package to investigate the environmental niche dynamics of an invasive species across its native and invaded ranges. We will develop the example of the spotted knapweed (*Centaurea stoebe)*, an Asteraceae native to Central and Eastern Europe. In Europe, both diploid and tetraploid cytotypes can be found, but only the tetraploids have spread to North America, where it is considered an invasive species, most presumably because of its ability to produce perennial polycarpic plants (Mráz et al. 2011; Treier et al. 2009). This species served as a study case upon which most of the methodological development on niche dynamics presented in this section have been based (Broennimann et al. 2007, 2012, 2014; Broennimann and Guisan 2008; Guisan et al. 2014; Petitpierre et al. 2012). Note however that, for practical reasons, the analyses presented here use occurrence and environmental data at a lower resolution than used in the original analyses, which can lead to slightly different results. The main idea of the niche dynamics analyses is to quantify realized environmental niches (sensu Hutchinson 1957; see Araújo and Guisan 2006, Broennimann et al. 2007) by investigating the occurrence density of species in a gridded climatic space allowing for direct pixel quantification and permutation tests (Broennimann et al. 2012).

2.2.1 Data Preparation

First, we need to load the R packages employed for the analyses. In addition to the package *ecospat*, the package *terra* is required for handling geo-referenced datasets, the package *geodata* to download the climatic dataset, the package *dplyr* to query, subset and summarize data, and the package *ade4* to perform the ordination analyses:

```
# Load the needed packages
library(ecospat)
library(terra)
library(tidyterra)
library(geodata)
library(dplyr)
library(ade4)
library(viridis)
```

Then, we download the climatic data, here the bioclim variables of the WorldClim v2.1 dataset at 10′ resolution using *worldclim_global* from the *geodata* package. For computational efficiency, we aggregate the climatic data at 30′ resolution (~50 km) using *aggregate* from *terra.* To delimit the study area for the analyses, we also download (and unzip) the geo-referenced polygons of the Terrestrial Ecoregions of the World distributed by the WWF:

We now load and prepare the data for the two regions of analysis, namely, Europe, where the species is native, and North America, where the species is invasive. The occurrence data are retrieved from the extdata folder included in the *ecospat* package and converted to a convenient vector format with *vect.* We thus use only occurrences of tetraploid cytotypes to ensure a fair comparison between ranges, as diploids are absent from the North American continent. We restrict the regions of analysis (background) to the biomes where the species is present in each realm using a pipe of *dyplr* functions. The biomes selected are temperate broadleaf forests (4), temperate coniferous forests (5), boreal forests (6), temperate grasslands (8), Mediterranean forests and woodlands (12), and deserts (13). We

```
## seed set for reproducible results
set.seed(123)
## create a randomly attributed temporary local folder to download the data
tempdir<-tempdir()

## download worlclim 2.1 climate data at 10' resolution and biomes data

clim.world<-
geodata::worldclim_global("bio",res=10,version="2.1",path=tempdir)
clim.world<-terra::aggregate(clim.world,3,fun=mean,na.rm=TRUE)
names(clim.world)<-sub("wc2.1_10m_","",names(clim.world))
names(clim.world)<-sub("_","",names(clim.world))

## download Terrestrial Ecoregions of the World dataset

teow.URL<-"https://storage.googleapis.com/teow2016/Ecoregions2017.zip"
# official website for manual download:
# https://files.worldwildlife.org/wwfcmsprod/files/Publication/file
# /6kcchn7e3u_official_teow.zip
download.file(teow.URL,destfile = paste0(tempdir,"/TEOW.zip"))
unzip(paste0(tempdir,"/TEOW.zip"),exdir = tempdir)
biomes<-
terra::vect(paste0(tempdir,"/Ecoregions2017.dbf"),crs="+proj=longlat")
```

further restrict the background to areas distant of less than 500 km from existing occurrences (this avoids including large regions of the Paleoarctic realm where the species does not occur in the native range). The background vector can then be used to *crop* and *mask* the climatic data. Note that the choice of the background area has important consequences on niche quantification (Broennimann et al. 2012; Rödder and Engler 2011), especially in the similarity tests presented below. The general advice in selecting the background is to include areas that are, or have been accessible to the species through dispersal during its evolutionary life (Barve et al. 2011). We recommend performing a sensitivity analysis on the parameters chosen for the background selection to ensure the stability and reliability of results. The plots show the background area and overlaid occurrences to visualize the final data to be analyzed (Figs. 8 and 9).

2.2.2 Niche Quantification

The niche quantification will be performed in two-dimensional space represented by the two first components (PC1 and PC2) of a principal component analysis (PCA). We chose to use two niche axes that maximize the variation contained in the original 19 bioclim variables from the WorldCim dataset, but the user could as well choose two specific variables relevant for the ecology of the species based on expert assessment. Niche quantification in one dimension is presented in a following section. To calibrate the PCA on the full extent of climatic values present in both background areas, we first join the European and North American datasets using *merge* from *terra*. We then extract the climatic values for each pixel of the global background (*clim.glob*), for each continental background (*clim.glob.EU*, *clim.glob.NAM*), and for each set of continental occurrences (*clim.occ.EU*, *clim.occ.NAM*) using *extract*

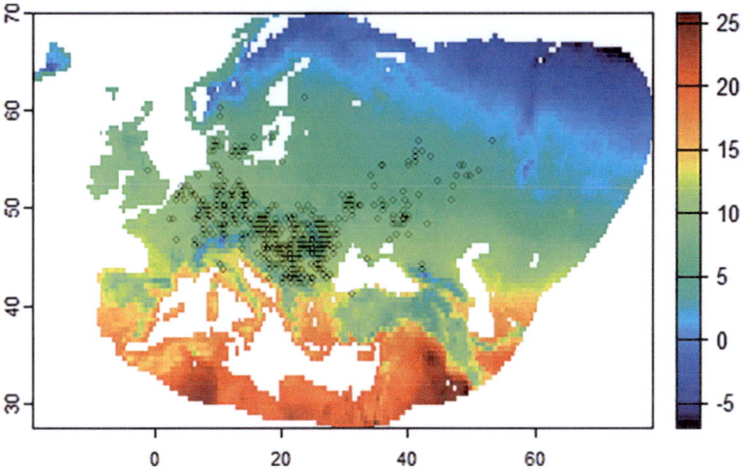

Fig. 8 European background and occurrences. Occurrences are shown with open circles. The color gradient corresponds to mean annual temperatures (bio1) over the background area

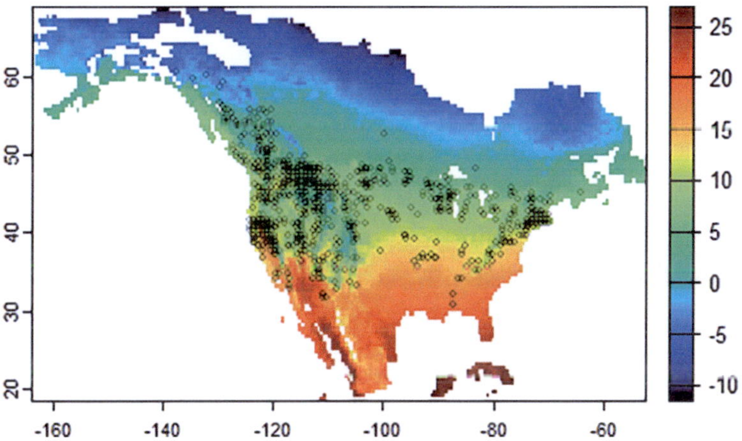

Fig. 9 North American background and occurrences. Occurrences are shown with open circles. The color gradient corresponds to mean annual temperatures (bio1) over the background area

from *terra*. The PCA is calibrated on the values of the global background using *dudi.pca* from *ade4*. The function *ecospat.plot.contrib* allows to plot the correlation circle (Fig. 10) indicating the correlation of original variables with the principal component axes with arrows. Here PC1 is mostly (but not exclusively) correlated with precipitation variables (bio12 to bio 19), while PC2 is more correlated with temperature variables (bio1 to bio11). The inertia of each principal component (similar to the variance explained in a regression model context) is also indicated. Here the combined contribution of PC1 and PC2 explains almost 80% of the climatic variation in the original dataset. The calculation of PCA scores (*li* vectors

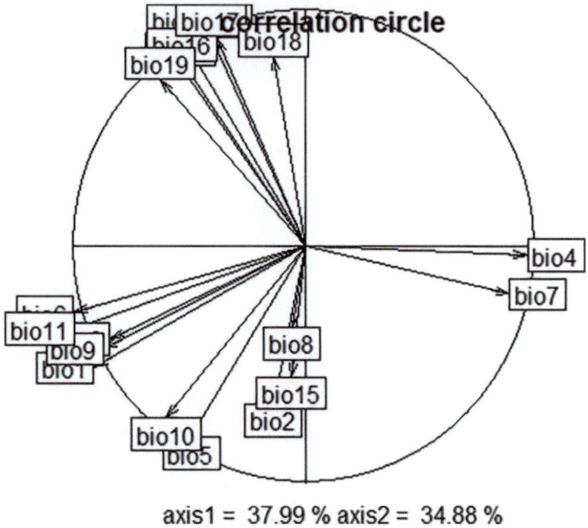

Fig. 10 Correlation circle showing the importance of original variables in the PCA

```
## data for Europe
# occurrences

occ.EU.df<-read.delim(system.file("extdata/Csto4xEU.txt", package="ecospat"))
occ.EU<-vect(occ.EU.df,geom=c("x", "y"),crs="+proj=longlat")

# biomes

buf<-buffer(occ.EU,1500000) %>% aggregate() # 1500km buffer around
occurrences
bkg<-biomes  %>% filter(BIOME_NUM %in% c(4,5,6,8,12,13)) %>% group_by(REALM)
%>%
     summarize() %>% crop(buf) # biomes in the realm where Centaurea is
present
clim.EU<-crop(clim.world,bkg)
clim.EU<-mask(clim.EU,bkg)
plot(clim.EU[[1]], col=viridis(100,option="turbo"))
    #plot annual temperature (bio1)
plot(occ.EU,cex=0.5,pch=21,alpha=0.4,add=TRUE)
```

in *dudi.pca* objects) for the pixels of the backgrounds (*scores.glob, scores.glob.EU, scores.glob.NAM*) and for the occurrences (*scores.occ.EU, scores.occ.NAM*) using *suprow* from *ade4* will be useful for the next step.

The following step is the core of the niche quantification analysis. The function *ecospat.grid.clim.dyn* first creates a gridded two-dimensional climatic space of RxR pixels bounded by the minimum and maximum values of the PCA scores of the background points (*score.glob*). It then uses kernel density estimation functions to calculate the density of occurrence in each pixel. The result is a grid of occurrence densities in a two-dimensional climatic space, with high values indicating the core of species niche, low values indicating the margin of the niche, and zero values indicating conditions outside of the niche. For comparison between ranges, occurrences

```
## data for North America
# occurrences

occ.NAM.df<-read.delim(system.file("extdata/Csto4xNAM.txt",
package="ecospat"))
occ.NAM<-vect(occ.NAM.df,geom=c("x", "y"),crs="+proj=longlat")

# biomes

buf<-buffer(occ.NAM,1500000) %>% aggregate()
bkg<-biomes %>% filter(BIOME_NUM %in% c(2,4,5,6,8,12,13)) %>% group_by(REALM)
%>%
    summarize() %>% crop(buf)
clim.NAM<-crop(clim.world,bkg)
clim.NAM<-mask(clim.NAM,bkg)
plot(clim.NAM[[1]], col=viridis(100,option="turbo"))
plot(occ.NAM,cex=0.5,pch=21,alpha=0.4,add=TRUE)
```

```
# extraction of climate values

clim<-terra::merge(clim.EU,clim.NAM)
clim.glob<-na.omit(terra::values(clim))
clim.glob.EU<-na.omit(terra::values(clim.EU))
clim.glob.NAM<-na.omit(terra::values(clim.NAM))
clim.occ.EU<-terra::extract(clim.EU,occ.EU,ID=FALSE)
clim.occ.NAM<-terra::extract(clim.NAM,occ.NAM,ID=FALSE)

## PCA calibration
pca.env<-dudi.pca(clim.glob,scannf=FALSE,nf=2)
ecospat.plot.contrib(contrib=pca.env$co,eigen=pca.env$eig)
```

```
## PCA scores

scores.glob<-pca.env$li
scores.glob.EU<-suprow(pca.env,clim.glob.EU)$li
scores.glob.NAM<-suprow(pca.env,clim.glob.NAM)$li
scores.occ.EU<-suprow(pca.env,clim.occ.EU)$li
scores.occ.NAM<-suprow(pca.env,clim.occ.NAM)$li
```

densities are standardized between zero and one (*object z.uncor*). An alternative version of occurrences densities is also calculated by dividing in each pixel the occurrence density by the density of the corresponding climatic condition in the background (object *z.cor* return by *ecospat.grid.clim.dyn;* defined as occurrence occupancy in Broennimann et al. (2012)). The function *ecospat.plot.niche* allows to plot the calculated niche as a gradient of occurrence densities along the two first PCA axes (Fig. 11). Alternatively, the niches can be directly plotted with the *plot* function as *rast* objects (Fig. 12):

We can analyze the results of the niche quantification more in depth. First, we want to know how much the two niches overlap. This can be achieved with the *ecospat.niche.overlap.* which calculates the Schoener's D and Hellinger I overlap indices. The parameter *cor=TRUE* allows to assess the overlap when correcting

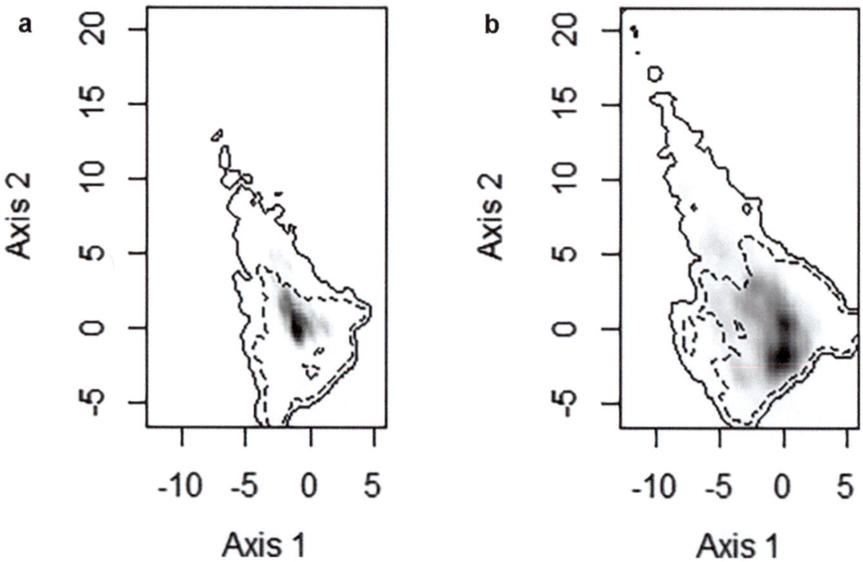

Fig. 11 Occurrence densities in (**a**) Europe and (**b**) North America. Shades of gray pixels indicate occurrence densities along the two first PCA axes. Solid lines show the contours of all climates, and dashed line show the 50% most common climates

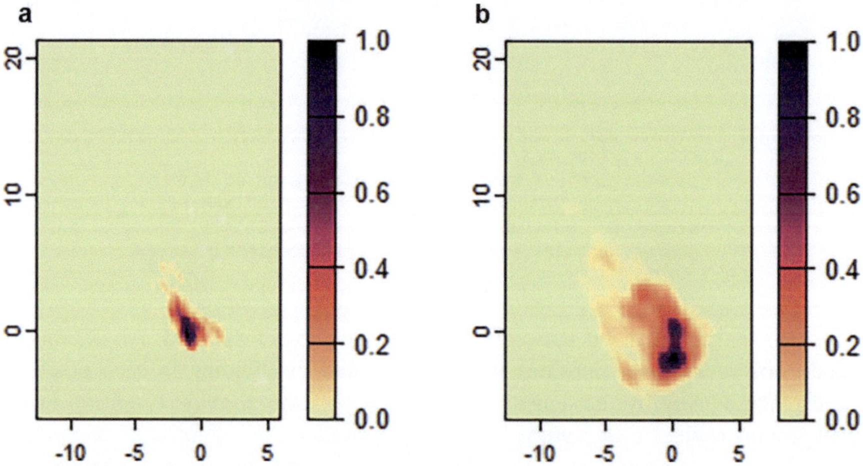

Fig. 12 Occurrence densities in (**a**) Europe and (**b**) North America plotted as rasters

for the density of environmental conditions in the background. In our case, using *cor=FALSE*, we detect an overlap D of 0.36:

We can ask the question whether the overlap measured is higher than random. This can be achieved with *ecospat.niche.equivalency.test* and *ecospat.niche.similarity.test* functions, based on Warren et al. (2008). The test of niche equivalency

```
z.EU  <- ecospat.grid.clim.dyn(glob=scores.glob,
                               glob1 = scores.glob.EU,
                               sp= scores.occ.EU,
                               R=100, th.sp=0,th.env=0,kernel.method = "ks")
z.NAM <- ecospat.grid.clim.dyn(glob=scores.glob,
                               glob1 = scores.glob.NAM,
                               sp= scores.occ.NAM,
                               R=100, th.sp=0,th.env=0,kernel.method = "ks")
par(mfrow=c(1,2))
ecospat.plot.niche(z.EU,title="a)")
ecospat.plot.niche(z.NAM,title="b)")

par(mfrow=c(1,2))
plot(z.EU$z.uncor,col=viridis(100,option="magma",direction = -1),main="a)")
plot(z.NAM$z.uncor,col=viridis(100,option="magma",direction = -1),main="b)")
```

```
ecospat.niche.overlap(z1=z.EU,z2=z.NAM,cor=FALSE)
```

```
## $D
## [1] 0.3287533
##
## $I
## [1] 0.5597273
```

asks whether the overlap observed between the two niches is significantly higher than the overlap that one would get if the occurrences between the two niches would be randomly reallocated. On the other hand, the test of similarity asks whether the overlap observed between the two niches is significantly higher than the overlap one would get if the two niches would be randomly moved in the background space. Both tests have been implemented in *ecospat* for correspondence with Warren et al. (2008), though here in environmental space (while in geographical space for Warren et al. 2008; see Broennimann et al. 2012), but the niche equivalency test almost always provides significant results even when the overlap is very low and is thus in general not informative (Collart et al. 2021; Peterson et al. 2011). We thus only run the niche similarity test, with parameters *rand.type* = 1 to randomize both niches in the background, *rep* = 100 to set the number of replications, *intersection* = 0 to perform the randomization in global background without removing marginal environments, and using *overlap.alternative* = "higher" as the alternative hypothesis for the test (i.e., a *p*-value <0.05 indicate that the observed overlap is significantly higher than the simulated overlap from randomized niches). Here, with a *p*-value of 0.25, the niche similarity test indicates that the niches of the spotted knapweed in Europe and in North America do not significantly depart from similarity.

We can further investigate whether some parts of the native niche are not occupied in the invaded range and, conversely, if the species occupies new conditions in the invaded range. This can be done with the *expansion*, *stability*, and *unfilling* indices returned by the *ecospat.niche.dyn.index* function (see COUE framework in Guisan et al. 2014). *Stability* and *expansion* correspond to the percentages of pixels of the invasive niche which conditions present, and not present in the native range,

```
sim.test<-ecospat.niche.similarity.test(z1=z.EU,z2=z.NAM,rand.type=1,rep=100,
                                        intersection=0,
overlap.alternative="higher")

sim.test$obs$D

## [1] 0.2537814
```

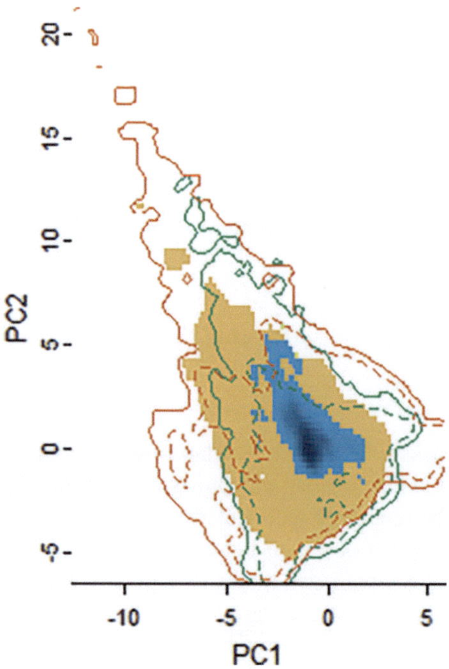

Fig. 13 Niche dynamics. The plot shows the niche stability in blue, niche expansion in orange, and niche unfilling in yellow. The solid green and red contour lines indicate the extent of environmental conditions that exist in the native and invaded ranges, respectively. The dotted contour lines indicate the quantile of the most abundant environment in both ranges (here 50% most abundant, with quant = 05). The densities of occurrences in the native range (argument interest = 1) are displayed using gray shading

respectively. *Unfilling* corresponds to the percentage of the native niche which is not present in the invaded range. The results indicate that the invasive niche of the spotted knapweed expands largely beyond conditions present in the native niche, while only a small fraction of the native niche is not occupied in the invaded ranges. It is possible to plot the *expansion*, *stability*, and *unfilling* indices in the environmental space using the function *ecospat.plot.niche.dyn* (Fig. 13).

Finally, we might wonder where are the areas predicted with a high density of occurrence by the niche quantification models. Indeed, every pixel in the geography corresponds to a predicted density of occurrence in the environmental space. The function *ecospat.niche.zProjGeo* allows doing that. It is possible to project the density of occurrence of one niche (*z* object) across its own range of calibration (e.g., for Europe using *zproj* = NULL and *env* = clim.EU; Fig. 14) but also to project the density of occurrence to another range, here, for example, to project the European niche of the spotted knapweed in North America (Fig. 15), which interestingly becomes then a form of SDM.

```
dyn<-ecospat.niche.dyn.index(z.EU,z.NAM,intersection=0)
dyn$dynamic.index.w

##     expansion     stability     unfilling
## 0.5719737854 0.4280262146 0.0007058004

# colorblind friendly palette, Wong, B. (2011). Nature Methods
pal<-c("#D55E00","#E69F00","#F0E442","#CC79A7","#009E73","#0072B2","#56B4E9")

ecospat.plot.niche.dyn(z.EU,z.NAM,quant=0.5,name.axis1 ="PC1",
             name.axis2="PC2",interest=1,col.unf="#F0E442", col.exp="#E69F00",
             col.stab="#0072B2", colZ1="#009E73", colZ2="#D55E00",transparency
= 40)
```

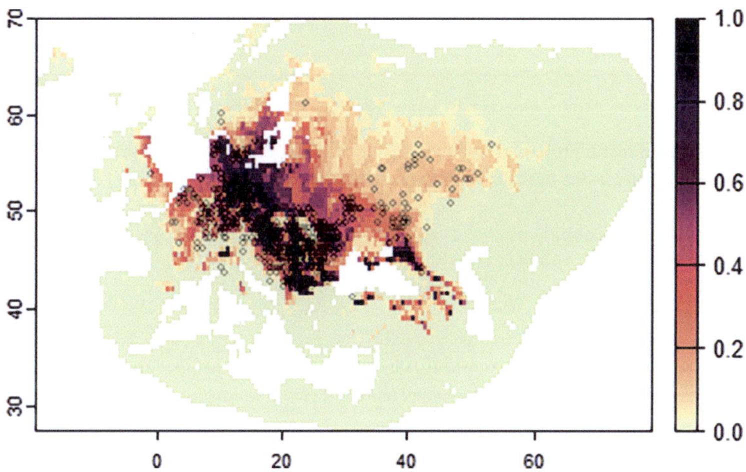

Fig. 14 European niche projected in Europe. Dark colors indicate predicted high density of occurrence. Open circles indicate know occurrences of the species

Similarly, we can also project niche dynamic indices in geography using the function *ecospat.niche.dynIndexProjGeo* (Fig. 16). The argument *proj* controls the range of projection (i.e., proj = 2 for the North American range), and *env* sets the environmental *rast* object used for the projection. The results here show that a large part of the North East and Midwest, as well as smaller fragmented areas of the Rocky Mountains in the West, have climatic conditions corresponding to the native niche of spotted knapweed, but that large areas, especially in the South, correspond to an expansion of the niche.

3 Discussion and Perspectives

We have illustrated the use of the ecospat R package to model a rare alpine plant species with ensemble of small models (ESMs; Lomba et al. 2010; Breiner et al. 2015) and to quantify and compare niches in the native and invaded ranges of

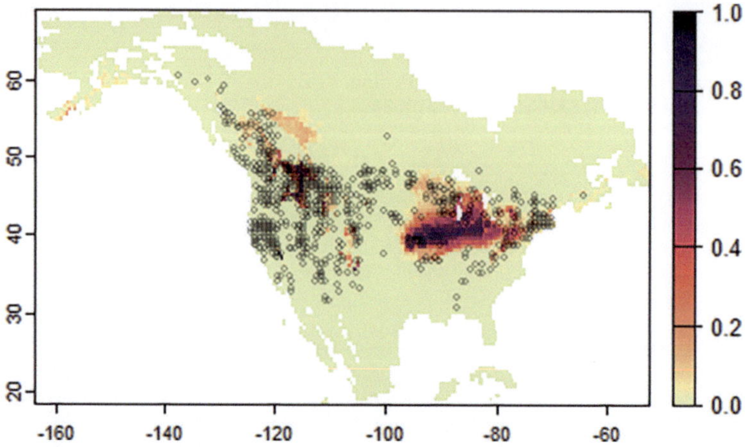

Fig. 15 European niche projected in North America. Dark colors indicate predicted high density of occurrence. Open circles indicate known occurrences of the species

```
geo.z.EU<-ecospat.niche.zProjGeo(z=z.EU,zproj=NULL,env=clim.EU)
plot(geo.z.EU,col=viridis(100,option="magma",direction =-1))
plot(occ.EU,add=TRUE,cex=0.5,pch=21,alpha=0.4)
```

```
geo.z.EUtoNAM<-ecospat.niche.zProjGeo(z=z.EU,zproj=z.NAM,env=clim.NAM)
plot(geo.z.EUtoNAM,col=viridis(100,option="magma",direction =-1))
plot(occ.NAM,add=TRUE,cex=0.5,pch=21,alpha=0.4)
```

Fig. 16 Niche dynamics of the European niche projected in North America. Colors indicate the type of niche dynamics in regard to the European niche

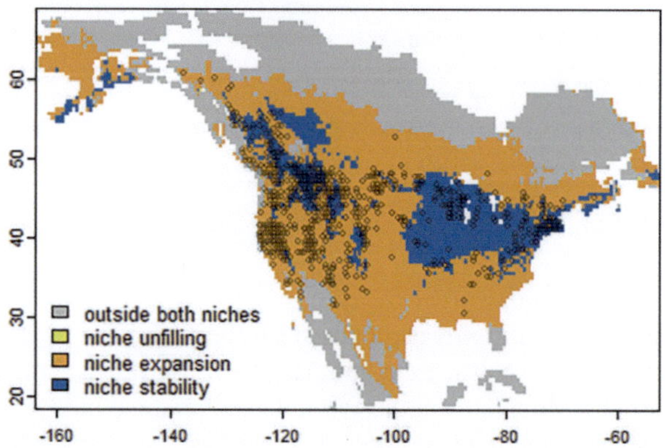

an invasive plant species with the COUE framework (Broennimann et al. 2012; Guisan et al. 2014). These two examples only used part of the functions available in ecospat. Other examples could have been developed for instance along the steps of the spatially explicit species assemblage modelling (SESAM) frameworks (Guisan and Rahbek 2011), including pre-modeling functions to analyze co-occurrences, biotic interactions, and assembly of species to form assemblages (D'Amen et al. 2018b; Scherrer et al. 2019b), post-modeling of assemblages and communities

```
geo.dyn<-ecospat.niche.dynIndexProjGeo(z.EU,z.NAM,proj=2,env=clim.NAM)

plot(geo.dyn,col = c("grey", "#F0E442", "#E69F00", "#0072B2"),
     plg=list(legend=c("outside both niches",
                       "niche expansion", "niche stability"),
x="bottomleft"))
plot(occ.NAM,add=TRUE,cex=0.5,pch=21,alpha=0.4)
```

from individual species predictions (D'Amen et al. 2015), or the pre- and post-analyses and modeling of spatial patterns of phylogenetic diversity, at the level of individual species (D'Amen et al. 2018a; Pio et al. 2014) or communities (Ndiribe et al. 2013). For an example of community modeling using these functions, we refer interested readers to Di Cola et al. (2017) (example 2) and to the related published references (some cited above or in D'Amen et al. 2015). By containing some unique features (COUE niche framework, ESMs, SESAM framework), the ecospat package is complementary to other R packages for SDMs and spatial analyses in ecology, biogeography, and conservation (see, e.g., Joo et al. 2019; Meynard et al. 2019; Sillero et al. 2023; Kass et al. 2025).

Furthermore, ecospat is complemented by three other packages initially developed in our ecospat group: migclim (Engler et al. 2012), covsel (Adde et al. 2023a), and n-sdm (Adde et al. 2023a). The migclim package allows integration of dispersal constraints into projections of species distribution models (see, e.g., Engler et al. 2009), while covsel allows for efficient selection of predictors in SDMs out of large sets of candidate covariates. The n-sdm package is very complementary to ecospat in the sense that it allows modeling large number of species on high-power computing platforms (HPCs; i.e., clusters) using three wishable properties:

(i) A nested hierarchical approach (the "n" in n.sdm) to avoid niche truncation (when the study area is too small to capture all environmental requirements of a species; Chevalier et al. 2021, 2022)
(ii) The covsel approach for an efficient selection of predictors (Adde et al. 2023b), allowing to consider also predictors that describe the environmental characteristics in the neighborhood of observations (i.e., focal windows analyses; as in Scherrer et al. 2019a), not only in the pixel contain the observation
(iii) The ESM approach described above to also include rare and infrequent species when modeling very large numbers of species, e.g., to assess how biodiversity might evolve under various environmental change scenarios at a national scale (Adde et al. in prep.)

Finally, ecospat is continuously evolving by increasing and adapting the number of functions to help researchers develop robust and efficient spatial analyses and models to predict the distribution of species and communities. For instance, the ESM functions are currently being optimized to run more efficiently and with less dependencies (Collart et al. in prep.), and the niche functions were expanded to include a "niche margin index" (NMI; Broennimann et al. 2021; Pearman,

Broennimann et al. in press) allowing to measure not only niche innerness (i.e., what SDM predictions do) but also niche outerness. Last, as the migclim package currently only runs under ancient versions of R, we foresee to integrate it into a next version of ecospat.

References

Adde A, Rey P-L, Brun P, Külling N, Fopp F, Altermatt F, Broennimann O, Lehmann A, Petitpierre B, Zimmermann NE, Pellissier L, Guisan A (2023a) N-SDM: a high-performance computing pipeline for nested species distribution modelling. Ecography 2023(6). https://doi.org/10.1111/ecog.06540

Adde A, Rey P-L, Fopp F, Petitpierre B, Schweiger AK, Broennimann O, Lehmann A, Zimmermann NE, Altermatt F, Pellissier L, Guisan A (2023b) Too many candidates: embedded covariate selection procedure for species distribution modelling with the covsel R package. Ecol Informatics 75:102080. https://doi.org/10.1016/j.ecoinf.2023.102080

Araújo MB, Guisan A (2006) Five (or so) challenges for species distribution modelling. J Biogeogr 33(10):1677–1688. https://doi.org/10.1111/j.1365-2699.2006.01584.x

Araújo MB, Anderson RP, Márcia Barbosa A, Beale CM, Dormann CF, Early R, Garcia RA, Guisan A, Maiorano L, Naimi B, O'Hara RB, Zimmermann NE, Rahbek C (2019) Standards for distribution models in biodiversity assessments. Sci Adv 5(1). https://doi.org/10.1126/sciadv.aat4858

Austin MP, Nicholls AO, Margules CR (1990) Measurement of the realized qualitative niche: environmental niches of five eucalyptus species. Ecol Monogr 60(2):161–177. https://doi.org/10.2307/1943043

Barbet-Massin M, Rome Q, Villemant C, Courchamp F (2018) Can species distribution models really predict the expansion of invasive species? PLoS One 13(3):e0193085. https://doi.org/10.1371/journal.pone.0193085

Barve N, Barve V, Jimenez-Valverde A, Lira-Noriega A, Maher SP, Peterson AT, Soberon J, Villalobos F (2011) The crucial role of the accessible area in ecological niche modeling and species distribution modeling. Ecol Model 222(11):1810–1819. https://doi.org/10.1016/j.ecolmodel.2011.02.011

Breiner FT, Guisan A, Bergamini A, Nobis MP (2015) Overcoming limitations of modelling rare species by using ensembles of small models. Methods Ecol Evol 6(10):1210–1218

Breiner FT, Nobis MP, Bergamini A, Guisan A (2018) Optimizing ensembles of small models for predicting the distribution of species with few occurrences. Methods Ecol Evol 9(4):802–808. https://doi.org/10.1111/2041-210x.12957

Broennimann O, Guisan A (2008) Predicting current and future biological invasions: both native and invaded ranges matter. Biol Lett 4(5):585–589. https://doi.org/10.1098/rsbl.2008.0254

Broennimann O, Thuiller W, Hughes G, Midgley GF, Alkemade JMR, Guisan A (2006) Do geographic distribution, niche property and life form explain plants' vulnerability to global change? Glob Change Biol 12(6):1079–1093. https://doi.org/10.1111/j.1365-2486.2006.01157.x

Broennimann O, Treier UA, Müller-Schärer H, Thuiller W, Peterson AT, Guisan A (2007) Evidence of climatic niche shift during biological invasion. Ecol Lett 10(8):701–709. https://doi.org/10.1111/j.1461-0248.2007.01060.x

Broennimann O, Fitzpatrick MC, Pearman PB, Petitpierre B, Pellissier L, Yoccoz NG, Thuiller W, Fortin MJ, Randin C, Zimmermann NE, Graham CH, Guisan A (2012) Measuring ecological niche overlap from occurrence and spatial environmental data. Glob Ecol Biogeogr 21(4):481–497. https://doi.org/10.1111/j.1466-8238.2011.00698.x

Broennimann O, Mráz P, Petitpierre B, Guisan A, Müller-Schärer H (2014) Contrasting spatio-temporal climatic niche dynamics during the eastern and western invasions of spotted knapweed in North America. J Biogeogr 41(6):1126–1136. https://doi.org/10.1111/jbi.12274

Broennimann O, Petitpierre B, Chevalier M, González-Suárez M, Jeschke JM, Rolland J, Gray SM, Bacher S, Guisan A (2021) Distance to native climatic niche margins explains establishment success of alien mammals. Nat Commun 12(1). https://doi.org/10.1038/s41467-021-22693-0

Broennimann O, Di Cola V, Guisan A (2023) Ecospat: spatial ecology miscellaneous methods. https://cran.r-project.org/web/packages/ecospat/ecospat.pdf

Chevalier M, Broennimann O, Cornuault J, Guisan A (2021) Data integration methods to account for spatial niche truncation effects in regional projections of species distribution. Ecol Appl 31(7). https://doi.org/10.1002/eap.2427

Chevalier M, Zarzo-Arias A, Guélat J, Mateo RG, Guisan A (2022) Accounting for niche truncation to improve spatial and temporal predictions of species distributions. Front Ecol Evol 10. https://doi.org/10.3389/fevo.2022.944116

Collart F, Guisan A (2023) Small to train, small to test: dealing with low sample size in model evaluation. Ecol Informatics 75:102106. https://doi.org/10.1016/j.ecoinf.2023.102106

Collart F, Hedenäs L, Broennimann O, Guisan A, Vanderpoorten A (2021) Intraspecific differentiation: implications for niche and distribution modelling. J Biogeogr 48(2):415–426. https://doi.org/10.1111/jbi.14009

D'Amen M, Rahbek C, Zimmermann NE, Guisan A (2015) Spatial predictions at the community level: from current approaches to future frameworks. Biol Rev 92(1):169–187. https://doi.org/10.1111/brv.12222

D'Amen M, Mateo RG, Pottier J, Thuiller W, Maiorano L, Pellissier L, Ndiribe C, Salamin N, Guisan A (2018a) Improving spatial predictions of taxonomic, functional and phylogenetic diversity. J Ecol 106(1):76–86. https://doi.org/10.1111/1365-2745.12801

D'Amen M, Mod HK, Gotelli NJ, Guisan A (2018b) Disentangling biotic interactions, environmental filters, and dispersal limitation as drivers of species co-occurrence. Ecography 41(8):1233–1244. https://doi.org/10.1111/ecog.03148

Di Cola V, Broennimann O, Petitpierre B, Breiner FT, D'Amen M, Randin C, Engler R, Pottier J, Pio D, Dubuis A, Pellissier L, Mateo RG, Hordijk W, Salamin N, Guisan A (2017) Ecospat: an R package to support spatial analyses and modeling of species niches and distributions. Ecography 40(6):774–787. https://doi.org/10.1111/ecog.02671

Di Febbraro M, Sallustio L, Vizzarri M, De Rosa D, De Lisio L, Loy A, Eichelberger BA, Marchetti M (2018) Expert-based and correlative models to map habitat quality: which gives better support to conservation planning? Glob Ecol Conserv 16:e00513. https://doi.org/10.1016/j.gecco.2018.e00513

Dubuis A, Pottier J, Rion V, Pellissier L, Theurillat J-P, Guisan A (2011) Predicting spatial patterns of plant species richness: a comparison of direct macroecological and species stacking modelling approaches. Divers Distrib 17(6):1122–1131. https://doi.org/10.1111/j.1472-4642.2011.00792.x

Elith J, Kearney M, Phillips S (2010) The art of modelling range-shifting species. Methods Ecol Evol 1:330–342. https://doi.org/10.1111/j.2041-210X.2010.00036.x

Elith J, Ferrier S, Huettmann F, Leathwick J (2005) The evaluation strip: new and robust method for plotting predicted responses from species distribution models. Ecol Model 186(3):280–289. https://doi.org/10.1016/j.ecolmodel.2004.12.007

Engler R, Randin CF, Vittoz P, Czáka T, Beniston M, Zimmermann NE, Guisan A (2009) Predicting future distributions of mountain plants under climate change: does dispersal capacity matter? Ecography 32(1):34–45. https://doi.org/10.1111/j.1600-0587.2009.05789.x

Engler R, Hordijk W, Guisan A (2012) The MIGCLIM R package seamless integration of dispersal constraints into projections of species distribution models. Ecography 35(10):872–878. https://doi.org/10.1111/j.1600-0587.2012.07608.x

Ferrarini A, Selvaggi A, Abeli T, Alatalo JM, Orsenigo S, Gentili R, Rossi G (2016) Planning for assisted colonization of plants in a warming world. Sci Rep 6(1). https://doi.org/10.1038/srep28542

Ferrier S, Guisan A (2006) Spatial modelling of biodiversity at the community level. J Appl Ecol 43(3):393–404. https://doi.org/10.1111/j.1365-2664.2006.01149.x

Fourcade Y, Engler JO, Besnard AG, Rödder D, Secondi J (2013) Confronting expert-based and modelled distributions for species with uncertain conservation status: a case study from the corncrake (Crex crex). Biol Conserv 167:161–171. https://doi.org/10.1016/j.biocon.2013.08.009

Franklin J (2010) Moving beyond static species distribution models in support of conservation biogeography. Divers Distrib 16(3):321–330. https://doi.org/10.1111/j.1472-4642.2010.00641.x

Guisan A, Rahbek C (2011) SESAM – a new framework integrating macroecological and species distribution models for predicting spatio-temporal patterns of species assemblages. J Biogeogr 38(8):1433–1444. https://doi.org/10.1111/j.1365-2699.2011.02550.x

Guisan A, Broennimann O, Engler R, Vust M, Yoccoz NG, Lehmann A, Zimmermann NE (2006) Using niche-based models to improve the sampling of rare species. Conserv Biol 20(2):501–511. https://doi.org/10.1111/j.1523-1739.2006.00354.x

Guisan A, Tingley R, Baumgartner JB, Naujokaitis-Lewis I, Sutcliffe PR, Tulloch AIT, Regan TJ, Brotons L, McDonald-Madden E, Mantyka-Pringle C, Martin TG, Rhodes MJR, Setterfield SA, Elith J, Schwartz MW, Wintle BA, Broennimann O, Austin M, Ferrier S et al (2013) Predicting species distributions for conservation decisions. Ecol Lett 16(12):1424–1435. https://doi.org/10.1111/ele.12189

Guisan A, Petitpierre B, Broennimann O, Daehler C, Kueffer C (2014) Unifying niche shift studies: insights from biological invasions. Trends Ecol Evol 29(5):260–269. https://doi.org/10.1016/j.tree.2014.02.009

Guisan A, Thuiller W, Zimmermann NE (2017) Habitat suitability and distribution models, with applications in R

Hirzel AH, Le Lay G, Helfer V, Randin C, Guisan A (2006) Evaluating the ability of habitat suitability models to predict species presences. Ecol Model 199(2):142–152. https://doi.org/10.1016/j.ecolmodel.2006.05.017

Hutchinson GE (1957) Concluding remarks. Cold Spring Harbor Symp Quant Biol 22(0):415–427. https://doi.org/10.1101/sqb.1957.022.01.039

IUCN (2012) IUCN red list categories and criteria: Version 3.1, 2nd edn. IUCN Species Survival Commission & The World Conservation Union

Jeliazkov A, Gavish Y, Marsh CJ, Geschke J, Brummitt N, Rocchini D, Haase P, Kunin WE, Henle K (2022) Sampling and modelling rare species: conceptual guidelines for the neglected majority. Glob Change Biol 28(12):3754–3777. https://doi.org/10.1111/gcb.16114

Jiménez-Valverde A (2020) Sample size for the evaluation of presence-absence models. Ecol Indic 114:106289. https://doi.org/10.1016/j.ecolind.2020.106289

Joo R, Boone ME, Clay TA, Patrick SC, Clusella-Trullas S, Basille M (2019) Navigating through the r packages for movement. J Anim Ecol 89(1):248–267. https://doi.org/10.1111/1365-2656.13116

Kass JM, Guénard B, Dudley KL, Jenkins CN, Azuma F, Fisher BL, Parr CL, Gibb H, Longino JT, Ward PS, Chao A, Lubertazzi D, Weiser M, Jetz W, Guralnick R, Blatrix R, Lauriers JD, Donoso DA, Georgiadis C et al (2022) The global distribution of known and undiscovered ant biodiversity. Sci Adv 8(31). https://doi.org/10.1126/sciadv.abp9908

Kass JM, Smith AB, Warren DL, Vignali S, Schmitt S, Aiello-Lammens ME, Arlé E, Márcia Barbosa A, Broennimann O, Cobos ME, Guéguen M, Guisan A, Merow C, Naimi B, Nobis MP, Ondo I, Osorio-Olvera L, Owens HL, Pinilla-Buitrago GE, Sánchez-Tapia A, Thuiller W, Valavi R, Velazco SJE, Zizka A, Zurell et al (2025) Achieving higher standards in species distribution modeling by leveraging the diversity of available software. Ecography 2025: e07346. https://doi.org/10.1111/ecog.07346

Lomba A, Pellissier L, Randin CR, Vicente J, Moreira F, Honrado J, Guisan A (2010) Overcoming the rare species modelling paradox: a novel hierarchical framework applied to an Iberian endemic plant. Biol Conserv 143(11):2647–2657. https://doi.org/10.1016/j.biocon.2010.07.007

Meynard CN, Leroy B, Kaplan DM (2019) Testing methods in species distribution modelling using virtual species: what have we learnt and what are we missing? Ecography 42(12):2021–2036. https://doi.org/10.1111/ecog.04385

Moilanen A, Lehtinen P, Kohonen I, Jalkanen J, Virtanen EA, Kujala H (2022) Novel methods for spatial prioritization with applications in conservation, land use planning and ecological impact avoidance. Methods Ecol Evol 13(5):1062–1072. https://doi.org/10.1111/2041-210x.13819

Mráz P, Bourchier RS, Treier UA, Schaffner U, Müller-Schärer H (2011) Polyploidy in phenotypic space and invasion context: a morphometric study of Centaurea stoebes. L. Int J Plant Sci 172(3):386–402. https://doi.org/10.1086/658151

Ndiribe C, Pellissier L, Antonelli S, Dubuis A, Pottier J, Vittoz P, Guisan A, Salamin N (2013) Phylogenetic plant community structure along elevation is lineage specific. Ecol Evol 3(15):4925–4939. https://doi.org/10.1002/ece3.868

Patiño J, Collart F, Vanderpoorten A, Martin-Esquivel JL, Naranjo-Cigala A, Mirolo S, Karger DN (2023) Spatial resolution impacts projected plant responses to climate change on topographically complex islands. Divers Distrib 29(10):1245–1262. https://doi.org/10.1111/ddi.13757

Peterson AT, Soberón J, Pearson RG, Anderson R, Martínez-Meyer E, Nakamura M, Araújo MP (2011) Ecological niches and geographic distributions. Princeton University Press

Petitpierre B, Kueffer C, Broennimann O, Randin C, Daehler C, Guisan A (2012) Climatic niche shifts are rare among terrestrial plant invaders. Science 335(6074):1344–1348. https://doi.org/10.1126/science.1215933

Pio DV, Engler R, Linder HP, Monadjem A, Cotterill FPD, Taylor PJ, Schoeman MC, Price BW, Villet MH, Eick G, Salamin N, Guisan A (2014) Climate change effects on animal and plant phylogenetic diversity in southern Africa. Glob Change Biol 20(5):1538–1549. https://doi.org/10.1111/gcb.12524

Pulliam HR (2000) On the relationship between niche and distribution. Ecol Lett 3(4):349–361. https://doi.org/10.1046/j.1461-0248.2000.00143.x

Rödder D, Engler JO (2011) Quantitative metrics of overlaps in Grinnellian niches: advances and possible drawbacks. Glob Ecol Biogeogr 20(6):915–927. https://doi.org/10.1111/j.1466-8238.2011.00659.x

Scherrer D, Christe P, Guisan A (2019a) Modelling bat distributions and diversity in a mountain landscape using focal predictors in ensemble of small models. Divers Distrib 25(5):770–782. https://doi.org/10.1111/ddi.12893

Scherrer D, Mod HK, Pottier J, Litsios-Dubuis A, Pellissier L, Vittoz P, Götzenberger L, Zobel M, Guisan A (2019b) Disentangling the processes driving plant assemblages in mountain grasslands across spatial scales and environmental gradients. J Ecol 107(1):265–278. https://doi.org/10.1111/1365-2745.13037

Schweiger O, Klotz S, Durka W, Kühn I (2008) A comparative test of phylogenetic diversity indices. Oecologia 157(3):485–495. https://doi.org/10.1007/s00442-008-1082-2

Shen T, Song L, Corlett RT, Guisan A, Wang J, Ma W Z, Mouton L, Vanderpoorten A, Collart F (2023) Disentangling the roles of chance, abiotic factors and biotic interactions among epiphytic bryophyte communities in a tropical rainforest (Yunnan, China). Plant Biol 25(6):880–891. https://doi.org/10.1111/plb.13570

Sillero N, Campos JC, Arenas-Castro S, Barbosa AM (2023) A curated list of R packages for ecological niche modelling. Ecol Model 476:110242. https://doi.org/10.1016/j.ecolmodel.2022.110242

Soberón J (2007) Grinnellian and Eltonian niches and geographic distributions of species. Ecol Lett 10(12):1115–1123. https://doi.org/10.1111/j.1461-0248.2007.01107.x

Thuiller W, Georges D, Gueguen M, Engler R, Breiner F, Lafourcade B, Patin R (2023) biomod2: Ensemble platform for species distribution modeling. https://CRAN.R-project.org/package=biomod2

Treier UA, Broennimann O, Normand S, Guisan A, Schaffner U, Steinger T, Müller-Schärer H (2009) Shift in cytotype frequency and niche space in the invasive plant Centaurea maculosa. Ecology 90(5):1366–1377. https://doi.org/10.1890/08-0420.1

van de Pol M, Bailey LD, McLean N, Rijsdijk L, Lawson CR, Brouwer L (2016) Identifying the best climatic predictors in ecology and evolution. Methods Ecol Evol 7(10):1246–1257. https://doi.org/10.1111/2041-210x.12590

van Proosdij ASJ, Sosef MSM, Wieringa JJ, Raes N (2015) Minimum required number of specimen records to develop accurate species distribution models. Ecography 39(6):542–552. https://doi.org/10.1111/ecog.01509

Warren DL, Glor RE, Turelli M (2008) Environmental niche equivalency versus conservatism: quantitative approaches to niche evolution. Evolution 62(11):2868–2883. https://doi.org/10.1111/j.1558-5646.2008.00482.x

Zurell D, Franklin J, König C, Bouchet PJ, Dormann CF, Elith J, Fandos G, Feng X, Guillera-Arroita G, Guisan A, Lahoz-Monfort JJ, Leitão PJ, Park DS, Peterson AT, Rapacciuolo G, Schmatz DR, Schröder S-DB, Thuiller W, Yates KL et al (2020) A standard protocol for reporting species distribution models. Ecography 43(9):1261–1277. https://doi.org/10.1111/ecog.04960

The `ecodist` Package for Dissimilarity-Based Analysis of Ecological Data

Sarah Goslee

Abstract

The concepts of similarity and difference have been intrinsic to community ecology since its early origins. Many ecological questions can only be expressed in terms of distances: spatial structure, but also community similarity/dissimilarity. The clearest and most direct approach to these kinds of questions requires working with these symmetrical dissimilarity matrices directly. This chapter presents a general workflow within the dissimilarity-based analysis framework. The key analyses—correlation, ordination, and spatial correlation—are demonstrated with R code using the ***ecodist*** package for a sample dataset from the Rocky Mountains. The likely driving variables in this region are geographic distance and elevation. Do changes in these variables relate to changes in plant community composition? Do sites that are farther apart in space have less similar vegetation? What about sites that differ in elevation? Key decision points within the workflow are highlighted, making it possible to adapt this framework for other systems and other research questions.

Keywords

Community ecology · Correlation · Dissimilarity · Distance · Ordination · Mantel test · Multivariate · Spatial ecology

S. Goslee (✉)
USDA Agricultural Research Service, Pasture Systems and Watershed Management Research Unit, University Park, PA, USA
e-mail: sarah.goslee@usda.gov

D. Rocchini (ed.), *R Coding for Ecology*, Use R!,
https://doi.org/10.1007/978-3-031-99665-8_4

1 Introduction

The concepts of similarity and difference have been intrinsic to community ecology since its early origins (see for instance Goodall 1986). Originally focused on classification of vegetation units, plant community ecology has grown to encompass many applications of community similarity, both applied and theoretical. The first attempts at community classification were qualitative, but with the advent of numerical ecology, quantifying similarity between vegetation samples became a standard approach. Although classification is still an important part of ecology, this chapter will focus on ordination: placing sampling units along a gradient rather than identifying discrete groups.

Similarity between communities reflects similarity in the environmental and historical processes that led to the development of those communities, and methods for identifying pattern and inferring process have become ever more sophisticated. In 1957, Bray and Curtis described an ordination method that starts with a community dissimilarity index, and could be constructed with pencil, paper, and ruler (Bray and Curtis 1957). While the details of the methods have changed, fueled by developments in ecological theory and in computing power, the basic principles described are foundational to numerical ecology.

Key ecological questions, such as those involving spatial distance and community similarity, can only be considered within a dissimilarity framework. The clearest and most direct approach to many of these sorts of questions requires working with distance values directly, most often in the form of symmetrical dissimilarity matrices. These have a very familiar form: a table of travel distances between cities is very much the same thing as a table of distances between plots, or of compositional dissimilarities between samples.

The objective of this chapter is to lay out a roadmap for deciding when dissimilarity-based analyses are appropriate and for performing common analyses, mainly correlations on dissimilarities, ordinations, and spatial correlations. This workflow is implemented in R using the ***ecodist*** package (Goslee and Urban 2007; Goslee and Urban 2023), and illustrated with a vegetation dataset collected in the Colorado mountains, USA (Lynn et al. 2014). The analysis examples show some of the most commonly used methods for analyzing community pattern and the relationship between community pattern and spatial and environmental gradients. I highlight the main points where you might wish to consider options more suited to your study system and research questions. Since this workflow lies at the interface between community and landscape ecology, texts on either might help to adapt the workflow to your own research. Good starting points include Legendre and Legendre (2012) and Turner and Gardner (2015).

1.1 Dissimilarity-Based Analyses

There has been substantial debate on the merits of analyzing dissimilarities compared to analyzing raw data. Two underlying issues drive these debates: what are the appropriate hypotheses and what are the proper statistics?

When is a dissimilarity-based analysis appropriate? When the questions of interest are dissimilarity-based! Somers and Jackson provide a review of the hypotheses proposed for the Mantel test (Table 1 in Somers and Jackson 2022). In all cases, the hypotheses are about whether sampling units that are farther apart in one way (often spatial distance) are farther apart in another way (often community dissimilarity). Questions about spatial, genetic, or community distance cannot be restated in terms of raw data, so these analyses are appropriate. The choice of analyses should be driven by the ecological questions (Goslee 2010; Legendre and Fortin 2010).

Most examinations of the power and error rates of dissimilarity-based analyses assume that these tests should perform identically to tests on raw data. However, the hypotheses are different, and a compound dissimilarity is a synthesis of the raw data, not a duplicate. One aspect of all correlation on dissimilarity analyses that confuses users is understanding the magnitude of the correlation coefficients. We are all accustomed to working with correlations on raw data, where a perfect relationship is 1 (or −1), and have established intuition about whether a particular correlation value is meaningful. Those intuitions do not necessarily apply to working with dissimilarities! The actual maximum value is determined by the number of variables comprising the dissimilarity, the internal covariance structure, any scaling applied, and the dissimilarity metric. This maximum value can be determined for scaled data and Euclidean distance (Goslee 2010), but has not been clearly articulated for other common combinations. It is important instead to rely on permutation testing to understand the likelihood of obtaining the observed value given the structure of that particular dataset. Permutation tests shuffle (permute) the values of one of the variables and compare the randomized test statistic to the original value to assess whether the original value could have been obtained by chance. For symmetric dissimilarity matrices, the design of the permutation test is particularly important, since not all random arrangements of the values preserve symmetry. The functions described in this chapter all preserve the structure of the data (Fig. 4.1).

The analytical workflow presented here is implemented in R using the *ecodist* package (at least version 2.1; Goslee and Urban 2007; Goslee and Urban 2023). The *igraph* package (Csárdi and Nepusz 2006; Csárdi et al. 2023) provides a shortest-path graph implementation (see section on extended dissimilarities). Additional packages were used to import and harmonize the example dataset, but these are not required if the supplementary data are downloaded from the GitHub repository. The *sf* package (Pebesma 2018; Pebesma 2023) was used to project the geographic coordinates into a more suitable coordinate system. If you wish to recreate the elevation map (optional), the *terra* package is also needed (Hijmans 2023), and the *maps* package was used to create the inset of Fig. 4.2 (Becker et al. 2022). The tables were created with *knitr* (Xie 2023) and *rmarkdown* (Allaire et al. 2023).

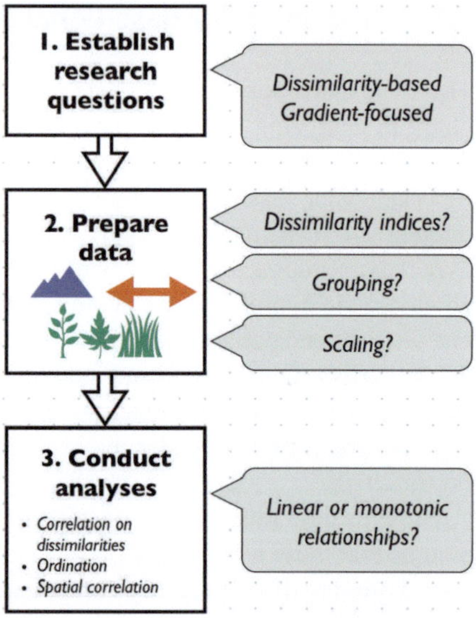

Fig. 1 Workflow for ecological dissimilarity analysis highlighting major decision points

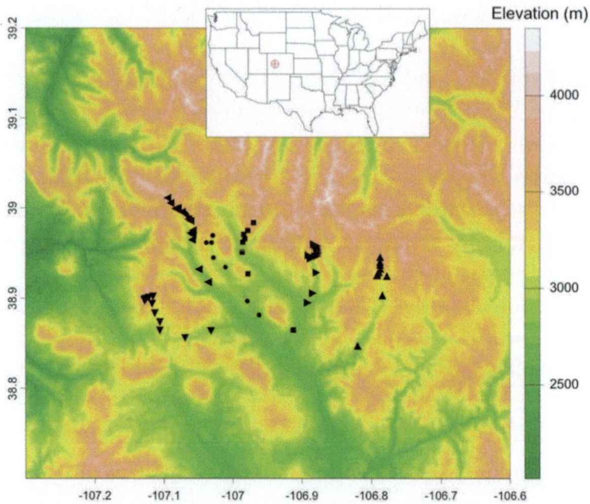

Fig. 2 Vegetation sampling sites along six elevation gradients in the Upper Gunnison Basin of the Colorado Rocky Mountains, USA (Lynn et al. 2014; US Geological Survey 2023)

The R code for the main sequence of analyses is embedded within this chapter. Where output is useful, it is included separately to facilitate copying code into your own R session. Because some of the calculations can be time-consuming, if the saved analyses are available, the calculations are not repeated. Some additional material is contained in supplementary code files, including the following:

- Downloading and formatting the example data.
- The code for creating the figures and tables.
- Custom functions *mprint()* and *mrmprint*() for prettier output.

- A custom function *ss()* to set the random seed before each analysis, so that anyone who runs the code gets exactly the same results as I did. This is not recommended for general analysis and is used here only to ensure repeatability.

All supplementary code files, as well as the harmonized data files, are available from https://github.com/RCodingForEcology.

2 Workflow Step 1

The first step in any analysis is to identify the research questions of interest (Fig. 4.1). The Gunnison plant community data, arranged along elevation transects in the Rocky Mountains, are likely to show strong elevation effects, and may show spatial effects within or between the separate gradients. Both community similarity and spatial separation can be analyzed effectively within the dissimilarity-based framework deployed here. Several research questions come to mind:

- What patterns of similarity/dissimilarity in community composition are evident?
- How do those patterns relate to location: are closer sites more similar in species composition?
- How do those patterns relate to elevation: are sites at the same elevation more similar than those at different elevations?
- Are the effects of location and elevation independent?

These questions all deal with pattern, which is a necessary precursor to understanding process. Carefully designed research questions and analyses can make it possible to eliminate some potential processes, and to develop deeper understandings of pattern that can help in the design of future field studies and analyses. It is a never-ending cycle, but we learn more with each pass.

2.1 Ecological Data

In 2014, Lynn and colleagues collected plant community composition data along six peak-to-valley gradients spanning 2700–4000 m in elevation in the Upper Gunnison Basin of the Colorado Rocky Mountains, USA (Lynn et al. 2014; Gray et al. 2018). Along each of the 6 gradients, sites were established every 100m in elevation, resulting in 9–13 sites per gradients for a total of 67 sites. For each site, species abundance was recorded in 33 quadrats (0.2 m × 0.2 m). Data were archived at the EDI Repository (Lynn et al. 2014). Further details on sampling methods are available there and in the cited references. For the analysis presented here, species abundances were averaged by site, resulting in 67 samples along six elevational gradients (Fig. 4.1).

These data have a clear spatial component with geographic distance nested within gradient, each potentially important, and elevation as a probable environmental

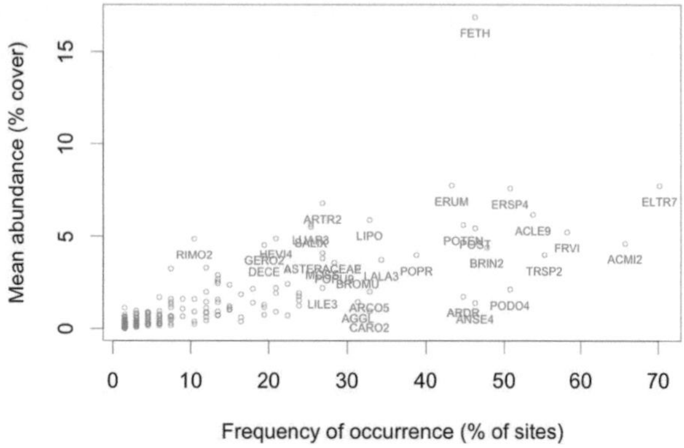

Fig. 3 Species frequency and abundance across all sites sampled. The labeled species include all those featured in ordination plots; more information is available in Table 4.1

driver (Fig. 4.2). The plant community comprises 281 species, with 7–39 species found at a site (mean = 25). The frequency-abundance curve in Fig. 4.3 highlights the dominant species, and details on those are given in Table 4.1.

Code Block 1. Workspace Configuration and Data Import

```
 # change to match preferred directory configuration
datadir <- “data”
codedir <- “code”
figdir  <- “figures”
outdir  <- “results”

# make figures and tables?
makefigs <- FALSE

library(ecodist)
# imports igraph

# custom functions
source(file.path(codedir, “0.functions.R”))

# custom color palettes
source(file.path(codedir, “1.colors.R”))

# data import
#  - veg: site by species table
#  - geo: spatial and environmental data
source(file.path(codedir, “2.gunnison.R”))

# focus on dominant and/or abundant species
veg.dom <- colSums(veg) > 250 |
  100 * colSums(veg > 0) / nrow(veg) > 25

### OPTIONAL ###
# create elevation map if desired
# both code files can stand alone
# loads additional libraries
# only used to produce Figure 1
# creates dem.tif in the data directory
# source(file.path(codedir, “3.dem.R”))
# source(file.path(codedir, “runlast.fig02.R”))

# load completed analyses if available
```

Table 1 Most abundant plant species in the Gunnison dataset. Frequency and abundance are shown in Fig. 4.3. Life form and life history information follow Lynn et al. 2014

Code	Family	Genus	Species	Lifeform	Life history
ACLE9	Poaceae	Achnatherum	lettermanii	Graminoid	Perennial
ACMI2	Asteraceae	Achillea	millefolium	Forb	Perennial
AGGL	Asteraceae	Agoseris	glauca	Forb	Perennial
ANSE4	Primulaceae	Androsace	septentrionalis	Forb	Annual
ARCO5	Caryophyllaceae	Arenaria	congesta	Forb	Perennial
ARDR	Brassicaceae	Boechera	stricta	Forb	Mixed
ARTR2	Asteraceae	Artemisia	tridentata	Shrub	Perennial
ASTERACEAE	Asteraceae	–	–	Forb	–
BRIN2	Poaceae	Bromus	inermis	Graminoid	Perennial
BROMU	Poaceae	Bromus	sp.	Graminoid	–
CARO2	Campanulaceae	Campanula	rotundifolia	Forb	Perennial
DECE	Poaceae	Deschampsia	caespitosa	Graminoid	Perennial
ELTR7	Poaceae	Elymus	trachycaulus	Graminoid	Perennial
ERSP4	Asteraceae	Erigeron	speciosus	Forb	Perennial
ERUM	Polygonaceae	Eriogonum	umbellatum	Forb/shrub	Perennial
FETH	Poaceae	Festuca	thurberi	Graminoid	Perennial
FRVI	Rosaceae	Fragaria	virginiana	Forb	Perennial
GERO2	Rosaceae	Geum	rossii	Forb	Perennial
HEVI4	Asteraceae	Heterotheca	villosa	Forb	Perennial
LALA3	Fabaceae	Lathyrus	leucanthus	Forb	Perennial
LILE3	Linaceae	Linum	lewisii	Forb	Perennial
LIPO	Apiaceae	Ligusticum	porteri	Forb	Perennial
LUAR3	Fabaceae	Lupinus	argenteus	Forb	Perennial
MOSS	–	–	–	Moss	–
PODO4	Polygonaceae	Polygonum	douglasii	Forb	Annual
POPR	Poaceae	Poa	pratensis	Graminoid	Perennial
POPU9	Rosaceae	Potentilla	pulcherrima	Forb	Perennial
POST	Poaceae	Poa	stenantha	Graminoid	Perennial
POTEN	Rosaceae	Potentilla	sp.	Forb	–
RIMO2	Grossulariaceae	Ribes	montigenum	Shrub	Perennial
SALIX	Salicaceae	Salix	sp.	Shrub	–
TRSP2	Poaceae	Trisetum	spicatum	Graminoid	Perennial

```
# otherwise recalculate everything
if(file.exists(file.path(outdir, "analyses.RDA")))
{
   load(file.path (outdir, "analyses.RDA"))
    runall <- FALSE
} else {
 runall <- TRUE
}
```

3 Workflow Step 2

There are many more distance and dissimilarity indices, each with its own strengths and weaknesses (see for instance Legendre and Legendre 2012). In this workflow, Euclidean distance is used for spatial and environmental data, and Bray-Curtis dissimilarity for community data, but a different choice might be more useful for your ecosystem and question.

But what is the difference between a distance and a dissimilarity? An index must meet several criteria to qualify as a metric index, a distance. Of these, the most relevant here is the triangle inequality, which requires that the distance from A to B plus the distance from A to C is always greater than the distance from A to C. The Euclidean distance, our familiar geographic measure, always meets this criterion, even in many dimensions. Unfortunately, it is not suitable for use with community data, because mathematically it treats mutually absent species as meaningful: two samples with no species in common could have a very low Euclidean distance because they are both missing many species.

The Bray-Curtis index is often used for community data, because it calculates the dissimilarity between species abundances of two samples while ignoring mutual absences. However, it does not follow the triangle inequality: the shortest distance between two samples can be along a path through third sample, rather than directly between the two samples themselves (Fig. 4.4). This index is described as semi-metric and, like all indices that do not fulfill all the criteria for being metric, is described as a dissimilarity rather than a distance. I use "dissimilarities" as the broader collective term: all distances are dissimilarities, but not all dissimilarities are distances. For dissimilarity indices like Bray-Curtis, which has 1 as a maximum value for samples with no species in common, community similarity can be expressed as *1-dissimilarity*. Sometimes it may be more satisfying to use one form or the other.

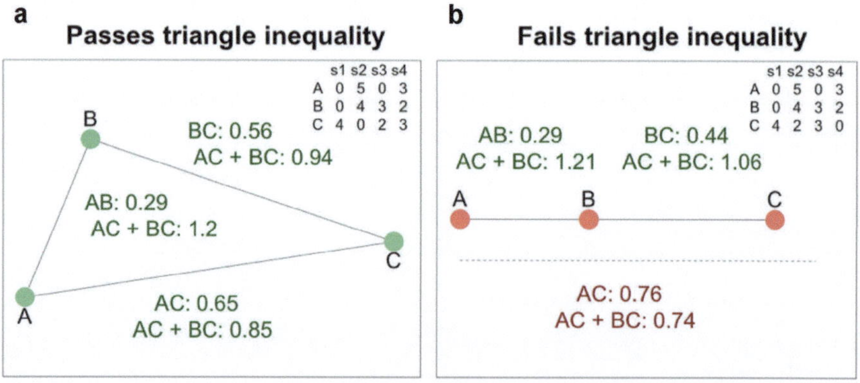

Fig. 4 Two small Bray-Curtis dissimilarity examples. In panel a, the dataset with four species and three sites passes the triangle inequality, but panel b shows a very similar dataset that fails the test. The dissimilarities in b cannot be plotted in a Euclidean (metric) space

3.1 Data Scaling

Besides the choice of index, there are several other things to consider. First is scaling the data. Environmental data should generally be standardized or scaled before calculating dissimilarities, unless (as for projected geographic data) the units are the same for all variables. Otherwise if elevation has a range of 0–10,000 but soil pH has a range of 4.5–7.5, patterns in the resulting distance matrix will be due entirely to elevation and not to pH simply because of the relative magnitudes. Scaling the data will give each variable equal weight.

Scaling choices for vegetation data are more nuanced, and there are various approaches. In this study, the raw abundance data have all values on the same scale (percentage), so the dissimilarity reflects the distribution of rare and abundant species, and is biased toward abundant species. Scaling the data, for instance to species maximum, instead gives rare and abundant species equal importance. For the Gunnison dataset, it makes little difference, with a correlation of 0.95 between Bray-Curtis dissimilarities on the raw and relativized data, but for other datasets or other questions, scaling may be needed.

Code Block 2. Data Scaling and Dissimilarity Options

```
# three different ways to conceptualize
# vegetation community composition

# raw % cover; emphasizes abundant species
veg.bc <- bcdist(veg)

# scaled to species maximum
# gives rare and abundant species the same weight
veg.rel <- relrange(veg)
veg.rel.bc <- bcdist(veg.rel)

# compares presences only
veg.jd <- distance(veg, method = "jaccard")

### compare methods

ss(); methods.mantel <- list(
    mantel(veg.bc ~ veg.rel.bc, nperm = 10000, nboot = 1000),
    mantel(veg.bc ~ veg.jd, nperm = 10000, nboot = 1000),
    mantel(veg.rel.bc ~ veg.jd, nperm = 10000, nboot = 1000))

# not continuing with these
rm(veg.rel.bc, veg.jd)

# same meaningful units in X and Y, so not scaled
geo.ed <- dist(subset(geo, select = c(X, Y)))

# single variable, so not scaled
elev.ed <- dist(geo$Elevation)
```

Another of the key decisions revolves around grouping variables for inclusion in a compound dissimilarity matrix. For this example, all plant species are grouped into one dissimilarity matrix, but for more complex questions, additional matrices might be needed. For instance, in a forested system, is difference in overstory species composition related to difference in seedling and sapling composition? What about difference in herbaceous species composition? The extremely flexible nature of the dissimilarity framework makes answering many different questions possible.

Similar decisions must be made for locational and environmental variables. Note that for location, it is necessary to project the coordinates; latitude and longitude are not distances (unless great circle calculations are used), and in nearly all cases, distance-based coordinates like the Albers Equal Area projection used here are required. I chose to use horizontal distance apart, including both x- and y-coordinates, but some applications might be better with x and y used separately.

To keep the analysis simple, elevation is the only environmental variable used here, but for more nuanced analyses, other environmental variables such as slope, aspect (decomposed into sine and cosine rather than as a circular variable), soils data, and so on might be relevant. Sorting these variables into thematic groups makes it possible to answer research questions like "Is difference in soil properties more important than difference in elevation or location for determining understory composition?"

4 Workflow Step 3

Finally we have made all the preliminary decisions, and are ready to begin the analyses. Data have been scaled, dissimilarities have been calculated for all the groups of variables of interest, and it is time to start producing results! Or, almost...

The methods presented here nearly all have two forms, a regular form that operates on the dissimilarities and a looser ranked form that uses ranked dissimilarities. This approach shows up in other kinds of statistical analyses as well: Pearson correlation on raw data, and Spearman correlation on ranked data, or the paired t-test and the Wilcoxon signed rank test. Just as for these examples, the ranked-distance methods have fewer assumptions about the form of the relationship: linear vs monotonic. Especially for semi-metric or nonmetric dissimilarities, the ranked versions may be more appropriate, but the choice will also depend on your research questions and understanding of the system you are studying. I will present both versions for the Gunnison dataset, but it would be more appropriate to select a version in advance based on your hypotheses.

4.1 Correlation

The Mantel test, a correlation on dissimilarities, provides a simple and flexible (yet controversial) framework for testing potentially complex relationships among dissimilarities. The basic Mantel test is a simple correlation, and various extensions provide partial correlations, ranked correlations that test monotonic rather than strictly linear relationships, and more. Mantel tests can even be used as a basis for variance partitioning, similar to path analysis on dissimilarities. As described above, there has a great deal of back-and-forth in the literature about the usefulness and validity of Mantel tests and dissimilarity tests in general, but much of this argument boils down to selecting appropriate statistical hypotheses; recognizing that analyses on raw data and on dissimilarities do not provide the same results; and using appropriate permutation procedures for dissimilarities (for instance, Goslee 2010).

Permutation tests of significance are crucial for dissimilarity-based statistics. These values do not necessarily follow particular predefined distributions, so probabilities cannot be determined that way. Instead, one or more components of the data being tested are randomly resorted many times. The original test statistic, for instance the Mantel correlation coefficient, is compared to the many test statistics that could be obtained from the same data in a different order. The ratio of the number of test statistics at least as extreme as the original to the total number of permutations results in a p-value describing the probability of obtaining the original result given the null hypothesis. Instead of looking up a p-value, we calculate it from scratch based on the actual distribution of the data, rather than a theoretical distribution. The more permutations, the better understanding of the distribution we have; I have used 10,000 in this chapter. You should use at least that many when preparing analyses for publication, although it is fine to use fewer when doing preliminary screening, because permutation tests can take a long time.

Code Block 3. Mantel Tests

```
if(runall)
 {
   ss();  veg.mantel <- list(
       mantel(veg.bc ~ geo.ed, nperm = 10000, nboot = 1000),
       mantel(veg.bc ~ geo.ed + elev.ed, nperm = 10000, nboot = 1000),
       mantel(veg.bc ~ elev.ed, nperm = 10000, nboot = 1000),
       mantel(veg.bc ~ elev.ed + geo.ed, nperm = 10000, nboot = 1000)
   )

   veg.mantel.labels <- c(
       "Veg ~ Space        ",
       "Veg ~ Space | Elev ",
       "Veg ~ Elev         ",
       "Veg ~ Elev | Space ")

   ###

   ss(); env.mantel <- mantel(elev.ed ~ geo.ed, nperm = 10000, nboot = 1000)

     ### compare rank-order tests

   ss(); veg.mantel.rank <- list(
       mantel(veg.bc ~ geo.ed, nperm = 10000, nboot = 1000, mrank = TRUE),
       mantel(veg.bc ~ geo.ed + elev.ed, nperm = 10000, nboot = 1000, mrank = TRUE),
       mantel(veg.bc ~ elev.ed, nperm = 10000, nboot = 1000, mrank = TRUE),
       mantel(veg.bc ~ elev.ed + geo.ed, nperm = 10000, nboot = 1000, mrank = TRUE)
   )
```

```
    veg.mantel.rank.labels <- c(
        "ranked Veg ~ Space          ",
        "ranked Veg ~ Space | Elev ",
        "ranked Veg ~ Elev           ",
        "ranked Veg ~ Elev | Space ")

    ###

    ss(); env.mantel.rank <- mantel(elev.ed ~ geo.ed, nperm = 10000, nboot = 1000, mrank = TRUE)

}
```

Results Block 1. Mantel Test Output

```
     # a. Linear relationships

> mprint(veg.mantel, veg.mantel.labels)

Veg ~ Space          r =  0.208, p = 0.0001, lim =  0.180 -  0.237
Veg ~ Space | Elev   r =  0.202, p = 0.0001, lim =  0.171 -  0.235
Veg ~ Elev           r =  0.581, p = 0.0001, lim =  0.549 -  0.615
Veg ~ Elev | Space   r =  0.580, p = 0.0001, lim =  0.547 -  0.616

> mprint(env.mantel, "Space ~ Elev")

Space ~ Elev         r =  0.077, p = 0.0206, lim =  0.045 -  0.106

    # b. Monotonic (rank-order) relationships

> mprint(veg.mantel.rank, veg.mantel.rank.labels)

ranked Veg ~ Space          r =  0.191, p = 0.0001, lim =  0.162 -  0.219
ranked Veg ~ Space | Elev   r =  0.180, p = 0.0001, lim =  0.151 -  0.213
ranked Veg ~ Elev           r =  0.605, p = 0.0001, lim =  0.569 -  0.643
ranked Veg ~ Elev | Space   r =  0.603, p = 0.0001, lim =  0.564 -  0.648

> mprint(env.mantel.rank, "ranked Space ~ Elev")

ranked Space ~ Elev         r =  0.080, p = 0.0103, lim =  0.048 -  0.107
```

The primary research question is on the relationship between plant community composition, location, and elevation, and the Mantel tests were designed to reflect that, testing the direct relationship between vegetation and each of the other variables, followed by partial Mantel tests examining the relationship with each after the other is partialled out, resulting in four tests. The fifth test tests whether the difference in elevation corresponds to the geographic distance. With 10,000 permutations, the smallest possible p-value is 0.0001, meaning that the observed Mantel r is greater than any of the permuted values. The output also provides the 95% confidence limits around r.

The Gunnison plant community composition showed a linear effect of space that was unlikely to be observed by chance, and a very strong elevation effect. While there was a relationship between geographic and elevational distances, r was small, and the partial Mantel results show that the effects of location and of elevation were largely independent: the partial r values were not much different from the simple values. For this dataset, the ranked relationships were about the same as the linear relationships.

4.2 Multiple Regression on Dissimilarities

Multiple regression on dissimilarities (MRM) offers a different view of the relationships (Lichstein 2007). Standard multiple regression functions cannot be used with dissimilarities because of the need for permutation tests of significance designed for symmetrical matrices. As for *mantel()*, *MRM()* offers options for ranked tests and for specifying the number of permutations. The regression framework is very similar to the Mantel framework, but allows separation of the independent variables. Multiple regression results are not the same as partial Mantel results conceptually, although for single variables the results are identical (given that *MRM()* returns r^2 and *mantel()* uses r). Partial Mantel tests quantify the relationship between dissimilarities once the effects of other variables have been partialled out, while MRM quantifies the contributions of each included variable jointly (Goslee 2010).

The MRM analysis of the Gunnison data supports the conclusion of the Mantel tests: that both geographic distance and difference in elevation are important determinants of plant community composition.

Code Block 4. Multiple Regression on Dissimilarities

```
if(runall)
{
  ss(); veg.bc.mrm <- MRM(veg.bc ~ geo.ed + elev.ed, nperm = 10000, mrank = FALSE)

  ss(); veg.bc.mrm.rank <- MRM(veg.bc ~ geo.ed + elev.ed, nperm = 10000, mrank = TRUE)

}
```

Results Block 2. MRM Output

```
> mrmprint(veg.bc.mrm)

$coef
              veg.bc    pval
Int      0.703179878 0.0001
geo.ed   0.000002652 0.0001
elev.ed  0.000243369 0.0001

$r.squared
    R2   pval
0.3649 0.0001

$F.test
       F   F.pval
634.3783   0.0001
```

4.3 Grouping Variables

A further extension of the correlation testing framework makes it possible to test whether groups are different by creating a dissimilarity matrix that has a value of 0 for sites in the same group, and 1 for sites in a different group (Legendre and Fortin 1989). The Gunnison data has obvious groups: the gradients. Are there differences between gradients that are not accounted for by geographic distance and elevation?

One advantage of this approach is that partial Mantel tests can address more complex questions, such as "Are there differences between gradients that cannot be explained by geographic distance?" That is, are the gradients more different that would be expected from their spatial distance? It is straightforward to calculate the group membership distance, as shown in the code below, but ***ecodist*** also provides a convenient *mgroup()* function that will create the membership distance and perform a simple Mantel test.

There is a significant difference between gradients (groups) in the Gunnison site, as shown by both the *mgroup()* test and the identical Mantel test. The partial Mantel tests, though, reveal that the differences between groups can be entirely explained by geographic distance, but is independent of elevation.

Code Block 5. Tests for Group Differences

```
if(runall)
{
  # convenience function to perform a test of group dissimilarity
   ss(); veg.bc.mg <- mgroup(veg.bc, geo$Gradient, nperm=10000)

   # duplicates mgroup: mantel(veg.bc ~ gradient.site.d)
   gradient.site.d <- dist(as.numeric(as.factor(sub("\\..*$", "", rownames(veg)))))
   gradient.site.d[gradient.site.d > 0] <- 1

   ss(); gradient.mantel <- list(
      mantel(veg.bc ~ gradient.site.d, nperm = 10000, nboot = 1000),
      mantel(veg.bc ~ gradient.site.d + geo.ed, nperm = 10000, nboot = 1000),
      mantel(veg.bc ~ gradient.site.d + elev.ed, nperm = 10000, nboot = 1000),
      mantel(veg.bc ~ gradient.site.d + geo.ed + elev.ed, nperm = 10000, nboot = 1000)
   )

   gradient.mantel.labels <- c(
      "Veg ~ Gradient              ",
      "Veg ~ Gradient | Space      ",
      "Veg ~ Gradient | Elev       ",
      "Veg ~ Gradient | Space + Elev  ")

  ###

   ss(); gradient.mantel.rank <- list(
      mantel(veg.bc ~ gradient.site.d, nperm = 10000, nboot = 1000, mrank = TRUE),
      mantel(veg.bc ~ gradient.site.d + geo.ed, nperm = 10000, nboot = 1000, mrank = TRUE),
      mantel(veg.bc ~ gradient.site.d + elev.ed, nperm = 10000, nboot = 1000, mrank = TRUE),
      mantel(veg.bc ~ gradient.site.d + elev.ed + geo.ed, nperm = 10000, nboot = 1000, mrank = TRUE)
   )

   gradient.mantel.rank.labels <- c(
      "ranked Veg ~ Gradient              ",
      "ranked Veg ~ Gradient | Space      ",
      "ranked Veg ~ Gradient | Elev       ",
```

```
        "ranked Veg ~ Gradient | Space + Elev  ")
  }
```

Results Block 3. Grouping Test Output

```
      # a. Mantel tests of group difference
  > mprint(gradient.mantel, gradient.mantel.labels)
  Veg ~ Gradient                  r =  0.062, p = 0.0054, lim =  0.041 -  0.085
  Veg ~ Gradient | Space          r = -0.055, p = 0.9938, lim = -0.073 - -0.034
  Veg ~ Gradient | Elev           r =  0.100, p = 0.0002, lim =  0.076 -  0.125
  Veg ~ Gradient | Space + Elev   r = -0.007, p = 0.5798, lim = -0.028 -  0.016

      # b. mgroup() test of group difference

  > cat(paste(capture.output(print(veg.bc.mg, digits = 4)), collapse = "\n"), "\n")

    nclust mantelr   pval
1        6 0.06228 0.0054
```

4.4 Ordination

Ordination methods are a very common way to reduce the dimensionality of a complex dataset so that major patterns can be identified and viewed in two (or sometimes three) dimensions and thus easily printed in publications. In its raw form, the Gunnison dataset comprises 67 points in 218 dimensions, one for each species: far too many for human brains to manage. The compositional dissimilarity matrix reduces the number considerably but is still 67-dimensional, so clearly further simplification is needed.

Principal coordinates analysis (PCO) is a metric ordination method that can be used with any dissimilarity index. The PCO algorithm tries to create a new configuration in Euclidean (metric) space such that the greatest variation in the data is shown along the first axes and the distances between points are preserved. The more familiar principal components analysis is very similar, but implicitly uses Euclidean distance so it is not as flexible (but fully metric!). When PCO is used with a nonmetric dissimilarity, the configuration cannot be fit exactly into a metric space, so the result will not exactly match the original data. PCO is still a useful technique, but it is important to be aware of its limitations. The *pco()* implementation in ***ecodist*** will return all of the possible dimensions, but usually the first ones are the most interesting.

Nonmetric multidimensional scaling (NMDS) instead tries to preserve the rank-order dissimilarities between samples, so it may be able to cope with semi- and nonmetric dissimilarities more ably. It is quite popular with community ecologists. Unlike PCO, the number of dimensions for NMDS has to be specified in advance. The algorithm can find local minima, so it is important to use many random starts to help ensure that a good global solution is identified. The *nmds()* function lets you specify a range of dimensions, and the number of starts to use. I have chosen to calculate 2–4 dimensions, and to use 100 random starts. It is also possible to use an initial configuration created by some other method as a starting point, instead of random configurations. Both r^2 and stress, a measure of goodness of fit, are saved for each of the random starts, and can be used to aid in choosing the dimensionality of the final ordination. The function *nmds.min()* will extract the best configuration of a given dimensionality. The two-dimensional solution is the easiest to work with, though, and is commonly used even in cases (like here) where additional dimensions improve the fit. For PCO, the eigenvalues provide a similar indication of the decline in important of each successive axis.

Code Block 6. Ordination

```
if(runall)
{
   ss(); veg.bc.nmds  <- nmds(veg.bc, mindim = 1, maxdim = 4, nits = 100)
   veg.bc.nmdsmin <- min(veg.bc.nmds)

   ss(); veg.bc.pco <- pco(veg.bc)

   ###

   # reconstructs the spatial pattern of the map, possibly rotated

   ss(); geo.nmds <- nmds(geo.ed, mindim = 2, maxdim = 2, nits = 100)
   geo.nmdsmin <- min(geo.nmds)

   ss(); geo.pco  <- pco(geo.ed)

   ###

   ss(); veg.bc.nmds.vf <- vf(veg.bc.nmdsmin, veg[, veg.dom], nperm = 10000)
   ss(); veg.bc.nmds.evf <- vf(veg.bc.nmdsmin, geo[, "Elevation", drop=FALSE], nperm = 10000)

   ###

   ss(); veg.bc.pco.vf <- vf(veg.bc.pco$vectors[, 1:2], veg[, veg.dom], nperm = 10000)
   ss(); veg.bc.pco.evf <- vf(veg.bc.pco$vectors[, 1:2], geo[, "Elevation", drop=FALSE], nperm = 10000)

}
```

Results Block 4. NMDS Ordination Output

```
 # NMDS fit metrics by dimension

> round(sapply(split(veg.bc.nmds$r2, rep(1:4, each = 100)), mean), 4)
      1      2      3      4
 0.0520 0.5581 0.6125 0.6556

> round(sapply(split(veg.bc.nmds$stress, rep(1:4, each = 100)), mean), 4)
      1      2      3      4
 0.5523 0.2673 0.2029 0.1622

# PCO axis variance indicated by eigenvalues

> head(veg.bc.pco$values)
 [1] 4.7407601 2.0543388 1.3998934 1.2148919 1.0325087 0.8903261
```

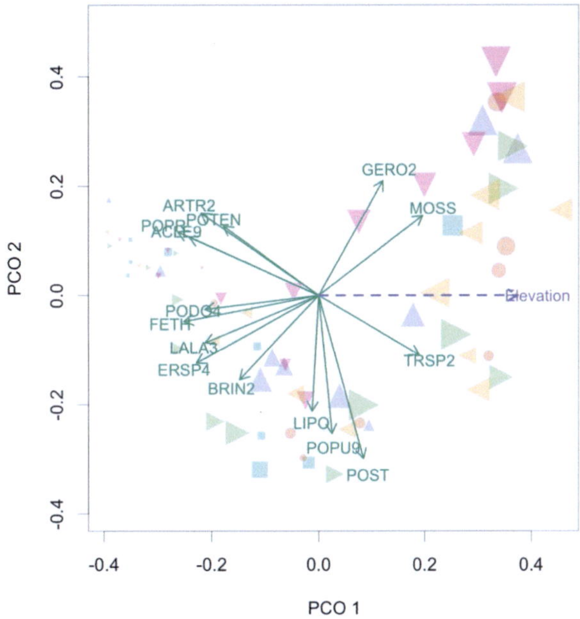

Fig. 5 PCO of Gunnison vegetation community data, with vector fitting used to place dominant species abundance and elevation onto the ordination. The ordination was rotated so elevation increases from left to right. Symbol shape and color indicate the elevation gradients (see map in Fig. 4.2), and symbol size increases with elevation

Both PCO and NMDS use a single dissimilarity matrix, here describing community composition, so if you want to overlay environmental or spatial information on the ordination a second step is required. This vector fitting of additional variables to an ordination configuration as implemented in *vf()* takes a two-dimensional ordination configuration and one or more raw data variables (not dissimilarities) and finds the direction of maximum increase of that variable within the ordination configuration. A permutation test is used to determine the probability of finding that result given the null hypothesis of no relationship. For the Gunnison example, I fitted dominant species abundances (those in Table 4.1), to provide insight into the community compositional differences producing the observed patterns, and elevation to better understand its relationship with community composition. Because *vf()* uses raw data, it is not possible to fit the spatial relationship. Figures 4.5 (PCO) and 4.6 (NMDS) both show strong patterns of vegetation community composition and elevation, and highlight similar but not identical sets of species as of particular importance.

4.5 Extended Dissimilarities

The Gunnison sites have 92 pairs of quadrats (out of 2211 pairs total) with no species in common. Those pairs all have the maximum Bray-Curtis dissimilarity value of 1. One consequence of this is that the linear relationships may be distorted, as can be clearly seen in the PCO ordination (Fig. 4.5), where the elevation gradient

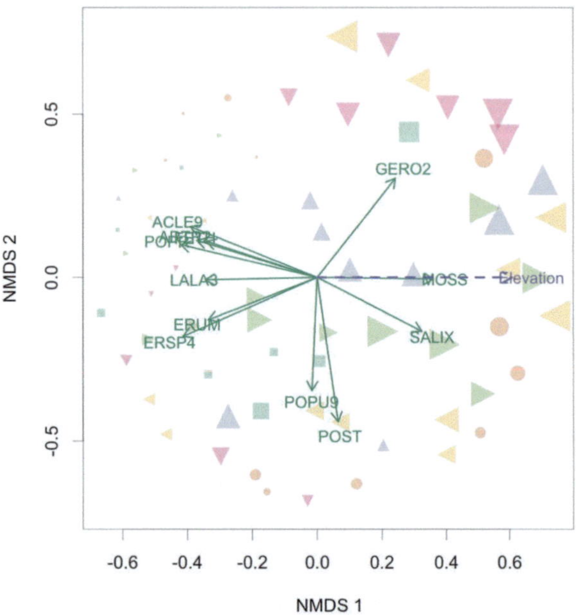

Fig. 6 NMDS of Gunnison vegetation community data, with vector fitting used to place dominant species abundance and elevation onto the ordination. The ordination was rotated so elevation increases from left to right. Symbol shape and color indicate the elevation gradients (see map in Fig. 4.2), and symbol size increases with elevation

as shown by the size of the symbols is bent into a horseshoe. This is perhaps a compelling argument to use NMDS, which is less affected by the presence of some sites with a maximum dissimilarity value (Fig. 4.6), but another approach is to estimate what the dissimilarity would be by adding together the dissimilarities for the two closest connected sites (Bradfield and Kenkel 1987, De'ath 1999). Referring back to Fig. 4.2, this would be the equivalent of determining the distance between sites A and C by adding AB and BC. The *pathdist()* function in ***ecodist*** performs this calculation using a graph-theoretic approach, and can estimate values for dissimilarities greater than a specified threshold (1 for Bray-Curtis). This approach is not always satisfying, since it results in the largest dissimilarity value for AC. The triangle inequality states that for metric distances, AC should be less than or equal to AB + BC; this method of estimation results in AC = AB + BC, creating in the discontinuity seen in Fig. 4.7. (Though note for semi-metric and nonmetric dissimilarities, AC could be larger than AB + BC.)

Using the shortest-path estimate for Bray-Curtis values equal to 1 (no species in common) changes the PCO configuration substantially (compare Figs. 4.5 and 4.8). The horseshoe effect is not eliminated, but the configuration is compressed. In this case, the shortest-path transformation does not improve the ordination. The NMDS ordination is almost unchanged (compare Figs. 4.6 and 4.9) because NMDS tries to preserve rank, rather than value. With a longer environmental gradient and more pairs of sites with no species in common, the transformation may be more valuable (Bradfield and Kenkel 1987). The extended dissimilarities can also be used in Mantel tests, or any other dissimilarity-based analysis.

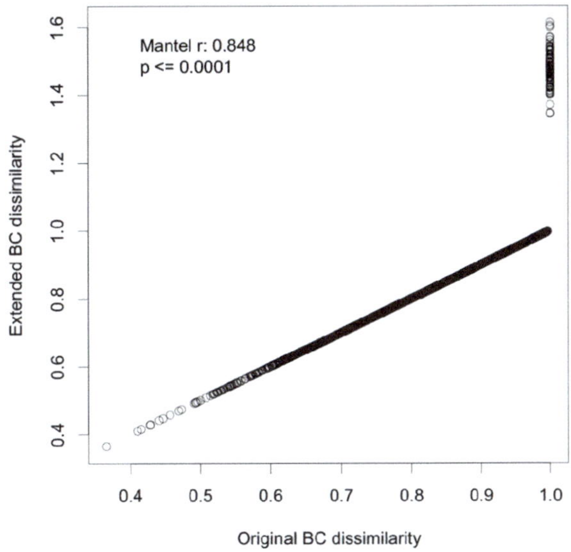

Fig. 7 Comparison of unmodified and extended dissimilarities for Gunnison vegetation community data

Fig. 8 PCO of extended dissimilarities for Gunnison vegetation community data, with vector fitting used to place dominant species abundance and elevation onto the ordination. The ordination was rotated so elevation increases from left to right. Symbol shape and color indicate the elevation gradients (see map in Fig. 4.2), and symbol size increases with elevation

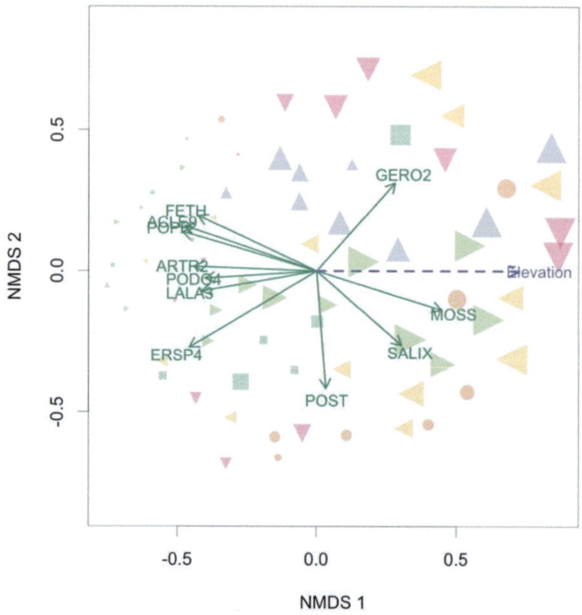

Fig. 9 NMDS of extended dissimilarities for Gunnison vegetation community data, with vector fitting used to place dominant species abundance and elevation onto the ordination. The ordination was rotated so elevation increases from left to right. Symbol shape and color indicate the elevation gradients (see map in Fig. 4.2), and symbol size increases with elevation

Code Block 7. Extended Dissimilarities

```
if(runall)
{
    veg.bc.path <- pathdist(veg.bc)

    ###

    ss(); veg.bc.nmds.path <- nmds(veg.bc.path, mindim = 2, maxdim = 2, nits = 100)
    veg.bc.nmdsmin.path <- min(veg.bc.nmds.path)

    ss(); veg.bc.nmds.vf.path <- vf(veg.bc.nmdsmin.path, veg[, veg.dom], nperm = 10000)
    ss(); veg.bc.nmds.evf.path <- vf(veg.bc.nmdsmin.path, geo[, “Elevation”, drop=FALSE], nperm = 10000)

    ###

    ss(); veg.bc.pco.path <- pco(veg.bc.path)

    ss(); veg.bc.pco.vf.path <- vf(veg.bc.pco.path$vectors[, 1:2], veg[, veg.dom], nperm = 10000)
    ss(); veg.bc.pco.evf.path <- vf(veg.bc.pco.path$vectors[, 1:2], geo[, “Elevation”, drop=FALSE], nperm = 10000)

}
```

4.6 Spatial Correlation

Mantel tests and MRM both assume linear relationships between community composition and related variables such as space. In complex landscapes, this may not be a reasonable assumption. Consider for instance a series of mountain peaks. Starting on a peak, the vegetation becomes more and more dissimilar as you travel into the valley, then more similar as you ascend the next peak, and so on. This is

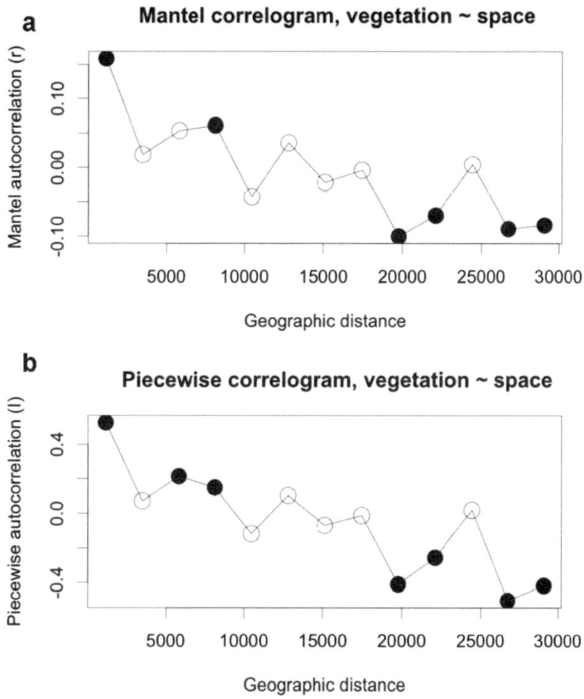

Fig. 10 Mantel correlogram and piecewise correlogram showing the spatial distances at which vegetation composition is more or less similar than would be expected to occur with random distribution. The solid symbols show values with $p <= 0.05$

contrary to the linearity assumption, or even the monotonicity of ranked tests. A logical extension would be to look at the correlation by distance classes: this creates a correlogram.

The standard Mantel correlogram calculated by *mgram()* tests the hypothesis that the mean compositional dissimilarity within a distance class differs from the mean of all the other distance classes combined. The *pmgram()* function instead produces a piecewise correlogram by testing the relationship between dissimilarities within each distance class on its own, without reference to relationships across other distance classes. This is a more satisfying hypothesis, and also makes it possible to incorporate partials into the analysis. In practice, the results of the two approaches are very similar (Fig. 4.10). The Mantel correlogram returns a Mantel correlation r; the partial correlogram uses a different statistic, comparable to Moran's I (Rossi et al. 1996), which is not bounded by -1 and 1. In both cases, permutation testing of significance is used.

The Gunnison data shows patterns of greater similarity than random at short distances, and again at around 8,000m distance, no relationship at intermediate distances, and two distances with less similarity than random at longer distances. This nonlinear relationship represents the spatial structure of the complex underlying topography and shows why the Mantel correlation value for the corresponding test was low. Much of the spatial complexity is eliminated by including elevation

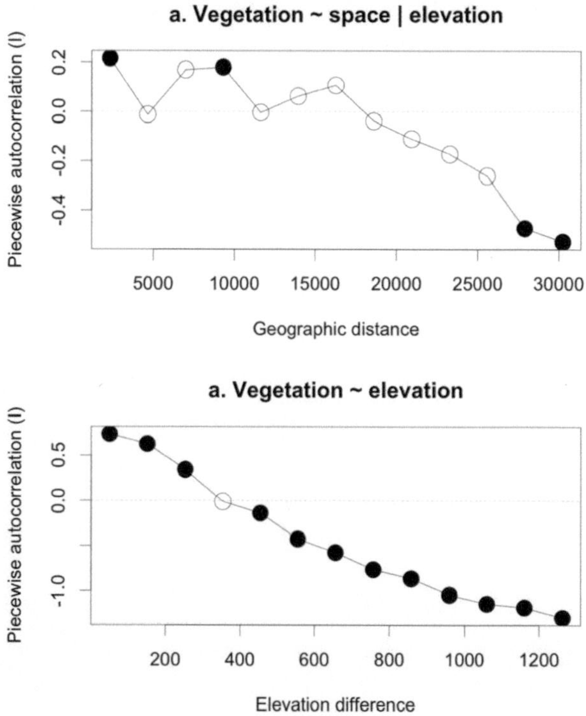

Fig. 11 Partial correlogram showing the spatial patterning of vegetation composition given elevation effects (**a**). Other distances than geographic can be used in correlograms; (**b**) shows the correlogram using elevation instead of location. The solid symbols show values with $p <= 0.05$

(Fig. 4.11a). Elevation class alone, used instead of location, shows a strongly linear pattern (Fig. 4.11b).

Code Block 8. Spatial Correlation

```
 if(runall)
{
    ss(); veg.1996.bc.pm <- pmgram(veg.bc, geo.ed, nperm = 10000)
    ss(); veg.bc.pme <- pmgram(veg.bc, geo.ed, elev.ed, nperm = 10000)

    ss(); veg.bc.epm <- pmgram(veg.bc, elev.ed, nperm = 10000)
    ss(); veg.bc.epmg <- pmgram(veg.bc, elev.ed, geo.ed, nperm = 10000)

    ss(); elev.pm <- pmgram(elev.ed, geo.ed, nperm = 10000)
    ###
    ss(); veg.bc.mm <- mgram(veg.bc, geo.ed, nperm = 10000)

}
```

After the analyses have been completed, the final steps are to save the results and to create any figures and tables, as in the code block below.

This chapter provides a general workflow for the analysis of community composition data in relation to location and environmental data, with specific examples of R code and analyses using data from a mountainous landscape. Some aspects of the data are linear, while others are better-described by more complex analyses. The dissimilarity-based framework is extremely flexible, and with careful attention to the design of the analysis can be adapted to a wide range of research questions with spatial and compositional data. All code is available from https://github.com/RCodingForEcology.

Code Block 9. Save Analyses, Figures, and Tables

```
if(runall)
{
 # save results
   f <- ls()
   f <- f[!(f %in% c("makefigs", "codedir", "col3.dark", "col3.lite", "col6.dark", "col6.dark.t", "col6.lite",
                   "sym6.filled", "datadir", "distance", "elev.ed", "figdir", "geo", "geo.ed",
                   "gradient.site.d", "mprint", "mrmprint", "pathdist", "outdir", "relrange", "rotate2d",
                   "runall", "species", "ss", "veg", "veg.bc", "veg.dom", "veg.rel", "f"))]
   save(list = f, file = "results/analyses.RDA")
   rm(f)

}

if(makefigs) Rossi et al. 1996(sapply(list.files(path = codedir, pattern = "^fig", full.names = TRUE),  source))
   rmarkdown::render(file.path(codedir, "table01.Rmd"), output_file = file.path("..", figdir, "table01_species.pdf"))

   # table 2 is incorporated into the chapter, but included
     # here to demonstrate how you might make a publication table
   rmarkdown::render(file.path(codedir, "table02.Rmd"), output_file = file.path("..", figdir, "table02_mantel.pdf"))

}
```

References

Allaire J, Xie Y, Dervieux C, McPherson J, Luraschi J, Ushey KA, Atkins H, Wickham J, Cheng W, Chang R, Iannone (2023) rmarkdown: Dynamic documents for R

Becker RA, Wilks AR, Brownrigg R, Minka TP, Deckmyn A (2022) maps: Draw Geographical Maps.

Bradfield GE, Kenkel NC (1987) Nonlinear ordination using flexible shortest path adjustment of ecological distances. Ecology 68:750–753. https://doi.org/10.2307/1938484

Bray JR, Curtis JT (1957) An ordination of the upland forest communities of Southern Wisconsin. Ecol Monogr 27:325–349. https://doi.org/10.2307/1942268

Csárdi G, Nepusz T (2006) The igraph software package for complex network research. Inter J Complex Syst:1695

Csárdi G, Nepusz T, Traag V, Horvát S, Zanini F, Noom D, Müller K (2023) igraph: Network Analysis and Visualization in R

De'ath G (1999) Extended dissimilarity: a method of robust estimation of ecological distances from high beta diversity data. Plant Ecol 144:191–199

Goodall DW (1986) Classification and ordination: their nature and role in taxonomy and community studies. Coenoses 1:3–9. https://doi.org/10.2307/43461230

Goslee SC (2010) Correlation analysis of dissimilarity matrices. Plant Ecol 206:279–286. https://doi.org/10.1007/s11258-009-9641-0

Goslee SC, Urban DL (2007) The ecodist package for dissimilarity-based analysis of ecological data. 22:1–19

Goslee SC, Urban DL (2023) ecodist: dissimilarity-based functions for ecological analysis

Gray M, Stansberry MJ, Lynn JS, Williams CF, White TE, Whitney KD (2018) Consistent shifts in pollinator-relevant floral coloration along Rocky Mountain elevation gradients. J Ecol 106:1910–1924. https://doi.org/10.1111/1365-2745.12948

Hijmans R (2023) terra: spatial data analysis

Legendre P, Fortin MJ (1989) Spatial pattern and ecological analysis. 80:107–138. https://doi.org/10.1007/BF00048036

Legendre P, Fortin M-J (2010) Comparison of the Mantel test and alternative approaches for detecting complex multivariate relationships in the spatial analysis of genetic data. Mol Ecol Resour 10:831–844. https://doi.org/10.1111/j.1755-0998.2010.02866.x

Legendre P, Legendre L (2012) Numerical Ecology. Elsevier

Lichstein JW (2007) Multiple regression on distance matrices: a multivariate spatial analysis tool. Plant Ecol 188:117–131. https://doi.org/10.1007/s11258-006-9126-3

Lynn JS, Kazenel MR, Kivlin SN, Rudgers JA (2014) Plant composition data from 67 grassland sites of the Upper Gunnison Basin, CO, USA, 2014 ver 1. Environ Data Initiat. https://doi.org/10.6073/pasta/f0050c1cfe11a5f78e7bd736c8d6f6ee

Pebesma E (2018) Simple features for R: standardized support for spatial vector data. The R Journal 10:439–446. https://doi.org/10.32614/RJ-2018-009

Pebesma E (2023) sf: Simple Features for R

Rossi JP, Lavelle P, Tondoh JE (1996) Statistical tool for soil biology. XI. Autocorrelogram and Mantel test. Eur J Soil Biol

Somers KM, Jackson DA (2022) Putting the Mantel test back together again. Ecology e3780. https://doi.org/10.1002/ecy.3780

Turner MG, Gardner RH (2015) Landscape ecology in theory and practice. Pattern and process. Springer, New York

US Geological Survey (2023) 3DEP 1 arc-second seamless digital elevation model

Xie Y (2023) knitr: a general-purpose package for dynamic report generation in R

Sampling Ecologically Meaningful Pseudo-Absences with the USE R Package

Daniele Da Re, Manuele Bazzichetto, and Enrico Tordoni

1 Introduction

Climate change (IPCC 2023), habitat loss and fragmentation (Newbold et al. 2015), and biological invasions (IPBES 2023) are profoundly transforming ecosystems, thereby increasing extinction risk and forcing species to move beyond their current ranges (Urban 2015; Newbold 2018). The necessity to anticipate future patterns of biodiversity has led to the advancement of quantitative methods for examining species distributions and their underlying drivers. These methods are fundamentally rooted in the concept of the "ecological niche."

The ecological niche of a species, as defined by Hutchinson (1957), can be thought of as the multidimensional space formed by the biotic and abiotic conditions where the species can survive and reproduce. Various, and sometimes contrasting, definitions and representations of the ecological niche have been proposed (see Sales et al. 2021 for a review), yet the concept is still pivotal to biogeographical studies (Soberón and Peterson 2020). The niche concept enables the identification of suitable environmental conditions for species and, using statistical tools, the mapping of these conditions from a multivariate environmental space to a geographical space defined by longitude and latitude coordinates.

D. Da Re (✉)
Research and Innovation Centre, Edmund Mach Foundation, San Michele all' Adige, Italy

Center Agriculture Food Environment, University of Trento, San Michele all' Adige, Italy
e-mail: daniele.dare@fmach.it

M. Bazzichetto
Faculty of Environmental Sciences, Department of Spatial Sciences, Czech University of Life Sciences Prague, Praha-Suchdol, Czech Republic

E. Tordoni
Institute of Ecology and Earth Sciences, University of Tartu, Tartu, Estonia

D. Rocchini (ed.), *R Coding for Ecology*, Use R!,
https://doi.org/10.1007/978-3-031-99665-8_5

A plethora of techniques have been developed to infer species' ecological niche and the associated species' geographical distribution using field-collected observations (e.g., species presence/absence or abundance), expert-based maps, and museum and herbarium records (Elith and Leathwick 2007; Erickson and Smith 2021). Among these techniques, species distribution models (SDMs) are probably the most widely used (Araújo et al. 2019; Franklin 2023). More specifically, correlative SDMs allow making inferences on the association between a species attribute observed in the geographical space (e.g., presence-absence, abundance) and a set of predictors typically representing the environmental conditions associated with the species occurrence. Although SDMs can be based on different statistical methods (like generalized linear models, random forest, and the maxent algorithm), they can all be used to generate geographical maps reporting a pixel-specific index of species' habitat suitability. High values of the suitability index are generally associated with areas featuring suitable environmental conditions (Guisan et al. 2017). At first glance, producing habitat suitability maps may appear to be merely a practical matter, namely, reducing the whole SDMs practice to selecting the correct input data (response variable and predictors) and algorithms. However, the reliability of SDM outputs, including suitability maps, depends on whether a set of important assumptions is met (Guisan et al. 2017). Three of these assumptions deserve particular attention:

1. The environmental variables selected for fitting SDMs are adequate to describe the ecological niche of the species.
2. The species is assumed to be at (quasi)equilibrium with the environment; in other words, it is assumed that the species has already colonized all suitable locations, and it is absent from unsuitable sites (Hattab et al. 2017).
3. The sample size and spatial coverage of the occurrence data (presence/absence or abundance) are adequate to estimate the underlying species-environment relationship (Guisan et al. 2017; Baker et al. 2022; Bazzichetto et al. 2023).

In this chapter, we will focus on the third assumption, examining its principles, implications, and impacts on SDMs.

While the information on species presence is generally available and accessible (e.g., from data platforms such as the Global Biodiversity Information Facility—GBIF), data on the species absence are more difficult to collect (Jiménez-Valverde et al. 2008) because their sampling would require time-consuming and expensive field campaigns (Hattab et al. 2017). One potential work-around is, for example, to generate pseudo-absences *in silico*. Here, we define pseudo-absences as locations within the geographical space that are likely to feature environmental conditions unsuitable for the species (Barbet-Massin et al. 2012; Sillero and Barbosa 2020). Based on this assumption, these locations are treated as absence data, even though no actual sampling was performed there. Usually, pseudo-absences are gathered in the geographical space by (i) generating a large number of points scattered across the entire study area (e.g., 10,000 points; Barbet-Massin et al. 2012) or sampling the points (ii) within or (iii) outside buffers created around presence locations

(VanDerWal et al. 2009; Bedia et al. 2013). These sampling procedures share two aspects: (1) points are located at random and (2) are generated in the geographical space. However, this aspect can introduce two detrimental effects, namely, sampling location bias and class overlap. Sampling location bias refers to the oversampling of the most prevalent habitat conditions within the study area (Tessarolo et al. 2014, 2021; Ronquillo et al. 2020), whereas class overlap indicates the degree of overlap between environmental conditions associated with species presences and absences, thus hindering the concept of pseudo-absences itself. In SDMs sampling location bias and class overlap can negatively affect the estimation of the species response to the environmental factors, especially in areas with high topographic heterogeneity or steep environmental gradients (Austin 2007; Hortal et al. 2008; Bazzichetto et al. 2023).

In this chapter, we deal with sampling bias and class overlap and propose a way to handle them in SDMs. First of all, we observe that the geographical space and environmental space are two communicating vessels: any species observation recorded in the geographical space (and therefore associated with a pair of longitude and latitude coordinates) also has a corresponding position within the multidimensional environmental space. In particular, the position of this observation within the environmental space would be determined by the environmental conditions at the location where the species was recorded in the geographical space. If we can map species presence from the geographical to the environmental space (and vice versa), we can also identify the portion of the environmental space associated with suitable conditions. This means that we can sample pseudo-absences directly within the environmental space, which may help overcome sampling location bias and class overlap. How? Regarding sampling location bias, this can be reduced by uniformly (or systematically) sampling pseudo-absences in the environmental space. In a nutshell, this is achieved by performing a principal component analysis (PCA) on the environmental variables used to define the environmental space, and then overlapping a grid to the obtained environmental space to collect pseudo-absences within each grid cell. Concerning class overlap, then, a kernel-based filter can be used to exploit the information on the environmental conditions associated with the presence data and delineate a portion of the environmental space which is less likely to include observations associated with suitable conditions for the species. This reduces the risk of including "false" (or low-quality) pseudo-absences in the datasets used to train and test the SDMs. We recently put the whole procedure described above in a unique framework, which we called the *uniform* approach, and also created the USE package to implement the *uniform* approach in R (Da Re et al. 2023).

As a result of systematically scanning the environmental space and using the kernel-based filter to exclude false pseudo-absences, the uniform approach proved to:

1. Effectively reduces sample location bias and class overlap.
2. Provides comparable predictive performance to sampling strategies carried out in the geographical space.
3. Ensures the gathering of pseudo-absences that accurately represent the variation in environmental conditions throughout the study area (Da Re et al. 2023).

In the following sections, we will explore a virtual species exercise to showcase the functionalities of the USE package and highlight its benefits. In practice, we:

1. Show how to collect pseudo-absences in the environmental space.
2. Show that the uniform approach, if compared to sampling strategies performed in the geographical space, allows for reducing sampling location bias and class overlap.
3. Provide a reproducible example of the application of the uniform approach to a typical SDM case study.

2 Virtual Species

The great advantage of using virtual species is that the ecological niche is known (i.e., it does not have to be estimated from the data), and, as a result, the geographic distribution of the virtual species is also known (Meynard et al. 2019). Here, we followed the approach used by Bazzichetto et al. (2023) and generated one virtual plant species. We focused on the species climatic niche, which we defined as a function of two bioclimatic variables: mean annual temperature (BIO1) and total annual precipitation (BIO12) (both bioclimatic variables were obtained from WorldClim2; Fick and Hijmans 2017). To lessen the computational effort of this example, the area of interest (AOI) was cropped to the spatial extent of the Italian peninsula, and the spatial resolution of the bioclimatic variables was aggregated to 0.083 degrees (~10 km at the Equator).

The virtual plant species had their thermal and precipitation optima around 8 ° C and above 1500 mm, respectively (Fig. 1a, b). This resulted in a species distributing primarily in the mountainous sectors of the Italian peninsula (Fig. 1c), following mostly a north-south gradient.

The (simulated) probability of species presence (p) across the AOI was computed as (Eq. 1).

$$\text{logit}\,(\text{pi}) = \alpha + \beta_{\text{tm}} \times \text{temperature} + \beta_{\text{tmq}} \times \text{temperature}^2 + \beta_{\text{pr}} \times \text{precipitation} \qquad (1)$$

where logit(•) is the natural logarithm of the odds pi /(1–pi), α is the model intercept, β_{tm} is the regression parameter for the linear term of temperature, β_{tmq} is the regression parameter of the quadratic term for temperature, and β_{pr} is the regression parameter for precipitation. Adding a quadratic term for temperature allowed the simulation of a unimodal response of the species to temperature. Regression

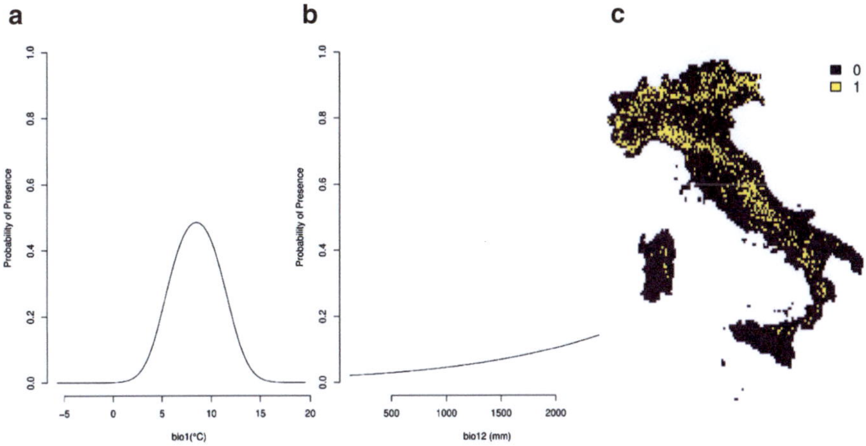

Fig. 1 Response curve of the virtual species in relation to (**a**) annual average temperature (BIO1, °C), and (**b**) total annual precipitation amount (BIO12, mm). Panel (**c**) shows the presence-absence (1 and 0 respectively) map of the virtual species for one iteration

parameters for the virtual plant species were set to −8 (α), 1 and 0.7 (β_{tm}), with the quadratic term equal to −0.1 (β_{tmq}), and 0.0009 (β_{pr}). Logit-transformed probabilities were mapped to the unit interval [0,1] using the logistic function. We then converted the species probability of presence (p) into a binary presence/absence layer, thereby mapping the species distribution into the geographical space. In particular, at each cell of the probability layer, we draw a random realization of a Bernoulli trial with probability p_i (see Eq. 1). Deriving presence/absence through the random draw from the Bernoulli distribution avoided selecting a fixed, arbitrary threshold to derive the presence/absence layer.

The procedure described above to generate the virtual species can be replicated in R using the function `USE::SpatialProba`, which is included in the USE package (from v0.1.6 available on the CRAN). Specifically, the argument `env.rast` requires a SpatRaster object with the environmental layers needed to create the layer of simulated probabilities for the virtual species (in our case the BIO1 and BIO12 raster layers obtained from WorldClim2 through the geodata package; Hijmans et al. 2024). The argument `coefs` of the `USE::SpatialProba` function requires a (numeric) vector with the values of the parameters that will be used to compute the probabilities of species presence (see Eq. 1). The name (character) of a quadratic term (`quadr_term`) can be provided in case of unimodal response curves for a single covariate. The function returns an object of the class list including a SpatRaster layer with the probability of species presence across the AOI. Additionally, the function returns plots showing the species response curves for the different environmental variables.

We utilized the `stats::rbinom` function to create a presence/absence layer based on the probability of species presence. This function simulates a Bernoulli process for each pixel, where a value of 1 indicates the presence of the species and a value of 0 indicates its absence.

```
library(sf)
library(terra)
library(geodata)
library(USE)
outdir <- getwd()
ita <- geodata::gadm(country="ITA", level=0, path = outdir)
BioData <- geodata::worldclim_global(var = 'bio',
                                     res = 5,
                                     path = outdir)
BioData <- terra::crop(BioData, ita, mask = TRUE )
names(BioData) <- paste0("bio", seq_along(names(BioData)))
myCoef <- c(-8, 1.7,  -0.1, 0.0009)
names(myCoef) <- c("intercept", "bio1", "quad_bio1", "bio12" )
VS_proba_layer <- USE::SpatialProba(coefs = myCoef,
                                    env.rast = BioData,
                                    quadr_term = "bio1",
                                    marginalPlots = TRUE)
# Convert to presence-absence using a random binomial draw
N_VS_layers <- terra::app(VS_proba_layer$rast, fun = function(x)
{ replicate(1, rbinom(n = length(x), 1, x)) })
```

3 Geographic Versus Environmental Sampling

3.1 Sampling of Pseudo-Absences in the Environmental and Geographical Space

The `USE::paSampling` function allows uniformly sample pseudo-absences within the environmental space through a two-step process:

- First, a PCA is performed on the environmental variables selected to define the environmental space. At this stage, it is worth reminding that each location in the geographical space has a corresponding position in the environmental space (i.e., it is a PC-score projected onto the space spanned by the first two PC components). Then, a kernel-based filter fitted on the presence points (in the environmental space) is used to exclude from the training (and, optionally, testing) dataset the observations likely associated with suitable conditions for the species. More specifically, all observations assigned a probability equal to or larger than a specified threshold (default set to 0.75) are considered to have suitable environmental conditions for the species and are thus excluded from the environmental space. This is achieved by drawing a convex hull that encompasses these observations.
- Second, a grid is overlaid to the environmental space, and it is systematically scanned to uniformly sample pseudo-absence from the observations left after the kernel-based filter (i.e., those lying outside the convex hull mentioned in the

previous step). Observations are randomly sampled from each cell of the grid. These observations will constitute the set of pseudo-absences that are less likely to be associated with the environmental conditions where the species was found.

From a practical point of view, our first step involves identifying the optimal resolution of the grid that will be used to sample pseudo-absences uniformly in the environmental space. To this end, we first perform a PCA to reduce the dimensionality of the environmental space using the `USE::rastPCA` function. The output of `USE::rastPCA` is then passed to the `USE::optimRes` function, which identifies, among a set of candidate resolutions (provided through the `grid.res` argument), the optimal resolution for the sampling grid.

```
rpc <- USE::rastPCA(BioData, stand = TRUE)
dt <- na.omit(as.data.frame(rpc$PCs[[c("PC1", "PC2")]], xy = TRUE))
dt <- sf::st_as_sf(dt, coords = c("PC1", "PC2"))
myRes <- USE::optimRes(sdf= dt,
                        grid.res= c(1:10),
                        perc.thr = 20,
                        showOpt = FALSE,
                        cr=5)

myRes
> myRes
$F_val
           Fval Res
 [1,] 3.8749856   1
 [2,] 2.1884022   2
 [3,] 1.5449909   3
 [4,] 1.0904999   4
 [5,] 0.9442570   5
 [6,] 0.8601195   6
 [7,] 0.6931168   7
 [8,] 0.6161875   8
 [9,] 0.5564388   9
[10,] 0.5014069  10

$Opt_res
[1] 4
```

The optimal resolution for this specific dataset is 4 (i.e., a 4 × 4 grid). Now, we pass an sf object (`myPres`), which includes the geographical coordinates of the presence data, to the `USE::paSampling`, specifically to its `pres` argument. We then set the `thres` parameter to a probability value bounded between 0 and 1 (0.75 is the default), to exclude those PC-scores likely associated with suitable conditions for the species. Finally, the `prev` parameter allows specifying a desired ratio between presences and pseudo-absences, in our example is kept to 1 (i.e., the same number of presences and pseudo-absences).

```
N_VS_layers.df <- as.data.frame(N_VS_layers, xy=TRUE, na.rm = TRUE)
names(N_VS_layers.df) <- c("x", "y", "PA")
pres <- terra::vect(subset(N_VS_layers.df, PA == 1),
                    geom = c("x", "y"),
                    crs = "EPSG:4326")

envpa <- USE::paSampling(env.rast = BioData, # environmental raster data
representing the variables of interest
                         pres = pres,
                         thres = 0.75,
                         grid.res = myRes$Opt_res,
                         prev = 1,
                         sub.ts = FALSE, plot_proc = FALSE)
```

We then use the random and buffer-out strategies to sample pseudo-absences within the geographical space. To implement the random strategy (Barbet-Massin et al. 2012), we simply generate random pseudo-absences throughout the AOI, setting prevalence equal to 1, using the `terra::spatSample`. We use the same settings for the buffer-out approach (Bedia et al. 2013). However, in this case, we first set a buffer of 15 km around each presence location and then mask out the areas included in the buffer from the random sampling.

```
# random
randpa <- terra::spatSample(BioData$bio1,
                            method= 'random',
                            size=nrow(pres),
                            xy=TRUE, as.df=TRUE,
                            na.rm=TRUE, values=FALSE)
randpa$PA <- 0

#buffer-out
buf.surface <- terra::mask(BioData$bio1,
                            terra::buffer(pres, width = 5),
                            inverse=TRUE)
buff.pa <- terra::spatSample(buf.surface,
                            method= 'random',
                            size=nrow(pres),
                            xy=TRUE, as.df=TRUE,
                            na.rm=TRUE, values=FALSE)
buff.pa$PA <- 0
```

3.2 Comparison of Sampling Location Bias and Class Overlap in Environmental and Geographical Approaches

To evaluate the degree of sampling location bias associated with each sampling strategy (i.e., uniform in the environmental space, random and buffer-out in the geographical space), we calculated the range covered by the two-component axes associated with the pseudo-absences obtained by the various sampling strategies, that is, the difference between the maximum and minimum value of the PC-scores associated with the axes. It should be noted that to do this for the random and buffer-out approaches, which are carried out in the geographical space, we first had to map the selected geographical locations onto the environmental space. Besides the observed range for each sampling strategy, we calculated the 95% confidence interval of the range through bootstrap ($n = 2000$) using the `smean.cl.boot` fuction of the Hmisc R package (Harrell Jr 2021). For simplicity, we here illustrate this procedure only for the pseudo-absences collected using the uniform approach.

```
unif.sample <- sf::st_drop_geometry(envpa) %>%
               dplyr::select(x, y, PA) %>%
               terra::vect(geom=c("x", "y"),
                           crs="EPSG:4326")
unif.sample <- unif.sample + pres
tmpVect<-na.omit(terra::extract(rpc$PCs, unif.sample, df=TRUE))
rangePC1 <- round(range(tmpVect$PC1)[2] - range(tmpVect$PC1)[1],3)
rangePC2 <- round(range(tmpVect$PC2)[2] - range(tmpVect$PC2)[1],3)
```

To quantify the degree of class overlap between presences and pseudo-absences, we computed their Gaussian hypervolumes in the environmental space using the hypervolumes R package (Blonder et al. 2014, 2022). First, we performed a PCA on the environmental conditions at presence and pseudo-absence locations for each sampling strategy, retaining the scores of the first two principal components. We then calculated the hypervolumes to quantify the multidimensional space occupied by absence (`hypAbs`) and presence (`hypPres`) data points in the PCA-transformed environmental space. Finally, we used the function `hypervolume::get_volume` to determine the overlap (`ovrlp`) between the hypervolumes of presences and pseudo-absences.

3.3 Sensitivity Analysis

For all tested sampling strategies, we kept the sample prevalence fixed at 1, which implies that an equal number of presences and pseudo-absences were sampled. To test how the different sampling approaches perform under varying sample prevalences in terms of sampling location bias and class overlap, we implemented a sensitivity analysis and repeated the workflow shown above, this time using the following prevalences: 0.25 and 0.5. In addition, only for the buffer-out approach,

```
pca.df <- data.frame(pa=unif.sample$PA,
                 terra::extract(x=BioData,
                                y=unif.sample,
                                ID=FALSE))
PCAenv <- data.frame(pa=pca.df$pa, princomp(scale(pca.df[,-1]))$scores[,1:2])
hypAbs <- hypervolume::hypervolume(PCAenv[PCAenv$pa==0,-1],verbose=FALSE)
hypPres <- hypervolume::hypervolume(PCAenv[PCAenv$pa==1,-1],verbose=FALSE)
ovrlp <- hypervolume::get_volume(hypervolume::hypervolume_set(hypPres, hypAbs,
check.memory=FALSE,verbose=FALSE))[[3]] #3 is intersection
```

we tested different buffer radii: 25 and 50 km. The entire workflow of the analyses, including the sensitivity analysis, is repeated 10 times the code of the analyses is available in the script "1_USE_chapter_workflow.R" on GitHub at https://github.com/RCodingForEcology/USE/.

3.4 Results

The uniform approach allowed sampling a wider portion of the environmental space than the geographical sampling strategies, especially along the first principal component (Fig. 2a). The uniform approach seemed less influenced by the sample prevalence than the two other sampling strategies, especially along the second principal component (Fig. 2b). However, the efficacy of the uniform approach in systematically sampling the first principal component seems to decrease at increasing prevalence (Fig. 2a).

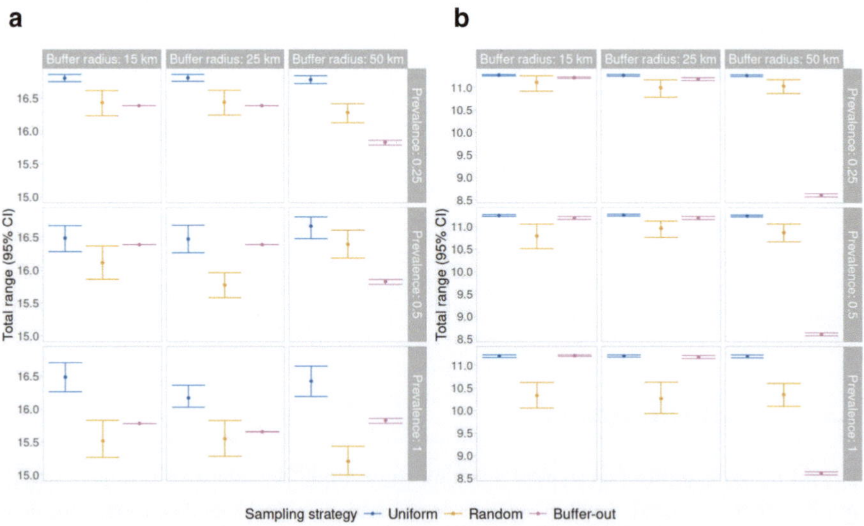

Fig. 2 Sampling location bias measured as the total range sampled by the three different sampling strategies in the (**a**) first principal component and (**b**) second principal component

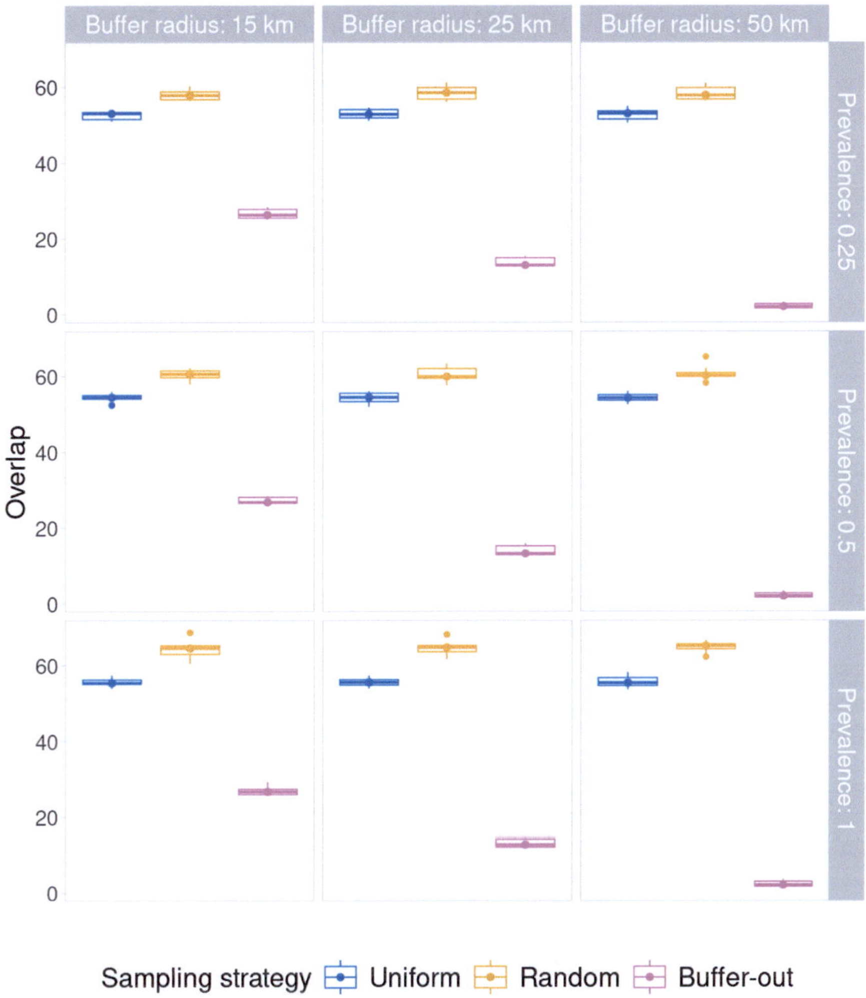

Fig. 3 Class overlap was measured using Gaussian hypervolume for the three different sampling strategies and their variations according to sample prevalence and buffer radius

Concerning class overlap, the three sample strategies seemed not to be affected by prevalence. The uniform approach always performs better than the random approach, i.e., the pseudo-absences sampled in the environmental space are less overlapping with the conditions of species' presence. On the contrary, the buffer-out approach performed better than the uniform and random approaches (Fig. 3).

4 Species Distribution Models with the Uniform Approach

In this section, we will delve into a detailed, real-world example of how to use the uniform approach for species distribution modelling. We will use *Acer pseudoplatanus*, also known as the sycamore maple, as our focal species. *Acer pseudoplatanus,* a deciduous tree native to central and southern Europe, often forms mixed stands with other broadleaf species like beech (*Fagus sylvatica*) and oaks (*Quercus* spp.). Also, it exhibits a wide ecological range, thriving in different habitats: from lowland forests to mountainous regions up to 1500 meters above sea level (Hein et al. 2009).

To begin, we download *A. pseudoplatanus* presence data from the Global Biodiversity Information Facility (GBIF) using the rgbif package (Chamberlain and Boettiger 2017), and we use the CoordinateCleaner package to refine the spatial accuracy of the data (Zizka et al. 2019). For clarity and to avoid redundancy, in the code example, we will use the already cleaned occurrence dataset (`occ object`). The R code to implement the postprocessing and cleaning of the GBIF data is not fully replicated in this chapter, but a thoroughly commented version of the script can be downloaded from the book's GitHub repository (https://github.com/RCodingForEcology/USE/script2_USE_chapter_GBIF.R).

```
library(sf)
library(terra)
library(geodata)
library(USE)
library(sdm)
# Load climatic data and Acer pseudoplatanus cleaned occurrences
bio <- geodata::worldclim_country(country = c("Italy"),
                                  var = c("bio"),
                                  path = getwd())
bio <- terra::aggregate(bio, fact = 10)
names(bio) <- paste0("bio", seq_along(names(bio)))
# load cleaned occurrences
occ <- readRDS("gbifClean.RDS")
```

As done before for the virtual species, we first find the optimal resolution of the sampling grid used to perform the uniform approach.

```
#perform PCA on the environmental space
PCA_ita <- USE::rastPCA(env.rast = bio, nPC = 2, stand = TRUE)
PCstack <- c(PCA_ita$PCs$PC1, PCA_ita$PCs$PC2)
#Get dataframe from PCstack
PCstack.df <- as.data.frame(PCstack, xy = T, na.rm = T)
#make the PCstack.df spatial
PCstack.sp <- st_as_sf(PCstack.df, coords = c("PC1", "PC2"))
#get optimal resolution of the grid for sampling European environmental
space
Optres_ita <- USE::optimRes(sdf = PCstack.sp,
                            grid.res = seq(1, 15),
                            cr = 5,
                            showOpt = FALSE)
```

At this point, we perform an environmental thinning (i.e., reducing the density of presence points in areas having similar environmental conditions) of the GBIF presence data using the function `USE::uniformSampling`. This approach is conceptually similar to the spatial thinning proposed by Aiello-Lammens et al. (2015), which aims at reducing the clustering of presences within the geographical space (Sillero and Barbosa 2020), except that here we apply it within the environmental space. This approach ensures that presences are evenly distributed across the environmental space, reducing sampling location bias and enhancing the environmental representativeness of the dataset. To this aim, we retrieve the PCA object `PCstack`, and we extract the PC-scores of the two first components associated with the presence locations of *A. pseudoplatanus*. The uniform sampling (`USE::uniformSampling`) is then employed to sample presences within the environmental space (Fig. 4).

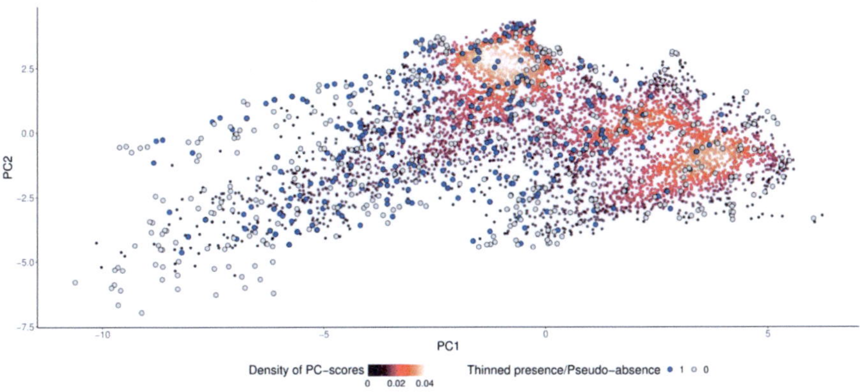

Fig. 4 Location of the thinned occurrences (blue dots) and pseudo-absences of *Acer pseudoplatanus* both sampled with the uniform approach in the environmental space

```
#extract PC-scores
occ <- occ %>%
  bind_cols(terra::extract(PCstack, occ,  xy=TRUE, ID=FALSE)) %>%
  relocate( geometry, .after = last_col())

occ.env <- occ %>%
  sf::st_drop_geometry() %>%
  sf::st_as_sf(coords=c("PC1", "PC2"))

# Perform uniform sampling within the environmental space
set.seed(17534)
occ.pres.tr.ts <- USE::uniformSampling(sdf = occ.env,
                                       grid.res = Optres_Eu$Opt_res,
                                       n.tr = 10,
                                       sub.ts = TRUE, n.ts = 10,
                                       plot_proc = TRUE)

# Convert training and testing datasets to spatial format
occ.pres.tr.ts <- lapply(occ.pres.tr.ts, function(x) {
  spdt <- x
  spdt$geometry <- NULL
  spdt.sp <- st_as_sf(spdt, coords = c("x", "y"))
  st_crs(spdt.sp) <- 4326
  return(spdt.sp)
})

# Check the number of presences in training and testing datasets
lapply(occ.pres.tr.ts, nrow)
```

Now that we have processed *A. pseudoplatanus* presence data, we can also generate pseudo-absences within the environmental space. This process involves using all cleaned species presence records (673 presences) obtained from GBIF to inform the kernel-based filter used to exclude as many "false" pseudo-absences as possible, i.e., pseudo-absences associated with suitable conditions for *A. pseudoplatanus*.

To achieve this, we adjust the prevalence of pseudo-absences (`myPrev`) to match that of the training set (`occ.pres.tr.ts$obs.tr`), ensuring that the prevalence ratio between presences and pseudo-absences remains consistent. This allows us to obtain a number of pseudo-absences comparable to the number of presences in the training set. We then sample the pseudo-absences using the `USE::paSampling` as demonstrated above for the virtual species (Fig. 4).

```
# Calculate prevalence for pseudo-absences
myPrev <- nrow(occ) / nrow(occ.pres.tr.ts$obs.tr)
myPrev

# Generate pseudo-absences using paSampling
set.seed(17893)
occ.abs <- USE::paSampling(env.rast = bio,
                           pres = occ, n.tr = 10,
                           grid.res = Optres_Eu$Opt_res,
                           prev = myPrev,
                           sub.ts = TRUE,
                           n.ts = 10, plot_proc = TRUE)

# Convert training and testing absence datasets to spatial format
abs.tr.ts.sp <- lapply(occ.abs, function(x) {
  spdt <- x
  spdt$geometry <- NULL
  spdt.sp <- st_as_sf(spdt, coords = c("x", "y"))
  st_crs(spdt.sp) <- 4326
  return(spdt.sp)
})

# Extract climatic data and join training and testing datasets
species_data <- Map(function(x, y) {
  x <- data.frame(PA = 1, st_coordinates(x))
  y <- data.frame(PA = 0, st_coordinates(y))
  colnames(x)[-1] <- colnames(y)[-1] <- c("long", "lat")
  df <- rbind(x, y)
  df <- data.frame(df, terra::extract(bio, df[c(2, 3)]))
  return(df)
}, occ.pres.tr.ts, abs.tr.ts.sp)

# Set names of the list including training and testing datasets
names(species_data) <- c("Tr_data", "Ts_data")

# Check number of presences and absences
table(species_data$Tr_data$PA)
table(species_data$Ts_data$PA)

# Ensure prevalence in the testing set equals 1 (i.e., reduce the number
of testing presences)
Subs.ts.pres <- species_data$Ts_data
Subs.ts.pres <- Subs.ts.pres[Subs.ts.pres$PA == 1, ]
Subs.ts.pres <- Subs.ts.pres[sample(x = nrow(Subs.ts.pres), size =
sum(species_data$Ts_data$PA == 0), replace = TRUE), ]


# Substitute testing dataset in species_data
species_data$Ts_data <- rbind(Subs.ts.pres,
species_data$Ts_data[species_data$Ts_data$PA == 0, ])

# Check updated prevalence
table(species_data$Ts_data$PA)

# Check for any missing values
sapply(species_data, anyNA)  # Should return FALSE for both datasets
```

In this last section, we showed how to process presence data for a real species, *A. pseudoplatanus*, for a typical SDM. This involved performing an environmental thinning of these data and sampling ecologically meaningful pseudo-absences. The current chapter provides the necessary preliminary steps to prepare presence and pseudo-absence data for accurate and reliable species distribution modeling. Here, we do not dive into the statistical aspects and best practices of species distribution modeling, as this topic is thoroughly covered in other chapters of the book (see chapter "The sdm R Package for Species Distribution Modeling"). However, we owe the readers a short demonstration, and therefore we will quickly build an SDM for *A. pseudoplatanus* using the sdm R package (Naimi and Araújo 2016). The whole procedure is available at the book's GitHub repository (https://github.com/RCodingForEcology/USE/script3_USE_chapter_SDM.R).

First, we set the percentage of data to be used for testing (`percTesting`), the number of model repetitions (`Nreps`), and the number of cores to be used for parallel processing (`myCores`). Using the `sdm::sdmData` function, we specify the response and predictor variables, as well as the training and testing datasets. Next, the sdm function is used to fit models using two methods: generalized linear models (GLM) and boosted regression trees (BRT). The models are replicated multiple times (`Nreps`), and the training data is partitioned into training and testing sets based on the specified percentage (`percTesting`). Parallel processing is enabled to speed up the computation. After fitting the models, the `sdm::rcurve` and `sdm::getVarImp` functions are used to inspect the response curves and variable importance for both GLM and BRT models. This provides insights into how each environmental variable influences the predicted presence (or absence) of *A. pseudoplatanus.* As a final step, we build an ensemble model by combining the predictions from both the GLM and BRT models using a weighted averaging approach, where the weights are determined based on each model's performance measured by the True Skill Statistic (TSS). This ensemble method leverages the strengths of both models, aiming to improve predictive accuracy by accounting for the individual performance of each model in capturing species distribution patterns. The final ensemble map, after converting suitability to presence-absence accordingly to a threshold that maximizes sensitivity and specificity, is displayed in Fig. 5, together with the chorological map of the species for comparison (Caudullo et al. 2017). All the codes used in this chapter are available for consultation at https://github.com/RCodingForEcology/USE

```
percTesting <- 30
Nreps <- 5
myCores <- 5

# extract environmental data
d <- sdm::sdmData(formula=PA~.  + coords(long+lat),
             train=species_data.f$Tr_data,
             test=species_data.f$Ts_data)
d

#model
m1 <- sdm::sdm(PA~.,
          data=d,
          methods=c('glm', 'brt'),
          replication='sub',
          test.percent=percTesting,
          n=Nreps,
          parallelSettings=list("parallel", ncore=myCores))

# Inspect the model
m1

# Response curves and variable importance
#glm
sdm::rcurve(m1, id=1:5)
sdm::getVarImp(m1, id=1:5)
#rf
sdm::rcurve(m1, id=6:10)
sdm::getVarImp(m1, id=6:10)

#ensemble model predictions based on a Weighted averaging that is
weighted using TSS statistic
e1 <- sdm::ensemble(m1, newdata = bio.rsp, #sdm not updated yet to terra
              overwrite = TRUE,
              setting = list(method='weighted',stat='tss'))
plot(e1)
```

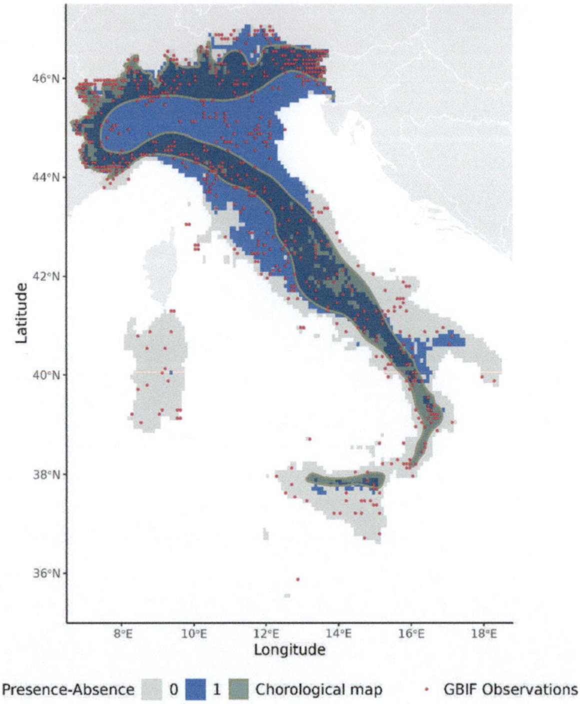

Fig. 5 Ensemble map of predicted presence-absence for *Acer pseudoplatanus*, converted from suitability using a threshold optimized for sensitivity and specificity. The chorological map of the species is overlaid for comparison

References

Aiello-Lammens ME, Boria RA, Radosavljevic A, Vilela B, Anderson RP (2015) spThin: an R package for spatial thinning of species occurrence records for use in ecological niche models. Ecography 38(5):541–545

Araújo MB, Anderson RP, Barbosa AM, Beale CM, Dormann CF, Early R, Garcia RA, Guisan A, Maiorano L, Naimi B, O'Hara RB (2019) Standards for distribution models in biodiversity assessments. Science. Advances 5(1):eaat4858

Austin M. (2007) Species distribution models and ecological theory: A critical assessment and some possible new approaches. Ecological Modelling 200(1-2):1–19. https://doi.org/10.1016/j.ecolmodel.2006.07.005

Baker DJ, Maclean IMD, Goodall M, Gaston KJ (2022) Correlations between spatial sampling biases and environmental niches affect species distribution models. Glob Ecol Biogeogr:1–13. https://doi.org/10.1111/geb.13491

Barbet-Massin M, Jiguet F, Albert CH, Thuiller W (2012) Selecting pseudo absences for species distribution models: how, where and how many? Methods Ecol Evol 3(2):327–338. https://doi.org/10.1111/j.2041-210X.2011.00172.x

Bazzichetto M, Lenoir J, Da Re D, Tordoni E, Rocchini D, Malavasi M, Barták V, Sperandii MG (2023) Sampling strategy matters to accurately estimate response curves' parameters in species distribution models. Glob Ecol Biogeogr 32:1717–1729

Bedia J, Herrera S, Gutiérrez JM (2013) Dangers of using global bioclimatic datasets for ecological niche modeling. Limitations for future climate projections. Glob Planet Chang 107:1–12. https://doi.org/10.1016/j.gloplacha.2013.04.00

Blonder B, Lamanna C, Violle C, Enquist BJ (2014) The n-dimensional hypervolume. Glob Ecol Biogeogr 23(5):595–609. https://doi.org/10.1111/geb.12146

Blonder B, Morrow CB, Harris DJ, Brown S, Butruille G, Laini A, Chen D (2022) Hypervolume: high dimensional geometry, set operations, projection, and inference using kernel density estimation, support vector machines, and convex hulls. R package version 3.0.4. https://CRAN.R-project.org/package=hypervolume

Caudullo G, Welk E, San-Miguel-Ayanz J (2017) Chorological maps for the main European woody species. Data Brief 12:662–666. https://doi.org/10.1016/j.dib.2017.05.007

Chamberlain SA, Boettiger C (2017) R Python, and Ruby clients for GBIF species occurrence data (No. e3304v1). PeerJ Preprints

Da Re D, Tordoni E, Lenoir J, Lembrechts JJ, Vanwambeke SO, Rocchini D, Bazzichetto M (2023) USE it: uniformly sampling pseudo-absences within the environmental space for applications in habitat suitability models. Methods Ecol Evol 14(11):2873–2887

Elith J, Leathwick J (2007) Predicting species distributions from museum and herbarium records using multiresponse models fitted with multivariate adaptive regression splines. Divers Distrib 13(3):265–275

Erickson KD, Smith AB (2021) Accounting for imperfect detection in data from museums and herbaria when modeling species distributions: combining and contrasting data-level versus model-level bias correction. Ecography 44(9):1341–1352

Fick SE, Hijmans RJ (2017) WorldClim 2: new 1-km spatial resolution climate surfaces for global land areas. Int J Climatol 37(12):4302–4315. https://doi.org/10.1002/joc.5086

Franklin J (2023) Species distribution modelling supports the study of past, present and future biogeographies. J Biogeogr 50:1533–1545. https://doi.org/10.1111/jbi.14617

Guisan A, Thuiller W, Zimmermann NE (2017) Habitat suitability and distribution models: with applications in R. Cambridge University Press

Harrell F Jr (2021) Hmisc: Harrell miscellaneous. R package version 4.6-0. https://CRAN.R-project.org/package=Hmisc

Hattab T, Garzón-López CX, Ewald M, Skowronek S, Aerts R, Horen H, Brasseur B, Gallet-Moron E, Spicher F, Decocq G, Feilhauer H, Honnay O, Kempeneers P, Schmidtlein S, Somers B, van de Kerchove R, Rocchini D, Lenoir J (2017) A unified framework to model the potential and realized distributions of invasive species within the invaded range. Divers Distrib 23(7):806–819. https://doi.org/10.1111/ddi.12566

Hein S, Collet C, Ammer C, Goff NL, Skovsgaard JP, Savill P (2009) A review of growth and stand dynamics of Acer pseudoplatanus L. in Europe: implications for silviculture. Forestry 82(4):361–385

Hijmans RJ, Barbosa M, Ghosh A, Mandel A (2024) _geodata: download geographic Data_. R package version 0.6-2. https://CRAN.R-project.org/package=geodata

Hortal J, Jiménez-Valverde A, Gómez JF, Lobo JM, Baselga A (2008) Historical bias in biodiversity inventories affects the observed environmental niche of the species. Oikos 117(6):847–858

Hutchinson G (1957) Concluding remarks cold Spring Harbor symposia on quantitative biology, vol 22. GS SEARCH, pp 415–427

IPBES (2023) In: Roy HE, Pauchard A, Stoett P, Renard Truong T, Bacher S, Galil BS, Hulme PE, Ikeda T, Sankaran KV, McGeoch MA, Meyerson LA, Nuñez MA, Ordonez A, Rahlao SJ, Schwindt E, Seebens H, Sheppard AW, Vandvik V (eds) Summary for policymakers of the thematic assessment report on invasive alien species and their control of the intergovernmental science-policy platform on biodiversity and ecosystem services. IPBES Secretariat, Bonn, Germany. https://doi.org/10.5281/zenodo.7430692

IPCC (2023) In: Core Writing Team, Lee H, Romero J (eds) Summary for policymakers. In: climate change 2023: synthesis report. Contribution of working groups I, II and III to the sixth assessment report of the intergovernmental panel on climate change. IPCC, Geneva, Switzerland, pp 1–34. https://doi.org/10.59327/IPCC/AR6-9789291691647.001

Jiménez-Valverde A, Lobo JM, Hortal J (2008) Not as good as they seem: the importance of concepts in species distribution modelling. Divers Distrib 14(6):885–890. https://doi.org/10.1111/j.1472-4642.2008.00496.x

Meynard CN, Leroy B, Kaplan DM (2019) Testing methods in species distribution modelling using virtual species: what have we learnt and what are we missing? Ecography 42(12):2021–2036

Naimi B, Araújo MB (2016) sdm: a reproducible and extensible R platform for species distribution modelling. Ecography 39(4):368–375

Newbold T (2018) Future effects of climate and land-use change on terrestrial vertebrate community diversity under different scenarios. Proc R Soc B Biol Sci 285(1881):20180792. https://doi.org/10.1098/rspb.2018.0792

Newbold T, Hudson L, Hill S et al (2015) Global effects of land use on local terrestrial biodiversity. Nature 520:45–50. https://doi.org/10.1038/nature14324

Ronquillo C, Alves-Martins F, Mazimpaka V, Sobral-Souza T, Vilela-Silva B, Medina NG, Hortal J (2020) Assessing spatial and temporal biases and gaps in the publicly available distributional information of Iberian mosses. Biodivers Data J 8. https://doi.org/10.3897/BDJ.8.e53474

Sales LP, Hayward MW, Loyola R (2021) What do you mean by "niche"? Modern ecological theories are not coherent on rhetoric about the niche concept. Acta Oecol 110:103701

Sillero N, Barbosa AM (2020) Common mistakes in ecological niche models. Int J Geogr Inf Sci 35:1–14

Soberón J, Peterson AT (2020) What is the shape of the fundamental Grinnellian niche? Theor Ecol 13(1):105–115

Tessarolo G, Rangel TF, Araújo MB, Hortal J (2014) Uncertainty associated with survey design in species distribution models. Diversity and Distributions 20(11):1258–1269. https://doi.org/10.1111/ddi.12236

Tessarolo G, Lobo JM, Rangel TF, Hortal J (2021) High uncertainty in the effects of data characteristics on the performance of species distribution models. Ecological Indicators, 121, 107147. https://doi.org/10.1016/j.ecolind.2020.107147

Urban MC (2015) Accelerating extinction risk from climate change. Science 348(6234):571–573

VanDerWal J, Shoo LP, Graham C, Williams SE (2009) Selecting pseudo-absence data for presence-only distribution modeling: how far should you stray from what you know? Ecol Model 220(4):589–594. https://doi.org/10.1016/j.ecolmodel.2008.11.01

Zizka A, Silvestro D, Andermann T, Azevedo J, Duarte Ritter C, Edler D et al (2019) CoordinateCleaner: standardized cleaning of occurrence records from biological collection databases. Methods Ecol Evol 10(5):744–751

Rarefaction Curves with the Rarefy Package

Elisa Thouverai, Sandrine Pavoine, Enrico Tordoni, Alessandro Chiarucci, Carlo Ricotta, Duccio Rocchini, and Giovanni Bacaro

1 Introduction

Rarefy is a new *R package* that offers a suite of new functions useful to perform rarefaction analysis based on any diversity metric. It can calculate the expected values of a given taxonomic, functional, or phylogenetic index for a reduced sampling size while considering a spatially constrained and distance-based ordering of the sampling units. Rarefy and the functions within this package offer an advanced solution for ecologists seeking to "rarefy" any diversity metric while taking into account the contributions of a distance-based ordination of sampling units. This approach provides more ecologically meaningful and unbiased estimates of expected diversities. In this chapter, we will explore the concept of rarefaction, its evolution to incorporate spatial autocorrelation, and its application with different types of metrics while examining the functions available in Rarefy.

```
install.packages("devtools")
library(devtools)
install_github("elith95/Rarefy")
```

E. Thouverai (✉)
Department of Life Sciences, University of Trieste, Trieste, Italy

BIOME Lab, Department of Biological, Geological and Environmental Sciences, Alma Mater Studiorum University of Bologna, Bologna, Italy
e-mail: elisa.thouverai2@unibo.it

S. Pavoine
Centre d'Ecologie et des Sciences de la Conservation (CESCO), Muséum national d'Histoire naturelle, CNRS, Sorbonne Université, Paris, France

E. Tordoni
Institute of Ecology and Earth Sciences, University of Tartu, Tartu, Estonia

D. Rocchini (ed.), *R Coding for Ecology*, Use R!,
https://doi.org/10.1007/978-3-031-99665-8_6

2 Classic Rarefaction Curves

The concept of rarefaction in ecology and biodiversity analysis has a long history and was not introduced by a single researcher. Instead, it is a fundamental concept that has evolved over time as ecologists and statisticians developed methods to compare and standardize species richness or diversity in ecological samples (see Chiarucci et al. 2008 for a literature review). Some of the early work related to rarefaction can be attributed to the pioneering ecologist Robert H. Whittaker. He, in his 1960 paper titled "Vegetation of the Siskiyou Mountains, Oregon and California", introduced the concept of "rarefied" species richness as a way to standardize species richness estimates when comparing ecological samples of different sizes. Later, an analytic formulation of rarefaction curve using species richness was firstly derived by Shinozaki (1963) and then independently rediscovered by a number of authors during the 1970s (Kobayashi 1974; Holthe 1975; Engen 1976; Smith et al. 1979) and in the last decades (Ugland et al. 2003; Colwell et al. 2004).

Rarefaction is a method used for constructing curves that depict the increase in detected species as the sample size grows. It is widely employed in ecology to elucidate species diversity patterns (Bacaro et al. 2012; Sanders 1968). Specifically, it is most commonly utilized to compare species richness in assemblages of varying sizes to a common level of sampling effort or sample completeness and to extrapolate species richness to a larger sampled area (Ricotta et al. 2020). Moreover, rarefaction can be applied to assess whether a site has been adequately sampled: a curve that approaches an asymptote suggests that few or no additional species are likely to be found in further sampling units, whereas a curve that continues to rise when all samples are considered indicates that many species remain unsampled (Bacaro et al. 2012).

In 1968, Sanders first used rarefaction methods to compare species richness in marine benthic communities sampled with different sample sizes. Sanders observed that a valid comparison of species richness between different sample sizes could be achieved by considering the same number of individuals after constructing

A. Chiarucci
BIOME Lab, Department of Biological, Geological and Environmental Sciences, Alma Mater Studiorum University of Bologna, Bologna, Italy

C. Ricotta
Department of Environmental Biology, University of Rome "La Sapienza", Rome, Italy

D. Rocchini
BIOME Lab, Department of Biological, Geological and Environmental Sciences, Alma Mater Studiorum University of Bologna, Bologna, Italy

Faculty of Environmental Sciences, Department of Applied Geoinformatics and Spatial Planning, Czech University of Life Sciences Prague, Praha, Suchdol, Czech Republic

G. Bacaro
Department of Life Sciences, University of Trieste, Trieste, Italy
e-mail: gbacaro@units.it

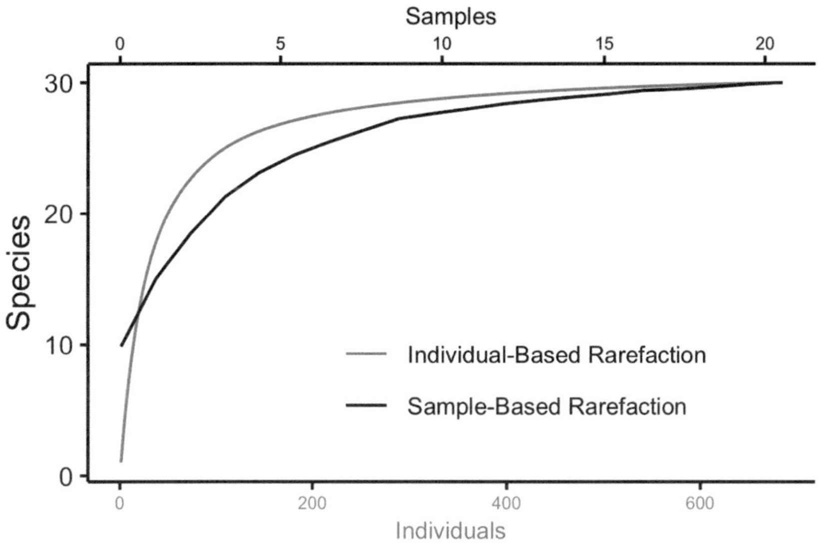

Fig. 1 Individual and sample-based rarefaction curves constructed with the dune dataset from vegan package (Oksanen et al. 2022)

the rarefaction curve. Consequently, a rarefaction curve for a collection of M individuals, plotted as the values of species richness against M, is defined as an individual-based rarefaction curve (Hubbell 2001). In 1971, John Hurlbert presented the concept of sample-based rarefaction for comparing species richness among different numbers of sampling units. As a result, a rarefaction curve for a collection of N sampling units, plotted as the values of species richness per sampling unit against N, is defined as a sample-based rarefaction curve (Gotelli and Colwell 2001). Therefore, as shown in Fig. 1, two types of rarefaction curves (individual-based and sample-based) can be distinguished by the type of accumulation unit utilized: an individual or a sampling unit.

Sample-based rarefaction curves have the advantage that only presence or absence of species needs to be detected or recorded, meanwhile individual-based rarefaction curves require a measure of abundance for each species, e.g., a number of individuals (Chiarucci et al. 2008). The new *R package* `Rarefy` was specifically developed to construct sample-based rarefaction curves.

From a methodological point of view, sampled-based rarefaction curves can be calculated using the two following methods:

- **Analytic formulation**: The formula calculates the expected number of species obtained by combining n equal-sized plots randomly and without replacement from a larger sample of N plots:

$$E[S_n] = \sum_{i=1}^{S} \left[1 - \frac{\binom{N-N_i}{n}}{\binom{N}{n}} \right] \quad (1)$$

where $E[S_n]$ is the expected species richness in the n sampling units, S is the total species richness in the N sampling units, and N_i is the number of sampling units that contain species i. $E[S_n]$ can be easily decomposed into additive terms to quantify the contribution of any subset of species to the expected species richness of the entire assemblage (Ricotta et al. 2010). The same function is also used for the calculation of individual-based rarefaction considering n and N as the number of individuals and the population size, respectively (Ricotta et al. 2012).

- **Accumulation curves**: This method involves estimating multiple accumulation curves by repeatedly conducting random, non-replacement resampling of the pool of N sampling units. Species richness is calculated for $1, 2, 3, ..., N$ sampling units to construct one curve. The average of the accumulation curves calculated represents the sample-based rarefaction curve for the given pool of sampling units (Bacaro et al. 2012). To construct individual-based rarefaction curves, the accumulation curves are generated by randomly resampling individual organisms, rather than sampling units (McGlinn et al. 2019).

A prerequisite for rarefaction analysis is that the spatial distribution of individuals in the environment is random, that sample sizes are sufficient, and that the assemblages being compared have been sampled using the same sampling design (Gotelli and Colwell 2001). It is essential to consider these assumptions when calculating a rarefaction curve to avoid potential issues. If sample sizes are insufficient, rarefaction will be unable to distinguish between different richness patterns. Additionally, if the sampling methods are not identical, various species may be either over- or under-represented in different samples (Gotelli and Colwell 2001). Furthermore, in sample-based rarefaction, any spatial or temporal autocorrelation between individuals can lead to nonrandom occurrences of sampling units, resulting in an overestimation of diversity in the study area (Gotelli and Colwell 2001).

3 From Classic to Spatially Explicit Rarefaction

Rarefaction to an equal number of sampling units is a common practice when comparing species richness in regions of different sizes (Ricotta et al. 2012). However, two regions with different spatial extent are not directly comparable by a similar number of sampling units (Chiarucci et al. 2009). The larger area is likely to contain more individuals and exhibit greater environmental heterogeneity, resulting in sampling units with more significant differences in species composition. This can lead to an overestimation of the mean number of species for a given sample size (Chiarucci et al. 2009). Consequently, to account for the impact of spatial extent, it is essential to consider spatial autocorrelation between sampling units when constructing a rarefaction curve. This adjustment cannot be achieved analytically

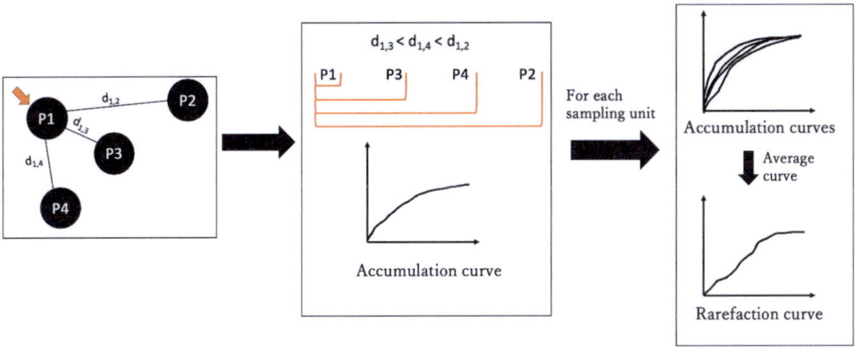

Fig. 2 Schematic representation of the process applied to calculate spatially explicit rarefaction curve. Specifically, one sampling unit is selected, and the other sampling units are sorted considering their spatial distance from the first one: One accumulation curve is constructed. The process is repeated for all the sampling units of the assemblage and the mean of the accumulation curves obtained is the spatially explicit rarefaction curve

using a formula, as explained earlier, but it necessitates adapting the way sampling units are resampled to estimate accumulation curves. Chiarucci et al. (2009) and Bacaro et al. (2012) have proposed a straightforward method for addressing this aspect during the construction of rarefaction curves, which is described as follows (Fig. 2):

- Picked one sampling unit randomly from a pool of N sampling units, the others are sampled in order of its spatial distance, and the accumulated species richness is calculated for $1, 2, 3, \ldots, N$ sampling units (as a term of reference, this corresponds to calculate a collector's curve based on the proximity of sampling units).
- The process is repeated for all the sampling units in the assemblage, and the rarefaction is then calculated as the average curve deriving from all the accumulation curves so obtained.

A rarefaction curve calculated using the described method is called spatially explicit rarefaction curve (SER) (Bacaro et al. 2016). As shown in Fig. 3 in Bacaro et al. (2016) which compared SER and classic rarefaction curves for two published vegetation datasets (Maremma Park and Migliarino San Rossore Masaciuccoli Park) that differ in both sampling strategies and vegetation type, SER curves increased less steeply than the classic ones resulting in lower estimates of species richness for a given extent.

`Rarefy` provides the function `directionalSAC()` for calculating spatially explicit rarefaction. It also calculates classic rarefaction using the analytic formulation (Eq. 1), allowing for result comparisons. To use the function, you need to provide the community matrix (sampling units x species) and a vector of numeric values, a matrix, a dataframe, or an object of class dist to define the order in

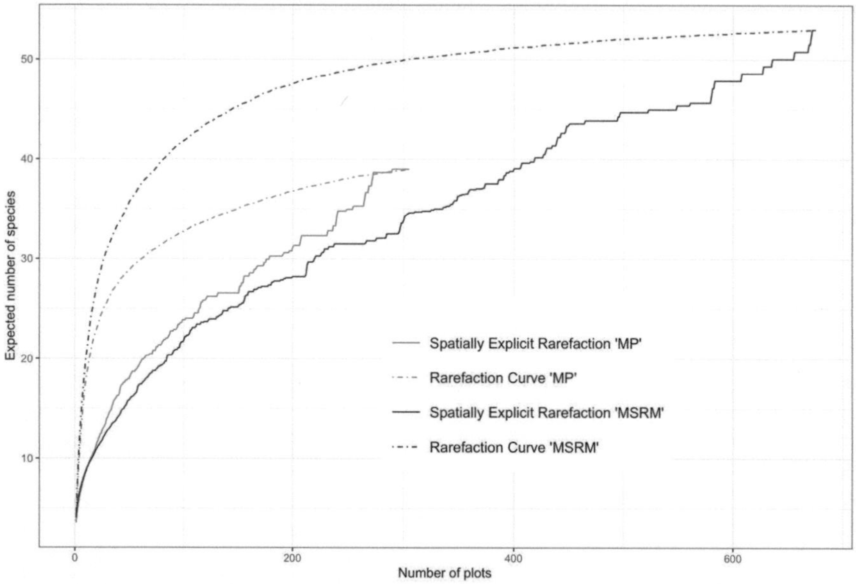

Fig. 3 Classic and spatially explicit rarefaction curves for two Tuscan coastal dune parks (Maremma Park and Migliarino San Rossore Masaciuccoli Park). For the SERs, the Maremma Park (MP) shows on average more species than those collected in an equally sampled area in the Migliarino San Rossore Masaciuccoli Park (MSRM), the reverse of the pattern of the RCs (Source: Bacaro et al. 2016)

which the sampling units should be accumulated. Here is an example using the data included in the package. The dataset is a list containing information about the abundance of native and alien vascular plants sampled in 128 16 m^2 squared plots in the Northern Adriatic Dunes, along with the geographic coordinates of the sampling units (Tordoni et al. 2019):

```
library(Rarefy)

#Sampling units x species dataset
data(duneFVG)

#Geographic coordinates of the sampling units
data(duneFVG.xy)

#Calculation of the sampling units distance matrix
dist_sp <- dist(duneFVG.xy$tot.xy)

#Spatially-explicit rarefaction
ser_rarefaction <- directionalSAC(duneFVG$total, dist_sp)
```

Table 1 Metrics included in `rare_alpha()`. The available names taken by the argument "`method`" of the function are indicated. In the formulas, S is the number of species, p_i is the relative abundance of species i, and q is the parameter that regulates the sensitivity to species abundance

Name	Reference	Formula
'HCDT'	HCDT entropy (Havrda and Charvát 1967; Daròczy 1970; Tsallis 1988)	$HCDT = \frac{\left(1-\sum_{i=1}^{S} p_i^q\right)}{(q-1)}$
'hill'	Hill numbers (Hill 1973)	${}^qD = \left(\sum_{i=1}^{S} p_i^q\right)^{1/(1-q)}$

Moreover, the `rare_alpha()` function offers additional metrics (see Table 1) for calculating taxonomic diversity beyond species richness. Users should provide the community matrix and an object of class dist for the calculation of spatially explicit rarefaction. The metric to be used for the rarefaction should also be specified. An example using Hill numbers is presented based on the same dataset introduced above:

```
ser_rare_hill <- rare_alpha(duneFVG$total,
                             dist_sp,
                             method = 'hill',
                             q = 2,
                             spatial = TRUE)
```

4 Extending Rarefaction to Other Components of Biological Diversity

Biodiversity encompasses several dimensions that complicate its characterization. Taxonomic diversity is the most commonly used measure, but current species diversity metrics treat all species as if they were functionally equivalent and evolutionarily independent. Consequently, they overlook the fact that the loss of certain species can have a more substantial impact on ecosystem functioning and evolutionary legacy (Arnan et al. 2017). To address this issue, diversity measures such as functional and phylogenetic diversities are necessary. These metrics can also be incorporated into the calculation of rarefaction curves.

4.1 Functional Rarefaction Curves

New methods have been proposed for the calculation of functional (Ricotta et al. 2010; Ricotta et al. 2012) rarefaction curves. Functional diversity is considered

Table 2 Metrics included in `ser_functional()`. The available names taken by the argument "`method`" of the function are indicated. In the formulas, S is the number of species. For the Chao index, $a_i(\tau)$ is the combined abundance of all functionally indistinct individuals from species i, $\nu_i(\tau)$ represents the attribute contribution of species i for a threshold level τ, n_+ is the total number of individuals in the community, and q is the parameter that determines the sensitivity of the measure to the relative abundance of the species. For the Rao index, p_i is the vector of relative species abundance within plot i, D is the matrix of functional dissimilarities among species, and d_{kl} is the functional dissimilarity between species k and l

Name	Reference	Formula
`'chao'`	Chao's functional β-diversity (Chao et al. 2019)	${}^{q}\mathrm{FD}(\Delta(\tau)) = \left(\sum_{i=1}^{S} \nu_i(\tau) \left(\frac{a_i(\tau)}{n_+} \right)^{(1/1-q)} \right)$
`'rao'`	Rao's quadratic entropy (Rao 1982)	$Q\,(p_i, D) = \sum_{k=1}^{S} \sum_{l=1}^{S} p_{kl_i} p_{kl_j} d_{kl}$

crucial for understanding ecosystem processes, responses to environmental stress, disturbance, and highlighting biogeographical patterns (Chao et al. 2019). It can be quantified using measures based on species' functional traits, which define species in terms of their ecological roles, combined with measures of their abundance (Chao et al. 2019). For instance, Rao's quadratic entropy Q (Rao 1982), which accounts for both functional pairwise distances between species and species' relative abundances, has been widely used to quantify functional diversity. This measure was used by Ricotta et al. (2012) to define a new type of rarefaction curve, known as "functional rarefaction." Functional rarefaction describes the increase in expected functional diversity as the sample size considered in the study area grows (Ricotta et al. 2012).

`Rarefy` presents the function `ser_functional()` for the computation of spatially explicit functional rarefaction. Users should provide the community matrix, a dist object containing species distances calculated based on their functional traits, and another dist object with distances between sampling units. An example using the Rao index is demonstrated (see Table 2 for the complete set of metrics). The data utilized is once again the dataset provided by the package, specifically, a list containing the functional traits of the sampled species:

```
library(ade4)

#Functional traits dataset
data(duneFVG.tr8)

#Calculation of the species pairwise distances
##Nominal traits
tr8_N <- duneFVG.tr8$traits.tot[,c(1,3,4)]
##Binary traits
tr8_D <- data.frame(duneFVG.tr8$traits.tot[,2])
##Quantitative traits
```

```
tr8_Q <- duneFVG.tr8$traits.tot[,5:15]
##Gower distance between species
tr8dist <- dist.ktab(ktab.list.df(list(tr8_N,
                                       tr8_D,
                                       tr8_Q)),
                     type = c('N','D','Q'))

#Spatially explicit functional rarefaction
ser_rare_rao <- ser_functional(duneFVG$total,
                               tr8dist,
                               dist_sp,
                               method = 'rao')
```

4.2 Phylogenetic Rarefaction Curves

Phylogenetic diversity is rooted in evolutionary histories, taxonomic classification, and well-supported phylogenetic trees of a community, shedding light on the role of species interactions and biogeographic histories in community structure and composition (Nipperess and Matsen 2013). The most widely used phylogenetic metric is Faith's phylogenetic diversity (PD), which is defined as the sum of the branch lengths of a phylogenetic tree connecting all species in the target assemblage (Faith 1992). This measure was employed by Chao et al. (2015) for the estimation and extrapolation of phylogenetic rarefaction curves. Phylogenetic rarefaction illustrates the increase in phylogenetic diversity with growing sample size in the study area (Chao et al. 2015).

Rarefy package provides the functions rare_phylo() for constructing phylogenetic rarefaction curves and ser_phylo for spatially explicit phylogenetic rarefaction curves. Users should provide the community matrix, a phylogenetic tree, and an object of class dist containing the distances between sampling units for spatially explicit rarefaction. An example is illustrated using Faith's index (see Table 3 for the complete set of metrics), utilizing the phylocom data provided by the picante package, which is designed for implementing ecophylogenetic approaches. A fake distance matrix between sampling units is also produced to show spatially explicit phylogenetic rarefaction:

```
library(picante)
data(phylocom)

#Fake sampling units coordinates
xy <- duneFVG.xy$tot.xy[1:6,]
rownames(xy) <- rownames(phylocom$sample)

#Sampling units distance matrix
dist_sp1 <- dist(xy)

#Phylogenetic rarefaction
rare_faith <- rare_phylo(phylocom$sample,
                         phylocom$phylo,
                         method = 'faith')
```

Table 3 Metrics included in `rare_phylo()` and `ser_phylo()`. The available names taken by the argument "`method`" of the function are indicated. In the formulas, B is the number of branches, L_i represents the length of branch i, p_i is the relative abundance of species i, and q is the scaling constant that weights the importance of rarity of the species. In the equation of Barker's PD, A_i represents the average abundance of the species sharing the branch i. In Pavoine's index, $H_{a,k}$ is the HCDT entropy computed for every evolutionary time period K and $t_K - t_{K-1}$ is the length of K

Name	Reference	Formula
'barker'	Barkerś weighted PD (Barker 2002)	$PDw = B \times \frac{\sum_i^B L_i A_i}{\sum_i^B A_i}$
'faith'	Faith's phylogenetic diversity (Faith 1992)	$PD = \sum_{i \in B} L_i$
'hill', 'tsallis', 'renyi'	Feature richness indexes (Pavoine and Ricotta 2019)	${}^q fdiv_{Hill} = \left[\sum_{i \in B} L_i (p_i)^q\right]^{\frac{1}{1-q}}$ ${}^q fdiv_{HCDT} = \frac{1-\sum_{i \in B} L_i (p_i)^q}{q-1}$ ${}^q fdiv_{Renyi} = \frac{1}{1-q} log\left[\sum_{i \in B} L_i (p_i)^q\right]$
'Ia'	Pavoine's index (Pavoine et al. 2009)	$I_a = \sum_{K=1}^{N} (t_K - t_{K-1})\, H_{a,K}$

```
#Spatially explicit phylogenetic rarefaction
ser_rare_faith <- ser_phylo(phylocom$sample,
                            phylocom$phylo,
                            dist_sp1,
                            method = 'faith')
```

4.3 Rarefaction of β Diversity

Spatially explicit or gradient-oriented (directional) turnover curves as a function of sampling effort have been firstly defined by Ricotta et al. (2019). This method provides insights into the use of rarefaction curves for analyzing β diversity, offering a significant contribution to summarizing multiple-site dissimilarity resulting from a species turnover process. This approach also incorporates spatial autocorrelation in species composition among plots along a spatial, temporal, or environmental gradient. `directionalSAC()` function can be used to calculate directional, normalized directional, nondirectional, normalized nondirectional β diversity as a function of sampling effort. A normalized measure of autocorrelation for directional β diversity calculated as the normalized difference between directional and nondirectional β is also available. An example of the directional rarefaction of β diversity is here provided using the mite dataset available in the `vegan` package (Oksanen et al. 2022). The same example is discussed in Appendix 1 in Ricotta et al. (2019):

Table 4 Metrics included in `rare_beta()`. The available names taken by the argument "`method`" of the function are indicated. In the Bray-Curtis dissimilarity equation, N is the number of plots, and x_i and x_j are the abundances of the species x in plots i and j. In the Cody index formula, $g(H)$ is the number of species gained along the habitat gradient H, while $l(H)$ is the number of lost species along the gradient. In the Jaccard index equation, a is the number of species in common between two plots, and α_1 and α_2 are the alpha diversities of the two plots. In the Whittaker species turnover formula, γ is the total species number of the study area, and α is the mean species number of the plots considered

Name	Reference	Formula
'bray'	Bray-Curtis dissimilarity (Bray and Curtis 1957)	$\beta_{bray} = \frac{\sum_{i,j}^{N}(x_i - x_j)}{\sum_{i,j}^{N} x_i + x_j}$
'cody'	Cody index (Cody 1975)	$\beta_c = \frac{[g(H)+l(H)]}{2}$
'jaccard'	Jaccard similarity coefficient (Jaccard 1901)	$\beta_j = \frac{a}{(\alpha_1+\alpha_2-a)}$
'whittaker'	Whittakerś species turnover (Whittaker 1960; Whittaker 1972)	$\beta_w = \frac{\gamma}{\alpha} - 1$

```
library(vegan)

#Community matrix
data(mite)

#Environmental data for the gradient
data(mite.env)

#Turnover curve
beta_directional <- directionalSAC(comm_matrix, mite.env$SubsDens)
```

In addition, the package `Rarefy` includes the function `rare_beta()`, which computes spatially explicit rarefaction exploiting different metrics of beta diversity (see Table 4):

```
ser_rare_bray <- rare_beta(duneFVG$total,
                           dist_sp,
                           method = "bray",
                           spatial = TRUE)
```

5 Customized Rarefaction

Finally, the `Rarefy` package provides users with the flexibility to choose a custom metric for calculating spatially and nonspatially explicit rarefaction through the function `design_rare()`. This flexibility allows users to tailor the rarefaction analysis to their specific needs. Users can select a personalized metric choosing between two methods:

- **"fun_div"**: A user-defined function will be used for the calculation of the rarefaction curve. The function must calculate the value of the chosen diversity index per plot and return a numeric vector with the values calculated.
- **"equation"**: A formula written by the user and passed to the function as a string will be used for the calculation of the rarefaction curve.

In the following example, the function `speciesdiv()` from `adiv` package (Pavoine 2020) is employed to perform spatially explicit rarefaction on the duneFVG dataset using the "`fun_div`" method.

Firstly, a list is created, which contains the values required for the `speciesdiv()`. In this list, the value NA should be provided in place of the community matrix, and the names of the elements within the list must correspond to the argument names expected by `speciesdiv()` function:

```
#List of the values for speciesdiv()
a <- list(NA,'Shannon')

#Assign to the list names matching the arguments of speciesdiv()
names(a) <- c('comm','method')

#Rarefaction
ser_rare_shannon <- design_rare(duneFVG$total,
                                dist_sp,
                                method = "fun_div",
                                args = a,
                                fun = "speciesdiv",
                                spatial = TRUE,
                                mean = TRUE)
```

The argument "`mean`" set TRUE means that the average instead of the sum of the species relative abundances is calculated during the accumulation of the sampling units. The reason is that Shannon is a concave index, so it can produce alteration to the final rarefaction curve due to its mathematical properties (see Ricotta et al. 2012 for more details).

In the second example, a rarefaction curve will be calculated using the "`equation`" method, employing the $FRic$ index, which is a measure of functional richness. The $FRic$ index is calculated as the sum of the product of species abundance, denoted as w_i, and the Euclidean distance of each species, denoted as d_i, from the centroid of the functional trait space:

$$FRic = \sum w_i \times d_i \tag{2}$$

where w_i represents the abundance of species i, and d_i is the Euclidean distance of species i from the centroid of the functional trait space.

```
#Quantitative traits
tr8_Q <- duneFVG.tr8$traits.tot[,5:15]
```

```
#Centroid coordinates calculated as mean
c <- colMeans(tr8_Q, na.rm = T)

#Distances from the centroid per species
d <- apply(tr8_Q, 1, function(r) dist(rbind(r, c)))

#Formula
#
eq <- "sum(unlist(lapply(1:length(w), function(i) w[i] * d[i])))"

#List of terms of the equation
#(w values are extracted from the community matrix)
a <- list(NA, d)

#Set the names of the list to match the formula terms
names(a) <- c("w", "d")

#Rarefaction
r <- design_rare1(duneFVG$total,
                  dist_sp,
                  method = "equation",
                  args = a,
                  formula = eq,
                  spatial = TRUE
                  )
```

6 Other Packages

Other packages and R functions are available that can calculate parts of the same or alternative methods to those provided by `Rarefy`. For example, nonspatially explicit rarefaction using species richness indices can be computed using the `iNEXT()` package's `iNEXT` version 3.0.0 function (Chao et al. 2014). Alternatively, the `vegan` package (Oksanen et al. 2022) offers the functions `specaccum()` and `rarefy()` for similar purposes. Nonspatially explicit functional rarefaction can be calculated using the `rare_Rao()` function included in the `adiv` package version 2.2 (Pavoine 2020).

7 Final Remarks

Spatial patterns are a fundamental aspect in the study of ecology, as the magnitude (and sometimes even the direction) of change in species diversity is significantly influenced by spatial scale (McGlinn et al. 2019) and the distribution of ecological and geographical measures. Consequently, there is a growing need for methods that can provide a comprehensive view of the biodiversity in a study area while accounting for spatial autocorrelation (Bacaro et al. 2016). Furthermore, it is increasingly recognized that to fully comprehend the structure, composition, and dynamics of natural communities, it is essential to measure various dimensions of biodiversity, including taxonomic diversity (species richness and turnover), phylogenetic diversity (the accumulated evolutionary history of a community), and functional diversity (diversity of morphological, physiological, and ecological traits) (Arnan et al. 2017).

In this context, as we saw in this chapter, `Rarefy` package offers the capability to calculate spatially explicit rarefaction using various metrics of functional (function `ser_functional()`), phylogenetic (function `ser_phylo()`), and taxonomic (functions `rare_alpha()` and `rare_beta()`) diversity. This allows for the exploration of different aspects of biodiversity, while taking into account spatial autocorrelation between sampling units. Other packages lack the versatility found in `Rarefy` when it comes to selecting measures. They also do not provide the option to choose a metric that is not already included in the package. Compared to other packages, `Rarefy` thus allows for the exploration of different aspects of biodiversity, while taking into account spatial autocorrelation between sampling units.

References

Arnan X, Cerdá X, Retana J (2017) Relationships among taxonomic, functional, and phylogenetic ant diversity across the biogeographic regions of Europe. Ecography 40:448–457

Bacaro G, Altobelli A, Cameletti M, Ciccarelli D, Martellos S, Palmer MW, Ricotta C, Rocchini D, Scheiner SM, Tordoni E, Chiarucci A (2016) Incorporating spatial autocorrelation in rarefaction methods: implications for ecologists and conservation biologists. Ecol Indic 69:233–238

Bacaro G, Rocchini D, Ghisla A, Marcantonio M, Neteler M, Chiarucci A (2012) The spatial domain matters: spatially constrained species rarefaction in a free and open source environment. Ecol Compl 12:63–69

Barker GM (2002) Phylogenetic diversity: a quantitative framework for measurement of priority and achievement in biodiversity conservation. Biol J Linnean Soc 76:165–194

Bray JR, Curtis JT (1957) An ordination of the upland forest communities of Southern Wisconsin. Ecol Monographs 27:325–349

Chao A, Gotelli NJ, Hsieh TC, Sander EL, Ma KH, Colwell RK, Ellison AM (2014) Rarefaction and extrapolation with Hill numbers: a framework for sampling and estimation in species diversity studies. Ecol Monographs 84:45–67

Chao A, Chiu CH, Hsieh TC, Davis T, Nipperess DA, Faith DP (2015) Rarefaction and extrapolation of phylogenetic diversity. Methods Ecol Evol 6:380–388

Chao A, Chiu C-H, Villéger S, Sun I-F, Thorn S, Lin Y-C, Chiang J-M, Sherwin WB (2019) An attribute-diversity approach to functional diversity, functional beta diversity, and related (dis)similarity measures. Ecol Monographs 89(2):e01343

Chiarucci A, Bacaro G, Rocchini D, Fattorini L (2008) Discovering and rediscovering the sample-based rarefaction formula in the ecological literature. Commun Ecol 9:121–123

Chiarucci A, Bacaro G, Rocchini D, Ricotta C, Palmer M, Scheiner S (2009) Spatially constrained rarefaction: incorporating the autocorrelated structure of biological communities into sample-based rarefaction. Commun Ecol 10(2):209–214

Cody ML (1975) Towards a theory of continental species diversities: bird distributions over Mediterranean habitat gradients. In: Cody ML, Diamond JM (eds) Ecology and Evolution of Communities. Belknap Press, Harvard, pp. 214–257

Colwell RK, Mao CX, Chang J, (2004) Interpolating, extrapolating, and comparing incidence-based species accumulation curves. Ecology 85:2717–2727

Daròczy Z (1970) Generalized information functions. Inf Control 16:36–51

Engen S (1976) A note on the estimation of the species–area curve. J du Conseil International Pour l'Exploration de la Mer 36:286–288

Faith DP (1992) Conservation evaluation and phylogenetic diversity. Biol Conserv 61:1–10

Gotelli NJ, Colwell RK (2001) Quantifying biodiversity: procedures and pitfalls in the measurement and comparison of species richness. Ecol Lett 4:379–391

Havrda J, Charvát F (1967) Quantification method of classification processes. Concept Struct-Entropy Kybernetika 3:30–35

Hill MO (1973) Diversity and evenness: a unifying notation and its consequences. Ecology 54:427–432

Holthe T (1975) A method for the calculation of ordinate values of the cumulative species–area curve. J du Conseil International Pour l'Exploration de la Mer 36:183–184

Hubbell S (2001) The unified neutral theory of biodiversity and bio geography. Princeton University Press, Princeton, NJ

Jaccard P (1901) Étude comparative de la distribution florale dans une portion des Alpes et des Jura. Bull de la Société vaudoise des sciences naturelles 37:547–579

Kobayashi S (1974) The species-area relation I. A model for discrete sampling. Res Popul Ecol 15:223–237

McGlinn DJ, Xiao X, May F, Gotelli NJ, Engel T, Blowes SA, Knight TM, Purschke O, Chase JM, McGill BJ (2019) Measurement of Biodiversity (MoB): a method to separate the scale-dependent effects of species abundance distribution, density, and aggregation on diversity change. Methods Ecol Evol 10:258–269

Nipperess DA, Matsen FA (2013) The mean and variance of phylogenetic diversity under rarefaction. Methods Ecol Evol 4:566–572

Oksanen J, Simpson G, Blanchet F, Kindt R, Legendre P, Minchin P, O'Hara R, Solymos P, Stevens M, Szoecs E, Wagner H, Barbour M, Bedward M, Bolker B, Borcard D, Carvalho G, Chirico M, De Caceres M, Durand S, Evangelista H, FitzJohn R, Friendly M, Furneaux B, Hannigan G, Hill M, Lahti L, McGlinn D, Ouellette M, Ribeiro Cunha E, Smith T, Stier A, Ter Braak C, Weedon J (2022). vegan: Community Ecology Package. *R package* version 2.6–4. https://CRAN.R-project.org/package=vegan

Pavoine S (2020) adiv: An r package to analyse biodiversity in ecology. Methods Ecol Evol 11:1106–1112. https://doi.org/10.1111/2041-210X.13430

Pavoine S, Ricotta C (2019) A simple translation from indices of species diversity to indices of phylogenetic diversity. Ecol Ind 101:552–561

Pavoine S, Love MS, Bonsall MB (2009) Hierarchical partitioning of evolutionary and ecological patterns in the organization of phylogenetically-structured species assemblages: application to rockfish (genus: Sebastes) in the Southern California Bight. Ecol Lett 12:898–908

Rao CR (1982) Diversity and dissimilarity coefficients: a unified approach. Theor Popul Biol 21:24–43

Ricotta C, Burrascano S, Blasi C (2010) Incorporating functional dissimilarities into sample-based rarefaction curves: from taxon resampling to functional resampling. J Veget Sci 21:280–286

Ricotta C, Pavoine S, Bacaro G, Acosta ATR (2012) Functional rarefaction for species abundance data. Methods Ecol Evol 3:519–525

Ricotta C, Acosta TRA, Bacaro G, Carboni M, Chiarucci A, Rocchini D, Pavoine S (2019) Rarefaction of beta diversity. Ecolol Ind 107:105606

Ricotta C, Laroche F, Szeidl L, Pavoine S (2020) From alpha to beta functional and phylogenetic redundancy. Methods Ecol Evol 11:487–493

Sanders HL (1968) Marine benthic diversity: a comparative study. Am Naturalist 102:243–282

Shinozaki K (1963) Note on the species area curve. In: Proceedings of the 10th Annual Meeting of the Ecological Society of Japan. (in Japanese), pp 1–5

Smith W, Grassle JF, Kravitz D (1979) Measures of diversity with unbiased estimates. In: Grassle JF, Patil GP, Smith W, Taille C (eds) Ecological Diversity in Theory and Practice. International Co-operative Publishing House, Fairland, pp. 177–191

Tordoni E, Petruzzellis F, Nardini A, Savi T, Bacaro G (2019) Make it simpler: Alien species decrease functional diversity of coastal plant communities. J Veg Sci 30:498–509

Tsallis C (1988) Possible generalization of Boltzmann-Gibbs statistics. J Stat Phys 52:479–487

Ugland KI, Gray JS, Ellingsen KE (2003) The species–accumulation curve and estimation of species richness. J Animal Ecol 72:888–897
Whittaker RH (1960) Vegetation of the Siskiyou mountains, Oregon and California. Ecol Monographs 30:279–338
Whittaker RH (1972) Evolution and measurement of species diversity. Taxon 21:213–251

Part III

Ecological Remote Sensing and Spatial Ecology

A Plant Trait-Based Approach to Map Forest Biodiversity Using Imaging Spectroscopy and Physical Modeling

J.-B. Féret, F. de Boissieu, and M. Lang

Abstract

This chapter provides a workflow taking advantage of imaging spectroscopy and physical modeling to assess a set of vegetation traits then used to feed remotely sensed (RS) diversity mapping techniques in the context of forest ecosystems. This approach intends to explicitly account for vegetation traits related to structural, compositional, and functional properties when computing diversity metrics. This study is based on three main R packages. The packages *prospect* and *prosail* are used to assess vegetation biophysical properties from imaging spectroscopy data: Leaf Area Index, leaf chlorophyll content, equivalent water thickness, and leaf mass per area. *Prosail* includes functions to run the radiative transfer model PROSAIL in both forward mode (simulation of canopy reflectance) and inverse mode (assessment of vegetation biophysical properties from canopy reflectance). The package *biodivMapR* is then used to compute various diversity metrics from these vegetation biophysical properties, including α- and β-diversity metrics usually obtained from species inventories in ecological applications, and a set of functional metrics. This workflow is illustrated with an application on imaging spectroscopy data acquired from a temperate forest site. Validation is based on inventories collected on a plot network. Results show consistent estimation of α-diversity, moderate agreement between observed, and RS β-diversity metrics. Further investigations are required for the validation of functional diversity metrics.

J.-B. Féret (✉) · F. de Boissieu
TETIS, INRAE, AgroParisTech, CIRAD, CNRS, Université Montpellier, Montpellier, France
e-mail: jean-baptiste.feret@inrae.fr; florian.deboissieu@inrae.fr

M. Lang
DYNAFOR, Université de Toulouse, INRAe, Toulouse, France
e-mail: marc.lang@toulouse-inp.fr

D. Rocchini (ed.), *R Coding for Ecology*, Use R!,
https://doi.org/10.1007/978-3-031-99665-8_7

Keywords

biodivMapR · Biophysical properties · PROSPECT · PROSAIL · Radiative transfer model inversion

1 Introduction

The biodiversity crisis currently occurring calls for improved monitoring of ecosystem structure, function, composition, and extent. Monitoring systems need to take advantage of ecological information obtained from ground observation, Earth observation data, environmental data, and modeling (Navarro et al. 2017; O'Connor et al. 2015). The integration of information obtained from Earth observation will contribute to produce biodiversity indicators based on frequent acquisitions of high-resolution imagery, bridging the data gap resulting from difficulties to collect sufficient ground observations (Jetz et al. 2019; Pereira et al. 2013). A variety of methods and concepts exist to assess various dimensions of biodiversity through remotely sensed (RS) diversity metrics (Wang and Gamon 2019). Optical data, including multispectral imagery and imaging spectroscopy, have been extensively used on that purpose, due to the link between vegetation biophysical and biochemical properties and canopy reflectance, which both reveal patterns of plant biodiversity (Asner et al. 2017; Asner and Martin 2009; Serbin and Townsend 2020). Most approaches use processing techniques for optical imagery based on spectral transformation (Féret and Asner 2014a; Laliberté et al. 2020) or spectral indices (Schneider et al. 2017), and few approaches attempt quantitative assessment of vegetation traits from remotely sensed information to derive RS diversity metrics.

The objectives of this chapter are to provide a reproducible workflow aiming at mapping RS diversity metrics to assess forest taxonomic and functional diversity, from a set of vegetation properties produced from physical modeling tools. This workflow uses a collection of R packages, including *prosail*,[1] which is used to assess vegetation biophysical variables from optical data using radiative transfer model (RTM) inversion, and *biodivMapR* (Féret and Boissieu 2020), which is used to map diversity metrics from raster data.

The method implemented in *biodivMapR* usually takes advantage of a data-driven spectral transformation (e.g., principal component analysis, PCA) followed by user-guided dimensionality reduction. This chapter introduces an alternative strategy, substituting the data-driven transformation for RTM inversion to produces a set of biophysical properties from imaging spectroscopy data. In the context of ecological analyses, our hypothesis is that accessing vegetation biophysical properties from RS data will increase consistency among sensors and generalization ability compared to statistically defined spectral features and spectral indices. Data-driven spectral features do not account for factors extrinsic to vegetation such as

[1] https://jbferet.gitlab.io/prosail/

the geometry of acquisition and soil background, which usually make these spectral features site- and sensor-dependent (Atzberger et al. 2011; Jacquemoud et al. 2009).

An open dataset is used to illustrate the diversity metrics produced with this method, related to the taxonomic dimension and to the functional dimension of biodiversity. This open dataset includes an airborne imaging spectroscopy mosaic acquired over a French temperate forest ecosystem, complemented with an inventoried plot network. The processing with free and open-source R packages ensures the full reproducibility of the analysis presented here.

2 Materials

2.1 Study Site and Ground Data Collection

The study area corresponds to the Fabas forest, located in the Haute-Garonne region, which covers an area of 831 hectares. This forest is characterized by a high diversity of tree species. A strong contribution of coniferous trees (douglas, diverse pines, spruce, larch, silver fir) to tree species diversity with mixed deciduous trees (oak and beech reserve) is reported from the forest-type geodatabase BD Forêt V2 (Institut National de l'Information Géographique et Forestière 2018). The local heterogeneity is enhanced in the surroundings of a reservoir bordering in the eastern part of the forest.

A field plot network including 44 plots was inventoried. Plot sampling was designed to cover the diversity of forest types hosted in the Fabas forest. Each plot corresponding to a 40 x 40 m square was accurately geolocated. Tree species inventories were performed: the relative abundance of each species was defined based on the surface emerging from the canopy in order to reduce the discrepancies between species abundance as seen from the ground and from remote sensing observations. Tree species contributing to less than 5% of the surface of the plot were not accounted for. The geolocated plot network and corresponding inventory are compiled in a geopackage file available online (Sheeren et al. 2023).

2.2 Imaging Spectroscopy Acquisition

Airborne imaging spectroscopy was acquired over the study area on September 15, 2015, at 10:07 (local time) at 1287 m above sea level by clear sky. A HySpex VNIR-1600 instrument (Norsk Elektro Optikk AS, Lørenskog, Norway) was used to acquire information in the visible to near-infrared spectral domain (VNIR, 400–1000 nm), and a HySpex SWIR-320 m instrument was used in the shortwave infrared (SWIR, 1000–2500 nm). The image spatial resolution of the instrument was 1 m in the VNIR and 2 m in the SWIR, and the full width at half maximum was 4.3 nm in the VNIR, and 9.0 nm in the SWIR. Flight lines were produced as georeferenced spectral radiances. The SWIR image was registered based on the VNIR image using the GeFolki algorithm (Brigot et al. 2016). The atmospheric

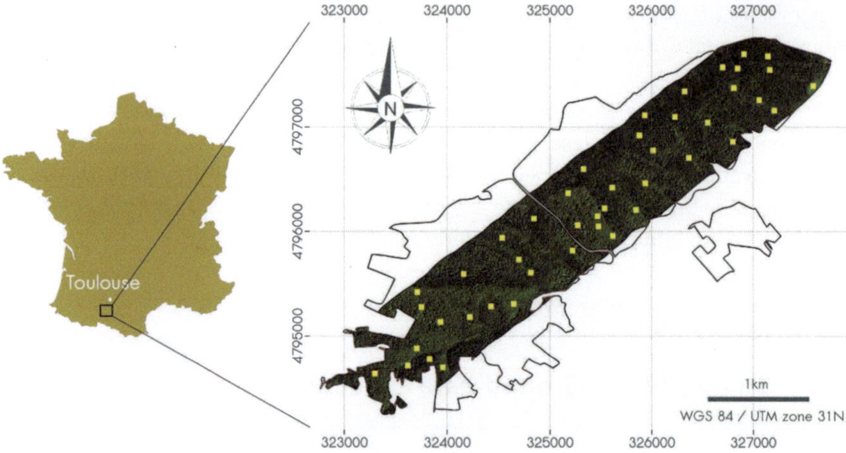

Fig. 1 RGB representation of the airborne acquisition corresponding to the study site of Fabas. The yellow squares represent the 40 m × 40 m field plots; the boundaries of the Fabas forest are delineated in gray

correction was applied to the resulting spectral radiance images in order to provide spectral reflectance using the Cochise atmospheric code (Poutier et al. 2002). The final reflectance product includes 409 spectral bands covering the reflective domain (400–2500 nm). Both VNIR and SWIR images were resampled to 4 m spatial resolution using a nearest neighborhood sampling, in order to reduce geometric distortions and preserve spectral information. Figure 1 shows a RGB representation of the airborne acquisition, with the location of the 44 field plots.

3 A Generic Framework to Compute Remotely Sensed Diversity Metrics

3.1 Description of the General Workflow

The R package *biodivMapR* aims at providing up-to-date implementation of the method described in Féret and Asner (2014a, b), which aimed at computing RS diversity metrics related to taxonomic diversity from imaging spectroscopy. Successive versions of the package introduced additional functionalities, including the following:

- The computation of functional diversity metrics (Villéger et al. 2008).
- Various optimizations to process large datasets, such as multiprocessing.
- Improved support for various remote sensing data sources (multispectral imagery and more) in addition to high spatial resolution imaging spectroscopy as initially used.

The method has been mainly used with optical data, including multispectral images such as Sentinel-2 data (Chraibi et al. 2021) and imaging spectroscopy (Féret and Asner 2014b; Kishore et al. 2023; Robertson et al. 2023) since the public release of the package.

Several data preprocessing steps were recommended in preparation for the computation of the RS diversity metrics in the original publications of the package, including the following:

- Spatial masking for irrelevant pixels corresponding to non-vegetated, shaded, and cloudy surfaces.
- Spectral masking for bands with low signal-to-noise ratio (e.g., water vapor absorption bands).
- Reflectance normalization with continuum removal.
- Spectral transformation with PCA or Minimum Noise Fraction (MNF) (Green et al. 1988).
- Feature selection based on visual examination of the operator.

The spectral transformation aims at enhancing spectral features from surface reflectance to increase discrimination among species or vegetation types. This spectral transformation prepares for feature selection, which aims at eliminating noisy components, and reducing dimensionality for computationally less demanding processing. Spectral transformation combined with component selection has been identified as particularly important to increase the correlation between RS diversity metrics based on spectral variance, and Shannon's H′ obtained from inventories (Badourdine et al. 2023). However, these data-driven transformations lack reproducibility and generalization ability in space and time, and the relative contribution of these different preprocessing steps on the estimation of biodiversity metrics is still not fully described. Spectral indicators more closely related to vegetation properties may contribute to improve the robustness of the link between remotely sensed information and ecological information on ecosystems. These spectral indicators include vegetation indices and biophysical properties, and examples of applications can be found in the literature. Schneider et al. (2017) and Helfenstein et al. (2022) computed vegetation indices from imaging spectroscopy data acquired over a temperate mixed forest, and used them as a proxy for vegetation traits, in order to assess and map functional diversity metrics. Hauser et al. (2021) applied RTM inversion to Sentinel-2 satellite data in order to compute a set of biophysical properties then used to assess functional diversity metrics in a (semi-)natural heterogeneous landscape located in Northern Portugal. Figure 2 summarizes the workflow followed by *biodivMapR* to compute RS diversity metrics. User can complement the main raster input with a mask in order to exclude unwanted regions of the image.

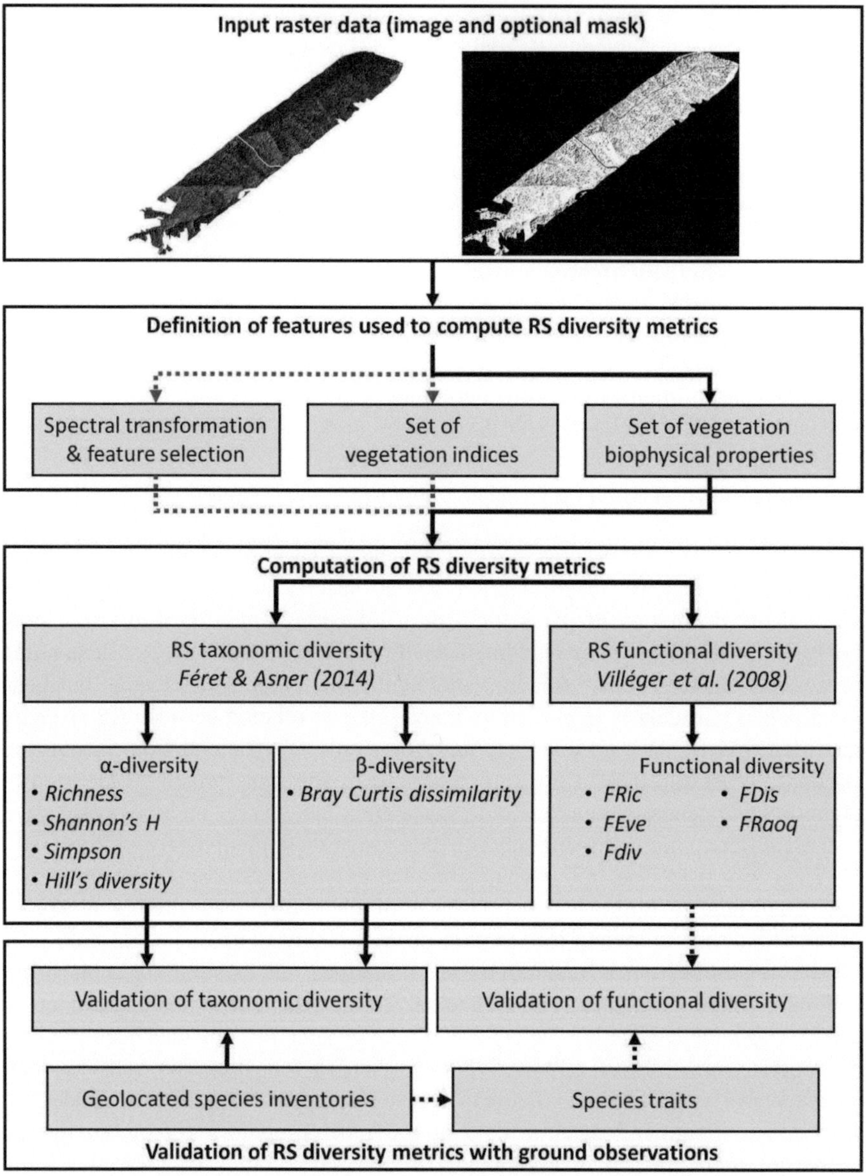

Fig. 2 General workflow applied to assess RS diversity metrics with *biodivMapR*. Plain arrows correspond to the workflow applied in this study; dotted arrows correspond to alternative options

3.2 Estimating Biophysical Properties with PROSAIL Inversion

Methods aiming at assessing vegetation properties are usually divided into two main categories: regression methods based on statistical/machine learning algorithms (ranging from simple vegetation indices to multivariate methods) and RTM model inversion (Verrelst et al. 2019). Data-driven regression methods are computationally efficient, but require intensive data collection to provide sufficient training data (remote sensing acquisition and corresponding ground measurement), which is often out of reach to ensure robust regression models. RTM accounts for extrinsic factors influencing surface reflectance (soil properties, geometry of acquisition), and their inversion does not require experimental training data. However, the inversion procedure is computationally more intensive than data-driven regression models when using iterative optimization or look-up table (LUT) inversions.

The combination of RTM with data-driven regression models (also commonly designated as hybrid inversion) takes advantage of the strengths of each approach: a RTM simulates canopy reflectance from a LUT including a large number of model parameter combinations. These parameters include leaf biophysical and chemical properties, canopy biophysical properties, and all factors extrinsic to vegetation accounted for by the RTM. The reflectance dataset and corresponding vegetation biophysical properties are then used as training dataset for a machine learning algorithm to produce regression models applicable to experimental data.

The design of the sampling strategy for the RTM input variables is key to determine the genericity or specialization of the regression models obtained at the end. Model genericity is usually favored, and the distribution and co-distributions of the input variables are then defined in order to cover at least the range observed over surfaces to process with random sampling. More refined methods based on active learning have also been implemented to improve retrieval accuracy by increasing specialization of the model, lower sampling rates, and increase computational efficiency (Verrelst et al. 2016).

The RTM inverted here is PROSAIL (Berger et al. 2018; Jacquemoud et al. 2009), implemented in the R package *prosail*. It results from the coupling of PROSPECT-D (Féret et al. 2017), implemented in the R package *prospect* (Féret and de Boissieu 2024), and SAIL (Verhoef et al. 2007). Readers are invited to refer to the original publications for detailed description of the input parameters of these models. The regression model used here is a Support Vector Regression (SVR) implemented in the R package *liquidSVM* (Steinwart and Thomann 2017). The distribution and co-distribution of the input parameters of PROSAIL was defined as described in the Algorithm Theoretical Basis Document (ATBD) of the Sentinel-2 toolbox (Weiss et al. 2020), except for LMA, which range was extended from 2 mg.cm-^{2} to 25 mg.cm-^{2}. The sun zenith angle was defined based on the location, day, and time of acquisition. The bidirectional reflectance factor (BRF) as measured by the sensor was obtained from simulations initially produced with 1 nm spectral sampling by applying the spectral response function (SRF) of the sensor. Finally, a 2% multiplicative Gaussian noise was applied on the BRF data.

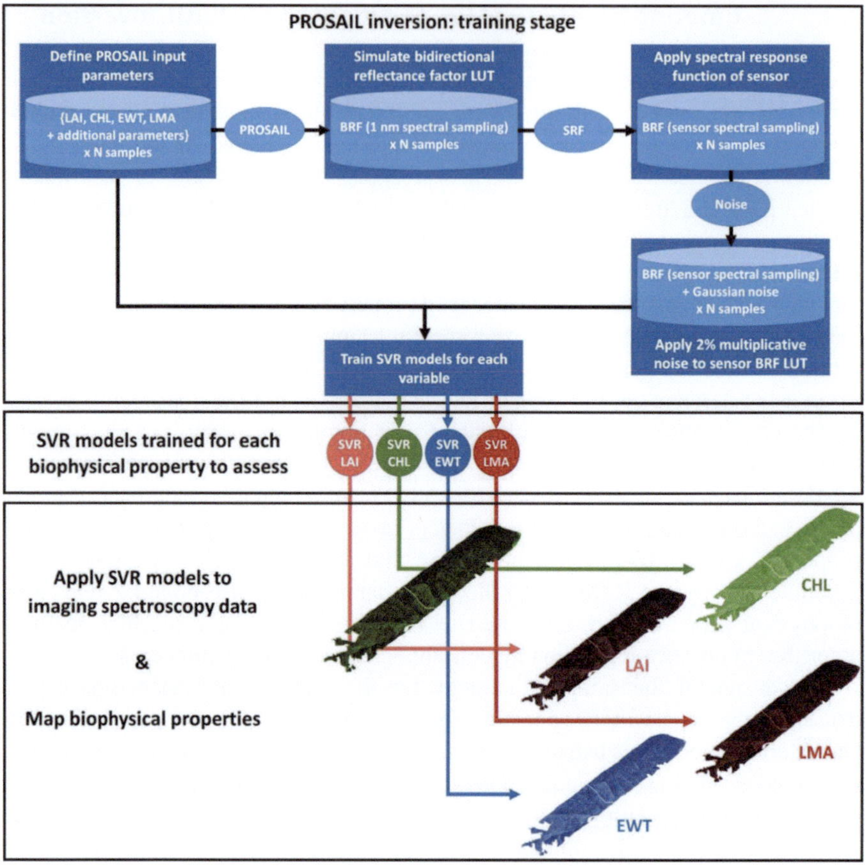

Fig. 3 Workflow for the hybrid inversion of PROSAIL applied to imaging spectroscopy data

The biophysical properties assessed with this inversion included Leaf Area Index (LAI), leaf chlorophyll a + b content (CHL), equivalent water thickness (EWT), and leaf mass per area (LMA). No ground measurements were performed for the biophysical properties assessed with this inversion. Therefore, the uncertainty associated with these properties is beyond the scope of this study. The workflow of the inversion is detailed in Fig. 3.

The scripts corresponding to the workflow described in Fig. 3 are detailed below. First, the required packages are loaded; data and results directories and files are defined. Imaging spectroscopy data is available on request from the authors. The corresponding atmospheric transmittance file computed with MODTRAN-4 (Berk et al. 1999) is available from an online repository and downloadable as described in the next example:

```
# load packages
library(biodivMapR)
library(prosail)
# data & results directories
data_dir <- '../01_DATA/raster_data'
prosail_dir <- '../03_RESULTS/PROSAIL_Inversion'
dir.create(prosail_dir, showWarnings = F, recursive = T)

# imaging spectroscopy data are available on request from the authors
# --> download and unzip data in 'data_dir'
# path for input image file
input_raster_path <-
  file.path(data_dir,'Fabas_HS_mosaic.tif')
HDR <- read_ENVI_header(get_HDR_name(input_raster_path))
SensorBands <- HDR$wavelength
SensorFWHM <- HDR$fwhm
# atmospheric transmittance file available on a repository
GitRepository <-
  'https://gitlab.com/jbferet/myshareddata/-/raw/master/Fabas_HS_AtmoT'
GitFile <- 'AtmoT_Fabas_Hyspex.csv'
url <- file.path(GitRepository, GitFile)
path_AtmoT <- file.path(data_dir,GitFile)
download.file(url = url, destfile = path_AtmoT)
```

Second, spatial and spectral filtering is performed. Spatial filtering produces a mask excluding pixels assumed to correspond to surfaces other than vegetation (NDVI thresholding, red band centered at 670 nm, near-infrared band centered at 835 nm) as well as shaded pixels (near-infrared thresholding) and cloudy pixels (blue thresholding, blue band centered at 480 nm). Spectral filtering excludes sensor bands showing low signal-to-noise ratio (SNR). These bands usually correspond to water absorption bands. Here we used information collected on atmospheric transmittance during the airborne campaign: sensor bands with atmospheric transmittance lower than 85% were discarded. An additional band was excluded after careful examination of spectral information and identification of poor quality:

```
##  SPATIAL FILTERING
Mask_File <- file.path(prosail_dir, 'Mask_FABAS')
NDVI_Thresh <- 0.5
Blue_Thresh <- 600
NIR_Thresh <- 1500
Mask_File <- radiometric_filtering(input_raster_path = input_raster_path,
                                   input_rast_wl = SensorBands,
                                   output_dir = prosail_dir,
                                   NDVI_Thresh = NDVI_Thresh,
                                   Blue_Thresh = Blue_Thresh,
                                   NIR_Thresh = NIR_Thresh)
##  SPECTRAL FILTERING
# discard bands with low atmospheric transmittance (<0.85)
Transmittance <- read.delim(path_AtmoT, header =TRUE)
LowTrans <- which(Transmittance$Transmittance<0.85)
# discard remaining noisy band
Noisyband <- which((SensorBands > 2200 & SensorBands < 2205))
# combine bands to discard
WLelim <- c(LowTrans, Noisyband)
```

PROSAIL inversion is then parameterized: the biophysical properties to be assessed are defined, along with the spectral domain used for the estimation of each of these properties. The simulation of a set of input PROSAIL variables is produced following specific distributions and co-distributions (Weiss et al. 2020). The level of the relative Gaussian noise applied on the BRF data is also defined for each biophysical property to assess (here, a unique value of 2% multiplicative Gaussian noise is defined for all properties):

```
# define parameters to assess
Parms2Estimate <- c('lai', 'chl', 'ewt', 'lma')
# define spectral domain and noise for each variable, discard WLelim
Bands2Select <- NoiseLevel <- list()
WLtrain <- data.frame(min = c('lai'=700, 'chl'=700, 'ewt'=1200, 'lma'=1200),
                      max = c('lai'=900, 'chl'=900, 'ewt'=2400, 'lma'=2400))

for(parm in Parms2Estimate){
  bandsel <- which(SensorBands > WLtrain[parm, 'min']&
                   SensorBands < WLtrain[parm, 'max'])
  Bands2Select[[parm]]<- bandsel[-na.exclude(match(WLelim, bandsel))]
  # 2% relative gaussian noise applied to BRF data
  NoiseLevel[[parm]]<- 0.02
}
# define geometry of acquisition based on location and time of acquisition
GeomAcq <- data.frame(min = c('tto' = 0, 'tts' = 50, 'psi' = 0),
                      max = c('tto' = 15, 'tts' = 70, 'psi' = 360))
# use same distribution as used in ATBD for Sentinel-2 data
InputPROSAIL <- get_InputPROSAIL(atbd =TRUE, GeomAcq = GeomAcq)
# extend range of LMA distribution compared to ATBD
InputPROSAIL$lma <- runif(n = length(InputPROSAIL[[1]]),
                          min = 0.002, max = 0.025)
```

Next, the SRF of the sensor is defined in order to simulate sensor BRF from the original 1 nm spectral sampling of the BRF simulated with PROSAIL:

```
# get sensor response function corresponding to the sensor
SensorName <- 'Hyspex'
SRF <- GetRadiometry(SensorName = SensorName,
                     SpectralProps = data.frame(wl = SensorBands,
                                                fwhm = SensorFWHM))
```

The SVR model is then trained, and the resulting regression models are applied to the BRF image:

```
# train a set of SVM regression models for each BP var to assess
modelSVR <- train_prosail_inversion(input_prosail = InputPROSAIL,
                                    parms_to_estimate = Parms2Estimate,
                                    geom_acq = GeomAcq,
                                    noise_level = NoiseLevel,
                                    srf = SRF,
                                    selected_bands = Bands2Select,
                                    Path_Results = prosail_dir)
```

```
# apply models on image
BPvars <- Apply_prosail_inversion(raster_path = input_raster_path,
                                  HybridModel = modelSVR,
                                  PathOut = prosail _dir,
                                  SelectedBands = Bands2Select,
                                  bandname = SensorBands,
                                  MaskRaster = Mask_File,
                                  MultiplyingFactor = 10000)
```

The variable BPvars includes the path corresponding to the biophysical properties obtained from PROSAIL inversion.

3.3 RS Diversity Metrics

biodivMapR computes two types of diversity metrics: functional diversity metrics and taxonomic diversity metrics. Functional diversity metrics are computed directly from a continuous spectral feature space defined by principal components, spectral indices, or assessed biophysical properties in our case. Taxonomic diversity metrics require a preliminary clustering step. This clustering aims at defining the "spectral species" from the spectral feature space. Spectral species are obtained from a k-means clustering. They are equivalent to the "optical types" defined by Ustin and Gamon (2010), who hypothesize that "functional types can be distinguished largely on the basis of optical properties detectable by remote sensing." In the context of high spatial resolution imaging applied to forest ecosystem observation, this hypothesis is transposed to the species level, as multiple studies evidenced that high spatial resolution optical information allows distinguishing among species (Baldeck et al. 2015; Féret and Asner 2013; Laybros et al. 2019).

Spectral species are conceptual proxy aiming at differentiating entities distinguishable from RS information, which may correspond to well-defined taxonomic entities such as species. However, multiple factors prevent from systematic discrimination among species, as reported by the non-negligible error rate reported by studies focusing on tree species identification. These factors include within-species variability of the RS features. The spectral properties of the sensor may be limiting as well, by poorly capturing features relevant for species discrimination. The spatial resolution of the sensor may also be limiting to capture information corresponding to pure species at pixel scale (Robertson et al. 2023). Therefore, grouping pixels based on a set of RS descriptors is assumed to capture a fraction of the taxonomic information only.

The default number of spectral species in *biodivMapR* is set to $k = 50$, as a trade-off between the accuracy of the assessed diversity metrics during validation over tropical sites and the computational efficiency. The temperate forest analyzed in this study shows lower diversity than tropical ecosystems; therefore, we decided to set the number of spectral species to $k = 25$. This parameter is expected to influence the RS taxonomic diversity metrics derived from imagery data. However, the search for an optimal number of spectral species is beyond the scope of this study.

The spectral diversity indices are computed from the input raster data once divided following a grid of square cells, which size is set by the parameter `window_size`. Each cell defines a spatial entity.

Starting from *biodivMapR v2.0*, all diversity metrics including functional diversity metrics and taxonomic diversity metrics are directly computed in a unique step:

- α-Diversity metrics are computed based on the spectral species inventory for each spatial entity. α-Diversity metrics implemented in *biodivMapR* include spectral species richness, Shannon's H′, Simpson index, and Hill diversity (with Hill order defined with dedicated input variable).
- β-Diversity is based on the Bray-Curtis (BC) dissimilarity among spatial entity. When computed from an image, the matrix corresponding to pairwise BC among all spatial entities is transformed into a 3D space based on a principal coordinate analysis (PCoA), a multidimensional scaling aiming at summarizing and representing inter-object (dis)similarity in a low-dimensional Euclidean space (Legendre et al. 1998). This allows three-color representation of the dissimilarity with the same spatial extent as the original raster.
- Functional diversity metrics correspond to functional richness (FRic), functional evenness (FEve), functional divergence (FDiv), functional dispersion (FDis), and Rao entropy (FRaoq), as defined by Villéger et al. (2008). These diversity metrics are computed with the package *fundiversity* (Gruson and Grenié 2022).

Intermediate data, including spectral species maps, are not stored in order to minimize requirements in terms of storage space for large images:

```
# define results directory
biodivMapR_dir <- '../03_RESULTS/biodivMapR'
# define biodivMapR parameters
window_size <- 10
nbclusters <- 25
nbCPU <- 4
maxRows <- 200
alphametrics <- c('richness', 'shannon', 'simpson', 'hill')
Hill_order <- 2
FDmetric <- c('FRic', 'FEve', 'FDiv', 'FDis', 'FRaoq')
## compute all diversity indices
ResDiv <- biodivMapR_full(input_raster_path = BPvars$mean,
                          input_mask_path = Mask_File,
                          output_dir = biodivMapR_dir,
                          window_size = window_size,
                          nb_clusters = nbclusters,
                          alphametrics = alphametrics,
                          FDmetric = FDmetric,
                          Hill_order = Hill_order,
                          maxRows = maxRows,
                          nbCPU = nbCPU)
```

3.4 Validation of the RS Diversity Metrics: Comparison with Forest Inventories

The validation of the RS diversity metrics computed with *biodivMapR* requires corresponding ground information. Here, the validation is performed for RS α-diversity and β-diversity metrics and their taxonomic counterpart only, as functional traits were not measured during the ground data collection, which does not allow computing functional diversity metrics from inventory data. The validation step requires defining polygons corresponding to the inventoried field plot network. By definition, RS diversity metrics reflect the diversity of information of a group of pixels; hence, the polygons corresponding to individual plots are expected to include a sufficient amount of pixels. We recommend a minimum of 25 pixels per plot to reduce the risk of discrepancies between RS diversity metrics and diversity observed in the field. A buffer can be applied on individual plots to reach a sufficient number of pixels per plot, if it can be assumed that the diversity corresponding to the area defined for a field plot can be compared to its surrounding area covered by the buffer. We also recommend defining plots covering areas of comparable surface, so that diversity metrics can actually be compared with each other. Here, each plot inventoried was defined with an individual vector file. No buffer was applied, as individual plots covered about 100 pixels.

The field diversity metrics are computed from the inventory data available in the *gpkg* vector file available from Sheeren et al. (2023). The following code example produces taxonomic diversity indices from the inventory data included in the vector file:

```
# data & results directories
data_dir <- '../01_DATA/vector_data'
# the validation plots & inventories can be downloaded from
# https://doi.org/10.57745/NHRNOR
# --> download and unzip data in 'data_dir'
# path for validation plot network (gpkg)
vect_path <- file.path(data_dir, 'forestPLOT_Fabas_2019.gpkg')
val_plots <- terra::vect(vect_path)
# get list of species found in inventoried plots
SpList <- unique(unlist(val_plots[[c('Species1', 'Species2',
                                     'Species3', 'Species4',
                                     'Species5')]]))
SpList <- sort(na.omit(SpList))
# initialize species abundance table from inventory data
Inventory <- data.frame(matrix(data = 0,
                               nrow = length(val_plots$ID),
                               ncol = length(SpList)))
colnames(Inventory)<- SpList
rownames(Inventory)<- val_plots$ID
# fill species abundance table
for(PlotID in val_plots$ID){
  whichPlot <- which(val_plots$ID == PlotID)
for(sp in seq_len(5)){
    spi <- paste0('Species',sp)
```

```
    wsp <- which(SpList == val_plots[[spi]][[1]][whichPlot])
    if(length(wsp)>0){
      Abundspi <- paste0('%Cover_Species',sp)
      Inventory[whichPlot,SpList[wsp]]<-
val_plots[[Abundspi]][[1]][whichPlot]
    }
  }
}
# compute diversity metrics from abundance table
plotID <- val_plots$ID
shannon_obs <- vegan::diversity(Inventory, index = "shannon")
richness_obs <- vegan::specnumber(Inventory)
bc_obs <- vegan::vegdist(Inventory, method = "bray")
```

The validation is then performed with the function `get_diversity_from_plots`, which computes the RS diversity metrics available from *biodivMapR* directly on the exact spatial extent of polygons corresponding to validation plots. Information on k-means clustering and beta diversity previously computed on the full raster data are also used as input data:

```
# compute spectral diversity metrics based on raster data and
# field plot network
load(file.path(biodivMapR_dir, 'Kmeans_info.RData'))
load(file.path(biodivMapR_dir, 'Beta_info.RData'))
input_rast <- lapply(X = BPvars$mean, FUN = terra::rast)
input_mask <- terra::rast(Mask_File)
validation <- get_diversity_from_plots(input_rast = input_rast,
                                       input_mask = input_mask,
                                       validation_vect = val_plots,
                                       Kmeans_info = Kmeans_info,
                                       Beta_info = Beta_info,
                                       alphametrics = alphametrics,
                                       Functional = FDmetric)
# RICHNESS
predicted <- validation$specdiv$richness_mean
df_richness <- data.frame('measured' = richness_obs,
                          'predicted' = predicted)
# SHANNON
predicted <- validation$specdiv$shannon_mean
df_shannon <- data.frame('measured' = shannon_obs,
                         'predicted' = predicted)
# Bray-Curtis dissimilarity
BCdissmat <- c(as.dist(validation$BC_dissimilarity, diag =FALSE))
Field_Beta <- c(as.dist(bc_obs, diag =FALSE))
df_beta_estimated <- data.frame('measured' = Field_Beta,
                                'predicted' = BCdissmat)
```

The comparison between measured and predicted values of species richness, Shannon's H′, and BC is performed with the code example below, and the resulting figure is displayed in Fig. 4. The Spearman correlation obtained between measured and predicted diversity metrics was strong for the Shannon's H, as well as for species richness, to a lesser extent. Lower correlation was obtained between measured and assessed BC:

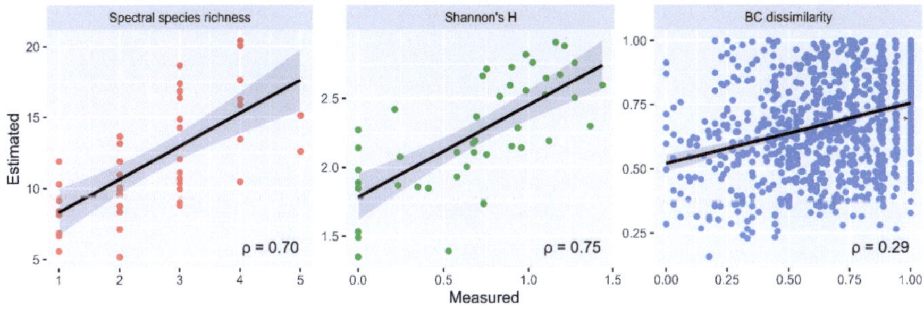

Fig. 4 Validation of the RS diversity metrics corresponding to (**a**) species richness, (**b**) Shannon's H′, and (**c**) Bray-Curtis dissimilarity. The Spearman correlation coefficient is displayed at the bottom for each metric

```
library(data.table)
library(ggplot2)
library(ggpmisc)

plot_data <- data.table::rbindlist(list(
  `Spectral species richness` = df_richness,
  `Shannon's H` = df_shannon,
  `BC dissimilarity` = df_beta_estimated), idcol = 'Metric')
plot_data <- dplyr::mutate(plot_data,
                           Metric = factor(Metric,
                                           levels = unique(Metric),
                                           labels = unique(Metric)))

plotval <- ggplot(plot_data, aes(measured, predicted, color=Metric))+
  geom_point()+
  geom_smooth(method="lm", color="black")+
  ggplot2::facet_wrap(~Metric, scales="free", ncol=3)+
  ggpmisc::stat_correlation(method = "spearman",
                            label.x = "right",
                            label.y = "bottom",
                            color="black")+
  guides(color="none")+
  annotate("label", colour = "black", size = 10)

filename <- file.path(biodivMapR_dir, 'Validation_AlphaBeta.png')
ggsave(filename, plot = plotval, device = 'png', path =NULL,
       scale = 0.75, width = 12, height = 4, units = "in",
       dpi = 600, limitsize =TRUE)
```

3.5 Diversity Maps

The maps corresponding to the RS diversity metrics are displayed in Fig. 5. These diversity maps can then be compared with forest species community maps and forest management plans, when available.

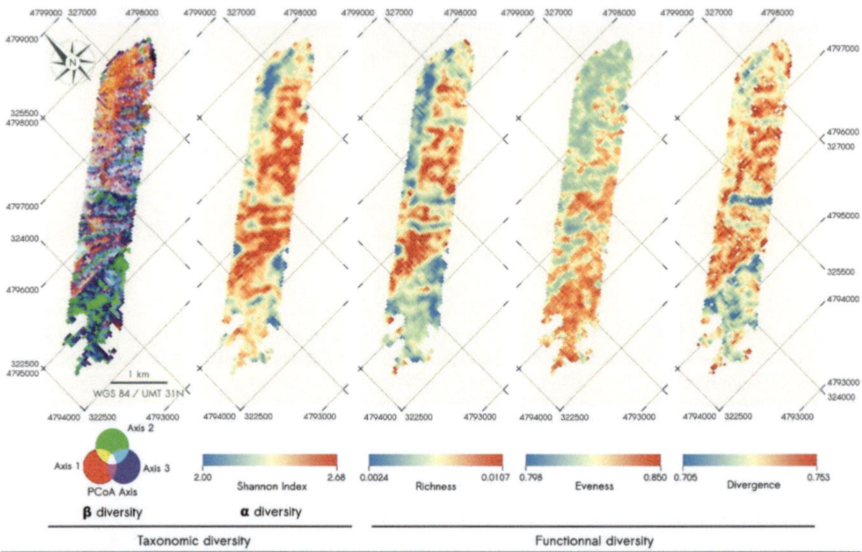

Fig. 5 RS diversity maps produced with *biodivMapR*. From left to right, these maps correspond to β-diversity (PCoA applied on a Bray-Curtis dissimilarity matrix), α-diversity (Shannon's H′), functional richness, evenness, and divergence

4 Discussion

4.1 Accuracy and Robustness of Estimated Biophysical Properties

The estimated values for the biophysical properties were not validated as no ground measurements were collected during the campaign. Here, we used the model PROSAIL to estimate these properties, which is based on hypotheses of homogeneity of the canopy. This hypothesis does not hold in the context of forested areas, and our experiment may diverge from the domain of validity of the model. Therefore, we can expect a strong uncertainty and possible bias in the biophysical properties assessed (Brown et al. 2019). Alternative RTM such as INFORM (Schlerf and Atzberger 2006) which accounts for canopy geometry, or DART (Gastellu-Etchegorry et al. 2015) which can take advantage of explicit 3D description of the canopy and terrain derived from LiDAR data (Ebengo et al. 2021), would reduce these uncertainties and improve the robustness of the estimated vegetation properties.

Alternative regression models such as partial least squares regression (PLSR) models (Asner et al. 2011; Serbin and Townsend 2020) or Gaussian processes (Verrelst et al. 2012) may be used instead of SVR used in our study to produce a set of vegetation traits. However, the differences among multivariate algorithms used to adjust regression models may not lead to differences in results as strong as those obtained for different RTM. Simple spectral indices can also be used instead of biophysical properties for the sake of computational efficiency, but the influence of factors extrinsic to vegetation would increase the uncertainty when attempting to relate the spectral indices to actual vegetation properties.

4.2 Definition of an Optimal Set of Biophysical Properties for the Estimation of Biodiversity

The set of biophysical properties estimated here was limited to four variables: LAI, CHL, EWT, and LMA. Additional variables possibly contributing to improve the estimation of diversity metrics are accessible from RTM inversion. These additional variables include pigments (carotenoids, anthocyanins, and brown pigments), proteins and carbon-based constituent if using PROSPECT-PRO instead of PROSPECT-D (Berger et al. 2020; Féret et al. 2021; Verrelst et al. 2021), and the fraction of absorbed photosynthetically active radiation (fAPAR). Variables obtained from other sensors such as LiDAR data may also contribute to estimate diversity metrics (Torresani et al. 2020). However, each of these estimated variables may be associated with its own uncertainty, and a more comprehensive sensitivity analysis and feature selection is needed to identify which combination of biophysical properties leads to the best relationship between measured and estimated biodiversity. The importance of each variable is expected to be ecosystem dependent and to vary in space and time.

4.3 Optimal Parametrization of biodivMapR

biodivMapR includes multiple parameters which may influence the diversity metrics estimated from remotely sensed data. The selection of spectral features used to estimate these metrics is certainly the most important factor. The standard implementation of *biodivMapR* is based on feature selection from a data-driven spectral transformation, which is limiting when attempting to compare diversity metrics obtained over sites corresponding to different acquisitions over time and space. The application of *biodivMapR* on biophysical properties may reduce the sensitivity expected from data-driven methods to the spatial scale of analysis (extent and grain size). Further analyses are still needed to explore this hypothesis. One strong interest of biophysical properties is the possibility to extend the analysis of diversity patterns by analyzing the contribution of each property to the estimated diversity metrics, and providing improved explanatory power and physical basis compared to spectral components resulting from linear or nonlinear combinations of multiple spectral bands.

The number of spectral species defined in a scene is also an important factor influencing the estimated diversity metrics. The hypothesis that unsupervised spectral clustering may lead to discrimination among taxonomically or functionally distinct groups of pixels should also be investigated.

5 Conclusions and Perspectives

This study introduces an original workflow to map RS diversity metrics from optical images, which combines a RTM inversion aiming at producing a set of biophysical variables relevant for the estimation of diversity metrics, combining the R package *prosail* and the latest version of the R package *biodivMapR*. RS α- and β-diversity metrics related to taxonomic diversity and a set of RS functional diversity metrics are computed.

The validation of the estimated biophysical properties was possible for α- and β-diversity metrics only, as functional traits were not measured in the field. This validation showed consistency between measured and estimated values of the Shannon's H′ and for species richness. The estimation of BC dissimilarity showed a weaker relationship with field inventories.

More efforts are needed to better identify the optimal parameterization of the method implemented in *biodivMapR*, in particular the selection of an optimal number of spectral species and the selection of the most relevant RS features for the estimation of diversity metrics. However, this study provides promising results to define a physically based and more robust approach for biodiversity monitoring.

Such methods are particularly relevant in the context of hyperspectral satellite missions, including operational missions such as EnMAP and PRISMA, as well as missions in preparation such as SBG and CHIME (Candela et al. 2016; Cawse-Nicholson et al. 2021; Guanter et al. 2015; Rast et al. 2021). This approach combining vegetation biophysical properties and RS diversity may also contribute to an improved characterization of ecosystem extent if combined with expert analysis on spatio-temporal variations of vegetation traits within and across ecosystems.

6 Data and Software Resources

The updated code corresponding to this chapter is stored at https://github.com/RCodingForEcology. Imaging spectroscopy data used for this study are available online on request from the authors. The plot network spatial extent and corresponding inventory data used for this study are available online: https://doi.org/10.57745/NHRNOR. The documentation and installation procedure for the R packages *prospect*, *prosail*, and *biodivMapR* are available online: https://jbferet.gitlab.io/prospect/, https://jbferet.gitlab.io/prosail/, and https://jbferet.github.io/biodivMapR/. the latest version (v3.0.0) of prosail is currently in the development branch "dev_joss" and will soon be released in the "master" branch.

Acknowledgments The authors acknowledge financial support from Agence Nationale de la Recherche (BioCop project—ANR-17-CE32-0001) and TOSCA program grant from the French Space Agency (CNES) (HyperBio project). The acquisition of hyperspectral images used in this work was funded in the frame of the NAOMI project between ONERA and TotalEnergies. The authors gratefully acknowledge Philippe Déliot, Laurent Poutier, and Sophie Fabre for their implications in image acquisitions and preprocessing. Forest plot inventories were collected by

David Sheeren, Marvin Graham-Fonseca, and Jérôme Willm. We warmly thank all airborne and ground data providers for giving public access to these datasets. We also thank Xavier Briottet for his contribution to the project HyperBIO.

References

Asner GP, Martin RE (2009) Airborne spectranomics: mapping canopy chemical and taxonomic diversity in tropical forests. Front Ecol Environ 7:269–276. https://doi.org/10.1890/070152

Asner GP, Martin RE, Knapp DE, Tupayachi R, Anderson C, Carranza L, Martinez P, Houcheime M, Sinca F, Weiss P (2011) Spectroscopy of canopy chemicals in humid tropical forests. Remote Sens Environ 115:3587–3598

Asner GP, Martin RE, Knapp DE, Tupayachi R, Anderson CB, Sinca F, Vaughn NR, Llactayo W (2017) Airborne laser-guided imaging spectroscopy to map forest trait diversity and guide conservation. Science 355:385–389. https://doi.org/10.1126/science.aaj1987

Atzberger C, Richter K, Vuolo F, Darvishzadeh R, Schlerf M (2011) Why confining to vegetation indices? Exploiting the potential of improved spectral observations using radiative transfer models. In: Neale CMU, Maltese A (eds) Presented at the SPIE remote sensing, Prague, Czech Republic, p 81740Q. https://doi.org/10.1117/12.898479

Badourdine C, Féret J, Pélissier R, Vincent G (2023) Exploring the link between spectral variance and upper canopy taxonomic diversity in a tropical forest: influence of spectral processing and feature selection. Remote Sens Ecol Conserv 9:235–250. https://doi.org/10.1002/rse2.306

Baldeck CA, Asner GP, Martin RE, Anderson CB, Knapp DE, Kellner JR, Wright SJ (2015) Operational tree species mapping in a diverse tropical Forest with airborne imaging spectroscopy. PLoS One 10:e0118403. https://doi.org/10.1371/journal.pone.0118403

Berger K, Atzberger C, Danner M, D'Urso G, Mauser W, Vuolo F, Hank T (2018) Evaluation of the PROSAIL model capabilities for future hyperspectral model environments: a review study. Remote Sens 10:85. https://doi.org/10.3390/rs10010085

Berger K, Verrelst J, Féret J-B, Hank T, Wocher M, Mauser W, Camps-Valls G (2020) Retrieval of aboveground crop nitrogen content with a hybrid machine learning method. Int J Appl Earth Obs Geoinformation 92:102174. https://doi.org/10.1016/j.jag.2020.102174

Berk A, Anderson GP, Bernstein LS, Acharya PK, Dothe H, Matthew MW, Adler-Golden SM, Chetwynd JH Jr, Richtsmeier SC, Pukall B, Allred CL, Jeong LS, Hoke ML (1999) MODTRAN4 radiative transfer modeling for atmospheric correction. In: Larar AM (ed) Presented at the SPIE'S international symposium on optical science, engineering, and instrumentation, Denver, CO, p 348. https://doi.org/10.1117/12.366388

Brigot G, Colin-Koeniguer E, Plyer A, Janez F (2016) Adaptation and evaluation of an optical flow method applied to Coregistration of Forest remote sensing images. IEEE J Sel Top Appl Earth Obs Remote Sens 9:2923–2939. https://doi.org/10.1109/JSTARS.2016.2578362

Brown LA, Ogutu BO, Dash J (2019) Estimating Forest leaf area index and canopy chlorophyll content with Sentinel-2: an evaluation of two hybrid retrieval algorithms. Remote Sens 11:1752. https://doi.org/10.3390/rs11151752

Candela L, Formaro R, Guarini R, Loizzo R, Longo F, Varacalli G (2016) The PRISMA mission. In: 2016 IEEE International Geoscience and Remote Sensing Symposium (IGARSS). Presented at the IGARSS 2016–2016 IEEE International Geoscience and Remote Sensing Symposium. IEEE, Beijing, China, pp 253–256. https://doi.org/10.1109/IGARSS.2016.7729057

Cawse-Nicholson K, Townsend PA, Schimel D, Assiri AM, Blake PL, Buongiorno MF, Campbell P, Carmon N, Casey KA, Correa-Pabón RE, Dahlin KM, Dashti H, Dennison PE, Dierssen H, Erickson A, Fisher JB, Frouin R, Gatebe CK, Gholizadeh H, Gierach M, Glenn NF, Goodman JA, Griffith DM, Guild L, Hakkenberg CR, Hochberg EJ, Holmes TRH, Hu C, Hulley G, Huemmrich KF, Kudela RM, Kokaly RF, Lee CM, Martin R, Miller CE, Moses WJ, Muller-Karger FE, Ortiz JD, Otis DB, Pahlevan N, Painter TH, Pavlick R, Poulter B, Qi Y, Realmuto VJ, Roberts D, Schaepman ME, Schneider FD, Schwandner FM, Serbin SP, Shiklomanov AN,

Stavros EN, Thompson DR, Torres-Perez JL, Turpie KR, Tzortziou M, Ustin S, Yu Q, Yusup Y, Zhang Q (2021) NASA'S surface biology and geology designated observable: a perspective on surface imaging algorithms. Remote Sens Environ 257:112349. https://doi.org/10.1016/j.rse.2021.112349

Chraibi E, Arnold H, Luque S, Deacon A, Magurran A, Féret J-B (2021) A remote sensing approach to understanding patterns of secondary succession in tropical Forest. Remote Sens 13:2148. https://doi.org/10.3390/rs13112148

Ebengo DM, de Boissieu F, Vincent G, Weber C, Féret J-B (2021) Simulating imaging spectroscopy in tropical Forest with 3D radiative transfer modeling. Remote Sens 13:2120. https://doi.org/10.3390/rs13112120

Féret J-B, Asner GP (2013) Tree species discrimination in tropical forests using airborne imaging spectroscopy. IEEE Trans Geosci Remote Sens 51:73–84. https://doi.org/10.1109/TGRS.2012.2199323

Féret J-B, Asner GP (2014a) Mapping tropical forest canopy diversity using high-fidelity imaging spectroscopy. Ecol Appl 24:1289–1296. https://doi.org/10.1890/13-1824.1

Féret J-B, Asner GP (2014b) Microtopographic controls on lowland Amazonian canopy diversity from imaging spectroscopy. Ecol Appl 24:1297–1310. https://doi.org/10.1890/13-1896.1

Féret J, Boissieu F (2020) biodivMapR: an R package for α- and β-diversity mapping using remotely sensed images. Methods Ecol Evol 11:64–70. https://doi.org/10.1111/2041-210X.13310

Féret J-B, de Boissieu F (2024) Prospect: an R package to link leaf optical properties with their chemical and structural properties with the leaf model PROSPECT. J Open Source Softw 9:6027. https://doi.org/10.21105/joss.06027

Féret J-B, Gitelson AA, Noble SD, Jacquemoud S (2017) PROSPECT-D: towards modeling leaf optical properties through a complete lifecycle. Remote Sens Environ 193:204–215. https://doi.org/10.1016/j.rse.2017.03.004

Féret J-B, Berger K, de Boissieu F, Malenovský Z (2021) PROSPECT-PRO for estimating content of nitrogen-containing leaf proteins and other carbon-based constituents. Remote Sens Environ 252:112173. https://doi.org/10.1016/j.rse.2020.112173

Gastellu-Etchegorry J-P, Yin T, Lauret N, Cajgfinger T, Gregoire T, Grau E, Féret J-B, Lopes M, Guilleux J, Dedieu G, Malenovský Z, Cook B, Morton D, Rubio J, Durrieu S, Cazanave G, Martin E, Ristorcelli T (2015) Discrete anisotropic radiative transfer (DART 5) for modeling airborne and satellite spectroradiometer and LIDAR acquisitions of natural and urban landscapes. Remote Sens 7:1667–1701. https://doi.org/10.3390/rs70201667

Green AA, Berman M, Switzer P, Craig MD (1988) A transformation for ordering multispectral data in terms of image quality with implications for noise removal. IEEE Trans Geosci Remote Sens 26:65–74. https://doi.org/10.1109/36.3001

Gruson H, Grenié M (2022) Fundiversity: easy computation of functional diversity indices. https://doi.org/10.5281/ZENODO.4761754

Guanter L, Kaufmann H, Segl K, Foerster S, Rogass C, Chabrillat S, Kuester T, Hollstein A, Rossner G, Chlebek C, Straif C, Fischer S, Schrader S, Storch T, Heiden U, Mueller A, Bachmann M, Mühle H, Müller R, Habermeyer M, Ohndorf A, Hill J, Buddenbaum H, Hostert P, van der Linden S, Leitão P, Rabe A, Doerffer R, Krasemann H, Xi H, Mauser W, Hank T, Locherer M, Rast M, Staenz K, Sang B (2015) The EnMAP Spaceborne imaging spectroscopy mission for earth observation. Remote Sens 7:8830–8857. https://doi.org/10.3390/rs70708830

Hauser LT, Féret J-B, An Binh N, van der Windt N, Sil ÂF, Timmermans J, Soudzilovskaia NA, van Bodegom PM (2021) Towards scalable estimation of plant functional diversity from Sentinel-2: in-situ validation in a heterogeneous (semi-)natural landscape. Remote Sens Environ 262:112505. https://doi.org/10.1016/j.rse.2021.112505

Helfenstein IS, Schneider FD, Schaepman ME, Morsdorf F (2022) Assessing biodiversity from space: impact of spatial and spectral resolution on trait-based functional diversity. Remote Sens Environ 275:113024. https://doi.org/10.1016/j.rse.2022.113024

Institut National de l'Information Géographique et Forestière (2018) BD Forêt – version 2.0

Jacquemoud S, Verhoef W, Baret F, Bacour C, Zarco-Tejada PJ, Asner GP, François C, Ustin SL (2009) PROSPECT+ SAIL models: a review of use for vegetation characterization. Remote Sens Environ 113:S56–S66. https://doi.org/10.1016/j.rse.2008.01.026

Jetz W, McGeoch MA, Guralnick R, Ferrier S, Beck J, Costello MJ, Fernandez M, Geller GN, Keil P, Merow C, Meyer C, Muller-Karger FE, Pereira HM, Regan EC, Schmeller DS, Turak E (2019) Essential biodiversity variables for mapping and monitoring species populations. Nat Ecol Evol 3:539–551. https://doi.org/10.1038/s41559-019-0826-1

Kishore BSPC, Kumar A, Saikia P, Khan ML (2023) Alpha and beta diversity mapping in Indian tropical deciduous forests using high-fidelity imaging spectroscopy. Adv Space Res S027311772300159X. https://doi.org/10.1016/j.asr.2023.02.031

Laliberté E, Schweiger AK, Legendre P (2020) Partitioning plant spectral diversity into alpha and beta components. Ecol Lett 23:370–380. https://doi.org/10.1111/ele.13429

Laybros A, Schläpfer D, Féret J-B, Descroix L, Bedeau C, Lefevre M-J, Vincent G (2019) Across date species detection using airborne imaging spectroscopy. Remote Sens 11:789. https://doi.org/10.3390/rs11070789

Legendre P, Legendre L, Legendre L, Legendre L (1998) Numerical ecology, 2nd English ed. Developments in environmental modelling. Elsevier, Amsterdam/New York

Navarro LM, Fernández N, Guerra C, Guralnick R, Kissling WD, Londoño MC, Muller-Karger F, Turak E, Balvanera P, Costello MJ, Delavaud A, El Serafy G, Ferrier S, Geijzendorffer I, Geller GN, Jetz W, Kim E-S, Kim H, Martin CS, McGeoch MA, Mwampamba TH, Nel JL, Nicholson E, Pettorelli N, Schaepman ME, Skidmore A, Sousa Pinto I, Vergara S, Vihervaara P, Xu H, Yahara T, Gill M, Pereira HM (2017) Monitoring biodiversity change through effective global coordination. Curr Opin Environ Sustain 29:158–169. https://doi.org/10.1016/j.cosust.2018.02.005

O'Connor B, Secades C, Penner J, Sonnenschein R, Skidmore A, Burgess ND, Hutton JM (2015) Earth observation as a tool for tracking progress towards the Aichi biodiversity targets. Remote Sens Ecol Conserv 1:19–28. https://doi.org/10.1002/rse2.4

Pereira HM, Ferrier S, Walters M, Geller GN, Jongman RHG, Scholes RJ, Bruford MW, Brummitt N, Butchart SHM, Cardoso AC, Coops NC, Dulloo E, Faith DP, Freyhof J, Gregory RD, Heip C, Hoft R, Hurtt G, Jetz W, Karp DS, McGeoch MA, Obura D, Onoda Y, Pettorelli N, Reyers B, Sayre R, Scharlemann JPW, Stuart SN, Turak E, Walpole M, Wegmann M (2013) Essential biodiversity variables. Science 339:277–278. https://doi.org/10.1126/science.1229931

Poutier, L., Miesch, C., Lenot, X., Achard, V., Boucher, Y., 2002. COMANCHE and COCHISE: two reciprocal atmospheric codes for hyperspectral remote sensing

Rast M, Nieke J, Adams J, Isola C, Gascon F (2021) Copernicus hyperspectral imaging Mission for the environment (Chime). In: 2021 IEEE International Geoscience and Remote Sensing Symposium IGARSS. Presented at the IGARSS 2021–2021 IEEE International Geoscience and Remote Sensing Symposium. IEEE, Brussels, Belgium, pp 108–111. https://doi.org/10.1109/IGARSS47720.2021.9553319

Robertson KM, Simonson E, Ramirez-Bullon N, Poulter B, Carter R (2023) Effects of spatial resolution, mapping window size, and spectral species clustering on remote sensing of plant Beta diversity using biodivMapR and hyperspectral imagery. J Geophys Res Biogeosci 128:e2022JG007350. https://doi.org/10.1029/2022JG007350

Schlerf M, Atzberger C (2006) Inversion of a forest reflectance model to estimate structural canopy variables from hyperspectral remote sensing data. Remote Sens Environ 100:281–294. https://doi.org/10.1016/j.rse.2005.10.006

Schneider FD, Morsdorf F, Schmid B, Petchey OL, Hueni A, Schimel DS, Schaepman ME (2017) Mapping functional diversity from remotely sensed morphological and physiological forest traits. Nat Commun 8. https://doi.org/10.1038/s41467-017-01530-3

Serbin SP, Townsend PA (2020) Scaling functional traits from leaves to canopies. In: Cavender-Bares J, Gamon JA, Townsend PA (eds) Remote sensing of plant biodiversity. Springer International Publishing, Cham, pp 43–82. https://doi.org/10.1007/978-3-030-33157-3_3

Sheeren D, Graham Fonseca M, Willm J (2023) Fabas forest plots acquired during the HYPERBIO project. https://doi.org/10.57745/NHRNOR

Steinwart I, Thomann P (2017) liquidSVM: a fast and versatile SVM package. https://doi.org/10.48550/ARXIV.1702.06899

Torresani M, Rocchini D, Sonnenschein R, Zebisch M, Hauffe HC, Heym M, Pretzsch H, Tonon G (2020) Height variation hypothesis: a new approach for estimating forest species diversity with CHM LiDAR data. Ecol Indic 117:106520. https://doi.org/10.1016/j.ecolind.2020.106520

Ustin SL, Gamon JA (2010) Remote sensing of plant functional types. New Phytol 186:795–816

Verhoef W, Jia L, Xiao Q, Su Z (2007) Unified optical-thermal four-stream radiative transfer theory for homogeneous vegetation canopies. IEEE Trans Geosci Remote Sens 45:1808–1822. https://doi.org/10.1109/TGRS.2007.895844

Verrelst J, Alonso L, Camps-Valls G, Delegido J, Moreno J (2012) Retrieval of vegetation biophysical parameters using Gaussian process techniques. IEEE Trans Geosci Remote Sens 50:1832–1843. https://doi.org/10.1109/TGRS.2011.2168962

Verrelst J, Dethier S, Rivera JP, Munoz-Mari J, Camps-Valls G, Moreno J (2016) Active learning methods for efficient hybrid biophysical variable retrieval. IEEE Geosci Remote Sens Lett 13:1012–1016. https://doi.org/10.1109/LGRS.2016.2560799

Verrelst J, Malenovský Z, Van der Tol C, Camps-Valls G, Gastellu-Etchegorry J-P, Lewis P, North P, Moreno J (2019) Quantifying vegetation biophysical variables from imaging spectroscopy data: a review on retrieval methods. Surv Geophys 40:589–629. https://doi.org/10.1007/s10712-018-9478-y

Verrelst J, Rivera-Caicedo JP, Reyes-Muñoz P, Morata M, Amin E, Tagliabue G, Panigada C, Hank T, Berger K (2021) Mapping landscape canopy nitrogen content from space using PRISMA data. ISPRS J Photogramm Remote Sens 178:382–395. https://doi.org/10.1016/j.isprsjprs.2021.06.017

Villéger S, Mason NWH, Mouillot D (2008) New multidimensional functional diversity indices for a multifaceted framework in functional ecology. Ecology 89:2290–2301. https://doi.org/10.1890/07-1206.1

Wang R, Gamon JA (2019) Remote sensing of terrestrial plant biodiversity. Remote Sens Environ 231:111218. https://doi.org/10.1016/j.rse.2019.111218

Weiss M, Baret F, Jay S (2020) S2ToolBox level 2 products: LAI, FAPAR, FCOVER (ATBD no. 2.0), Sentinel2ToolBox Level2 products. INRA, Avignon

Bridging Remote Sensing and Movement Ecology

Ruben Remelgado

Abstract

Remote sensing data are valuable assets in movement ecology. They provide unique insights into habitat composition and change, enabling unique insights into the drivers of animal movements that depict their behavior. However, combining movement and remote sensing data is not linear. These data are typically collected at contrasting spatial and temporal resolutions, impairing their direct integration. Due to this limitation, we must carefully consider our choices of remote sensing data, and how their spatial and temporal resolutions impair, and potentially mislead, analyses of movement. The R package *rsMove* supports this assessment. This chapter discusses the challenges of linking animal movement and remote sensing data, and how *rsMove* can help navigate them.

Keywords

Animal movement · Remote sensing · Data integration · Multidisciplinary science

1 Introduction

Within their habitats, animals must adapt to the abundance and spatial distribution of food and shelter (Abrahms et al. 2021). This adaptation is driven by an animal's ability to access those resources, which is conditioned by the species' physiological traits (Wu and Seebacher 2022), human pressures (Passoni et al. 2021; Xu et al.

R. Remelgado (✉)
Agro-ecological Modeling Group, Institute of Crop Science and Resource Conservation, University of Bonn, Bonn, Germany
e-mail: ruben.remelgado@idiv.de

D. Rocchini (ed.), *R Coding for Ecology*, Use R!,
https://doi.org/10.1007/978-3-031-99665-8_8

2023), and natural barriers (Plazio et al. 2020; Russo et al. 2023). In addition, animals may have to dynamically adapt to changes to the quality of foraging grounds due to vegetation phenology (Abrahms et al. 2021), human and natural disturbances (Doherty et al. 2021; Nimmo et al. 2019), predator-prey dynamics (Keim et al. 2021), and animal territoriality (Schlichting et al. 2022). All of these factors condition an animal's dispersal capabilities, ultimately defining its habitat preferences and extents (Alzate and Onstein 2022) and its susceptibility to habitat loss (Chetcuti et al. 2022).

As animals respond to these factors, they develop unique movement patters. These movements reflect animal-environment interactions (Abrahms et al. 2021), and record behavioral adaptations to environmental change (Shaw 2020), and to interactions with prey (Hansen et al. 2020) and competing species (Schlichting et al. 2022). Consequently, data on animal movements, which are enabled by GPS technologies, become essential for biodiversity conservation and its effectiveness (Pekarsky et al. 2021). When combined with satellite-borne imagery used to describe concurrent land surface dynamics, movement data helps us build knowledge on animal's behavioral responses to environmental change (Neumann et al. 2015; Smith et al. 2023). This enables us to map and monitor the home range dynamics of individual species (Kays et al. 2023), their migratory patterns (Howard et al. 2024), and their interactions with other species (Russell et al. 2017), creating a basis for adaptative conservation action (Nightingale et al. 2023).

The literature on the combined use of remote sensing and animal movement data is vast. Datasets derived with satellite imagery, such as spectral indices (Brum-Bastos et al. 2020), atmospheric variables (Bradarić et al. 2024), and land cover (Balouch et al. 2022), are commonly used in movement ecology. Most recently, metrics on vegetation structure (Duporge et al. 2022) and fragmentation (Rus et al. 2021) have been exploited, and, although less common, "raw" spectral images have also been used (Geremia et al. 2019; Remelgado et al. 2018). However, although the combined use of movement and remote sensing data is common, these data are not immediately compatible (Neumann et al. 2015; Remelgado et al. 2018).

GPS technologies allow us to monitor movements at increasingly finer fine temporal scales (down to minutes and seconds), and remarkedly fine spatial resolutions (down to meters and centimeters). Oftentimes, these data depict daily movement patterns across multiple years and continents (Brodie et al. 2018; Shamoun-Baranes et al. 2014). However, the temporal and spatial resolution of remote sensing data is constrained by the orbit of the satellite enabling them. For instance, in order to sustain a persistent global coverage at high temporal resolutions, satellite programs used in atmospheric monitoring (e.g., EUMETSAT, NOAA, ASCAT) adopt a higher orbit, leading to lower spatial resolutions (in the order of several kilometers). In contrast, satellite programs directed at land surface monitoring (e.g., Landsat, MODIS, Sentinel) offer enable higher resolutions (in the order of several meters) by adopting lower orbits. Yet, lower orbits force satellites to monitor smaller spatial extents, leading to spatial inconsistencies in temporal coverages.

Due to these caveats, no single Earth Observation sensor can provide data at consistently high temporal and spatial resolutions with a global coverage.

Therefore, matching movement and remote sensing data is often indirect, relying on environmental datasets that fail to represent environmental dynamics at the spatial temporal scales at which an animal perceive them. Mobilizing "inadequate" remote sensing data, or failing to match animal movement data to the spatial and temporal scales at which remote sensing data are produced, may mislead our understanding of how animals respond to environmental dynamics.

However, choosing "appropriate" remote sensing data is equally challenging. Due to the advent of open-access planetary computing platforms, disciplines such as movement ecology can now benefit from a wealth of satellite-borne imagery to study animal behaviors. Simultaneously, access to these data and computational resources made producing remote sensing data products increasingly feasible. While this accelerated progress in remote sensing (Nagaraj et al. 2020), it has also made it more challenging for nonexperts to navigate an exponentially growing pool of open-access datasets.

Perceiving and tackling these limitations requires in-depth knowledge of both animal movement and remote sensing data. However, knowledge on these caveats is not universal. Movement ecologists typically lack in-depth experience with remote sensing data to anticipate technical challenges, and remote sensing experts are rarely accustomed to the dynamic nature of movement data, in terms of their varying—and sometimes inconsistent—spatial and temporal coverages. To cement these cross-disciplinary knowledge gaps, we require tools to help communicate data limitations, to guide optimal data integration strategies, and to support the definition of remote sensing data requirements.

2 rsMove: Movement Ecology Through the Lens of Remote Sensing

The rsMove R package *rsMove* (Remelgado et al. 2019) was designed to improve communication between movement ecologists and remote sensing experts. It provides several (Table 1) analyzing how the integrating user-provided movement data with remote sensing datasets is affected by the spatial and temporal resolutions of the latter. Furthermore, *rsMove* helps its users translate movement data into spatially and temporally constrained areas of interest (or "sites") where the different remote sensing solutions may be tested most effectively. This ultimately supports the communication of remote sensing data requirements, potentially supporting collaborations between movement ecologists and remote sensing experts.

In the following sections, we will demonstrate how the *rsMove* can guide the selection of remote sensing data and study sites. The package *rsMove* is available on GitHub, and can be installed as follows:

```
devtools::install_github("RRemelgado/rsMove")
```

Table 1 rsMove functions. List of functions discussed in this chapter

Purpose	Function	Description
Evaluation of the potential limitations of earth observation data	*sMoveRes*	Simulation of data losses due to the choice of earth observation data with different spatial resolutions
	tMoveRes	Simulation of data losses due to the choice of earth observation data with different temporal resolutions
	moveCloud	Extraction of cloud cover fractions at each movement data coordinate pair, at the corresponding acquisition date
Selection of test sites	*map_sites*	Detection of movement data clusters to distinguish regions of interest where to test remote sensing applications

3 Example Movement and Remote Sensing Dataset

To demonstrate the use of *rsMove*, the package offers data from two exemplary movement datasets. First, it includes data for a population of bald eagles (*Haliaeetus leucocephalus*) composed by GPS records from five individuals. These individuals were monitored between 2014 and 2021 at an average frequency of 30 minutes (Fig. 1a–c). Second, *rsMove* includes data from a population of white storks (*Ciconia ciconia*) from two individuals monitored between 2014 at an average frequency of 5 minutes. These data were obtained from a movement dataset originally composed by 92 individuals tagged between 2014 and 2024 (Fiedler et al. 2021; Fig. 1d, e). We only preserved a subset of these data given its large size. Both example datasets were developed by the Max Planck Institute for Ornithology and provided through MoveBank (Wikelski and Kays 2022). Processed versions can be accessed as follows:

```
bald_eagle_data = read.csv(system.file('extdata', 'baldEagle.csv', package="rsMove"))
white_stork_ data = read.csv(system.file('extdata', 'whiteStork.csv', package="rsMove"))
```

4 Spatial Compatibility of Movement and Remote Sensing Data

Because movement and remote sensing data are typically retrieved at different spatial and temporal scales, integrating them directly is often inadequate. It often leads to pseudo-replication while increasing spatial autocorrelation (Fig. 2a). For instance, the use of remote sensing data on the Normalized Difference Vegetation Index (NDVI) gained popularity in movement ecology for modeling the quality of foraging grounds (e.g., Johnson et al. 2018), step-selection dynamics (e.g., Stratmann et al. 2021), or even the propagation of diseases through infected animals (Dougherty et al. 2022). MODIS NDVI data from TERRA and/or AQUA, in

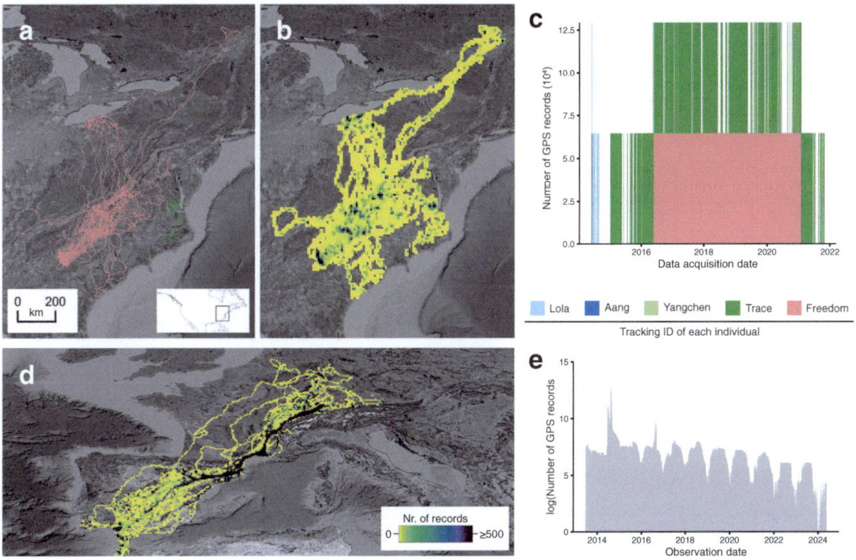

Fig. 1 Example movement data. Example movement dataset composed by GPS tracking data from five bald eagles (*Haliaeetus leucocephalus*). (**a**) Depiction of the area covered by the movement dataset, located in the eastern USA (area shown on the bottom right). Individuals are distinguished by different colors. (**b**) Frequency of GPS records per each 10-km pixel. (**c**) Temporal distribution of the frequency of GPS records per individual. (**d**) Spatial distribution and density of GPS coordinates recording the movement of 92 white storks (*Ciconia ciconia*) tagged originally in South-Eastern Germany. (**e**) Temporal distribution of the White Stork data. The data on the white stork (Fiedler et al. 2024) and bald eagle (Kays 2025) are available under a CC 1.0 Universal license

particular, have been frequently used (e.g., Dougherty et al. 2022; Johnson et al. 2018; Stratmann et al. 2021) due to their relatively high spatial and temporal resolutions of these data (250 m and 16 days), their simplicity, and their sensitivity to various environmental change processes. Due to the abundance of literature documenting the usefulness of MODIS-based NDVI data, one might feel compelled to use them in new movement studies. However, even if this choice is supported by literature, it may be inadequate for one's application.

The *sMoveRes()* function helps us revel how the choice of MODIS NDVI data may result in pseudo-replication and increased spatial autocorrelation (Fig. 2b). This function identifies which pixels in a regular spatial grid overlap with each data entry in a movement dataset. Then, it counts the number of unique pixels preserved at the given resolution to measure pseudo-replication, and the number of connected groups of unique pixels (or "regions"), detected with K-dimensional trees, to measure spatial autocorrelation.

The resolution of the grid used to assign GPS records to pixels corresponds to that of the target remote sensing dataset. However, *sMoveRes()* does not require us to provide the actual remote sensing data. It simulates the target remote sensing data

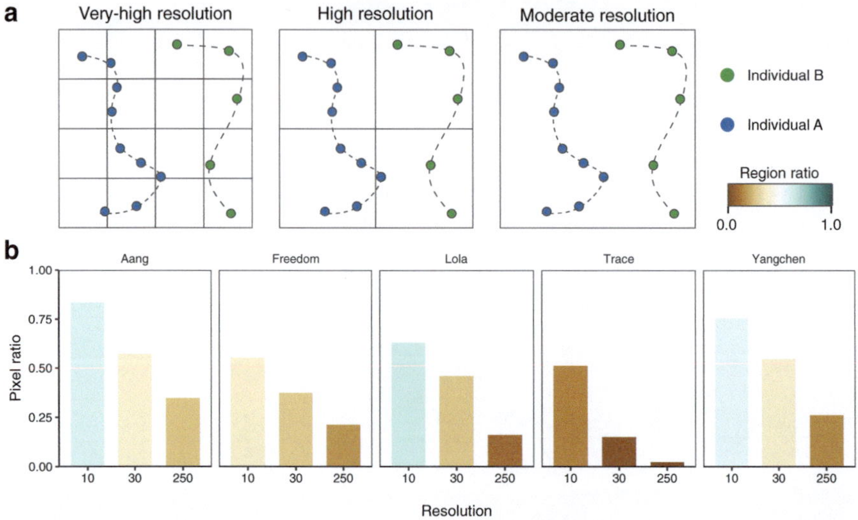

Fig. 2 Clustered movement data impacts the granularity of environmental data. (**a**) Conceptual diagram showing how decreasing spatial resolutions impact the data retrieved from a given environmental dataset (raster grid designated by white squares) over the GPS coordinates in a movement dataset composed by two bald eagles (dots and dashed lines; colors distinguish individuals). At a very-high spatial resolution (e.g., Sentinel), we already note that some pixels overlap with multiple GPS records. This implies that multiple environmental data entries will be replicated. As we move to a high resolution (e.g., Landsat 8–9) toward a moderate spatial resolution (e.g., MODIS), we see that the degree of pseudo-replication increases, until the inclusion of environmental data becomes redundant due to the lack of spatial variability. Furthermore, the spatial autocorrelation between the two movement tracks increases as they share larger numbers of pixels. (**b**) Evaluation of the quantity of unique pixels and pixel regions preserver at very-high, high, and moderate resolutions (i.e., Sentinel, Landsat, and MODIS, respectively) for each individual in the exemplary Bald Eagle movement dataset. The "pixel ratio" (measure on the y-axis) is the ratio between the number of unique pixels and the number of GPS tracking records. The "region ratio" (i.e., number of GPS coordinates divided by the number of regions with coordinates, shown by a gradient of colors in the bars) is the ratio between the number of unique pixel regions and the number of GPS tracking devices. As the values in these metrics decrease, they respectively show increasing degrees of pseudo-replication and spatial-autocorrelation. The Bald Eagle data are available under a CC 1.0 Universal license (Kays 2025)

based on a user-defined spatial resolution and the spatial extent of the movement data. Therefore, *sMoveRes()* provides a quick assessment of the limitations of acquiring an environmental dataset of a given spatial resolution before engaging in time-consuming downloads and/or data processing. To demonstrate this, we will apply *sMoveRes()* to the exemplary Bald Eagle data, as follows:

```
spatial_loss = sMoveRes(bald_eagle_data, c(10,30,250))
```

The code evaluates the implications of using NDVI data obtained with MODIS compared to NDVI data generated with Landsat 8–9 Sentinel 2, which have a spatial resolution of 30 and 10 m, respectively. *sMoveRes()* reveals that, with Sentinel data, we would have preserved 89,952 unique pixels and 61,214 unique pixel regions, approximately one-third more than enabled by Landsat data. In contrast, using MODIS data, we would have preserved only 34,353 unique pixels and 26,354 regions, representing a decrease of 38.1% and 43.0%, respectively, compared to the use of Sentinel data. This constitutes a substantial reduction in unique data given the original movement dataset contains 159,331 GPS records.

Furthermore, *sMoveRes()* demonstrates that the number of unique pixels and regions varies across individuals (Fig. 2b). For instance, using Sentinel data, the number of unique pixels varied between 44 and 82,608, compared to a range between 2 and 31,855 using MODIS data. This indicates that different individuals show varying degrees of dispersal, meaning that our choice of remote sensing data may need to be customized accordingly.

5 Temporal Compatibility of Movement and Remote Sensing Data

Pseudo-replication and autocorrelation also manifest in time. Because remote sensing data is typically provided at stable temporal intervals, movement data recorded in between two timesteps cannot be paired directly with Earth Observation measurements (Fig. 3). To measure potential data degradation due to temporal mismatches, we can use *tMoveRes()*. This function works similarly to *sMoveRes()*, but instead estimates the degree of pseudo-replication and spatial-autocorrelation resulting from the combination of movement data with remote sensing data of varying temporal resolutions, and a static spatial resolution.

As shown below, applying *tMoveRes()* to the Bald Eagle movement data shows varying degrees on pseudo-replication and pseudo-autocorrelation over time. Here, we assumed that the MODIS-based NDVI data remained the preferred choice, and therefore tested temporal resolutions of 1 day (i.e., MODIS daily surface reflectance products MOD09Q1 and MYD09Q1 from which the NDVI can be calculated) and 16 days (i.e., MODIS MOD13Q1 and MYD13Q1 NDVI products), both of which are available at a resolution of 250 m:

```
obs.dates = as.Date(bald_eage_data$timestamp)
temporal_loss = tMoveRes(bald_eagle_data[ind,], obs.dates[ind,], c(1, 16), 250)
```

The function highlights two important considerations (Fig. 3a). First, it shows us that the frequency of unique pixels is not constant over time (standard deviations of 44.8 and 492.1 for temporal resolutions of 1 and 16 days, respectively), neither is the frequency of unique pixel regions (standard deviations of 144.4 and 692.1). This is influenced by the daily dispersal and mobility of each tracked individual. Second, similarly to *sMoveRes()*, the function shows that a reduction in temporal resolution leads to a substantial decrease in unique pixels a pixel region relative to the amount of GPS coordinates. The ratio of pixel regions relative to the number of GPS tracking records dropped from an average of 0.38 at a 1-day resolution, to

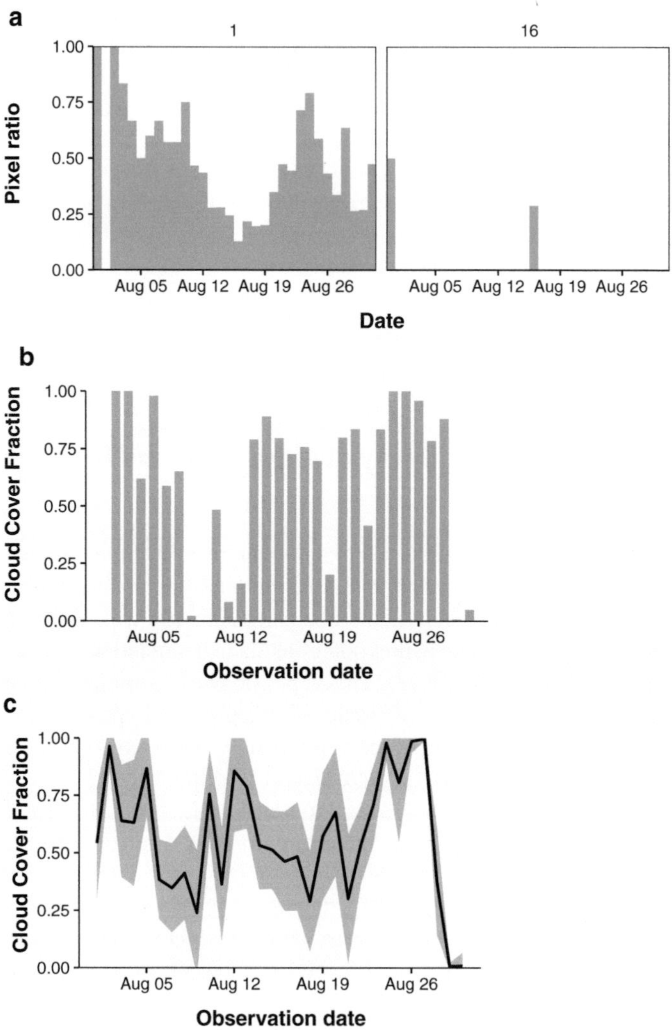

Fig. 3 Impact of varying temporal resolutions on data availability. (**a**) On the left, we see the variability in the pixel ratio (i.e., number of GPS coordinates divided by the number of unique pixels with coordinates) during the period between 2019-08-01 and 2019-08-30. This imagines the potential overlap of movement data with remote sensing data with a temporal resolution of 1 day. On the right, the same analysis with a temporal resolution of 16 days. Here, within each 16-day window, the GPS coordinates of the movement dataset that are contained within are were summarized into unique pixel coordinates given that no direct remote sensing information will be available before the end of that window. As a consequence, only two data entries can be found. (**b**) Temporal variation in the mean cloud cover fractions (CCFs) per each day with GPS coordinates. (**c**) Mean CCF across all GPS coordinates for each day for each CCF was acquired, represented by a black line. The gray buffer surrounding that line indicates the standard deviation

an average of 0.33 at a 16-day resolution. The ratio of pixel regions relative to the number of GPS tracking records dropped from 0.82 to 0.46.

Looking at the results generate by *tMoveRes()*, daily NDVI data is the most appealing. This would enable preserving the most amount of information. However, this comes with additional challenges. When data is obtained at such fine temporal resolutions, in addition to added preprocessing challenges (e.g., due to different in the swat on acquisitions, illumination angles), we must deal with variations in the amount of available data due to cloud cover, and due to the shadows cast by it. To help us perceive this challenge, we can use *moveCloud()* as follows:

```
start = as.Date("2019-08-01"),
end = as.Date("2019-08-30")
interval = 1
clouded_movements = moveCloud(bald_eagle_data, start, end, interval, "cloud_fraction")
```

We applied *moveCloud()* to the Bald Eagle data. Due to the extensive temporal coverage of these data, we focused on the period between 2019-08-01 and 2019-08-30 for exemplary purposes. With these parameters, *moveCloud()* retrieves data on cloud cover fractions (CCFs) within the defined temporal window on a daily basis. This information is then extracted for each pair of GPS coordinates in a movement dataset to inform us on the following: (i) the CCF at the dates when GPS data were recorded (Fig. 3b) and (ii) the daily mean and standard deviation of the CCF during the reference period estimated across the coordinates of all movement data records (Fig. 3c).

To acquire information on cloud cover fractions, *moveCloud()* accesses global layers on this variable provided by NASA's Earth Observations (NEO) platform at a resolution of 0.1 degrees (NEO 2024). These data are open-access, but they are not supported by a server able to process coordinate-specific data queries. For that reason, *moveCloud()* must download these data locally before subsequent analyses (here stored in a new directory named "cloud_fraction"). When the download is initiated, the function will report the amount of space required to store the data for all target dates. Once the data is downloaded, provided that the same directory is specified, the *moveCloud()* will not rerun the download if the target data is available locally.

The application of *moveCloud()* to the Bald Eagle data shows us how the use of daily remote sensing data will be impacted by cloud cover, which limits the number of usable pixels. In the proposed example, the CCF at the dates GPS coordinates were obtained was of 0.7 on average (±0.4). Across the reference period, cloud cover increased over time, from an average of 0.5 in the 1st of August (±0.2, indicating that half of the area in covered by clouds) to 1.0 in the 27th of August (±0.2, indicating the absence of usable data). These results suggest that both remote sensing and movement data should be handled at coarser temporal resolutions. By considering a larger temporal window, we can, for instance, create a mosaic of visible pixels to minimize occurrences of missing data across space, similarly to how existing MODIS NDVI data are produced. In parallel, several R packages are available to extract relevant information from movement data at similar temporal scales (Joo et al. 2020), such as by estimating the elapsed time at each pixel, or by distinguishing foraging and resting sites. Through this process, we would achieve a compromise between movement and remote sensing data that is sensitive to their respective strengths and weaknesses.

We should acknowledge that, should a finer temporal scale be necessary, it is possible to tackle limitations caused by cloud cover. Several methods have been proposed to general seemingly continuous time series by combining data from multiple optical satellite sensors (e.g., Claverie et al. 2018), or by combining data from satellites equipped with optical and synthetic aperture radar sensors (e.g., Tang et al. 2023). In addition, several methods exist to interpolate missing data using sophisticated regression and machine learning methods (Mo et al. 2023). However, these methods would at most enable daily analyses, meaning that movement data should be summarized at that scale. Nonetheless, the use of such advanced methods would naturally increase computational and data storage requirements. One must assess whether such an effort is both feasible and necessary and, if not, research objectives must be adjusted accordingly.

6 Study Site Selection

Although planetary computing platforms are today accessible to scientists of various disciplines, if processing requirements are extensive, they may still incur financial costs due to existing quotas, or unreasonable time investments. To design experiments using remote sensing adapted to such barriers, we must be able to benchmark our choices of methods and environmental datasets, and to carefully document how those choices interact with our target movement data, and how they may advance, or prevent, our research questions. To enable this, we must identify study sites, i.e., smaller geographical regions where movement data and representative and where handling multiple data are feasible with local computational resources.

However, improvements in GPS tracking technologies make it more challenging to reduce our focus to smaller areas. Even data from a single individual may contain millions of GPS records, and extend several countries or continents. Analyzing movement patterns recorded by such extensive datasets can demand vast and unmanageable amounts of remote sensing data. Here, we can use the function *map_sites()*.Using data from a movement dataset, the function identifies regions of pixels based on the clustering GPS coordinates and a user-defined spatial resolution. Then, at each pixel region, the function distinguishes temporal segments (i.e., sequences of days separated by 1-day intervals), and proceeds to report several metrics within those segments. These include counts of unique individuals, GPS records, and days, start and end dates with GPS observations, and mean frequencies of GPS recordings expressed in minutes. Furthermore, the function calculates metrics on the average proportion of a day that is covered by GPS records (i.e., "day cover"), and on the span of the day covered between the first and last GPS coordinates (i.e., "day overlap"). This information helps distinguish geographical regions where movement data is most abundant, and where these data occur for

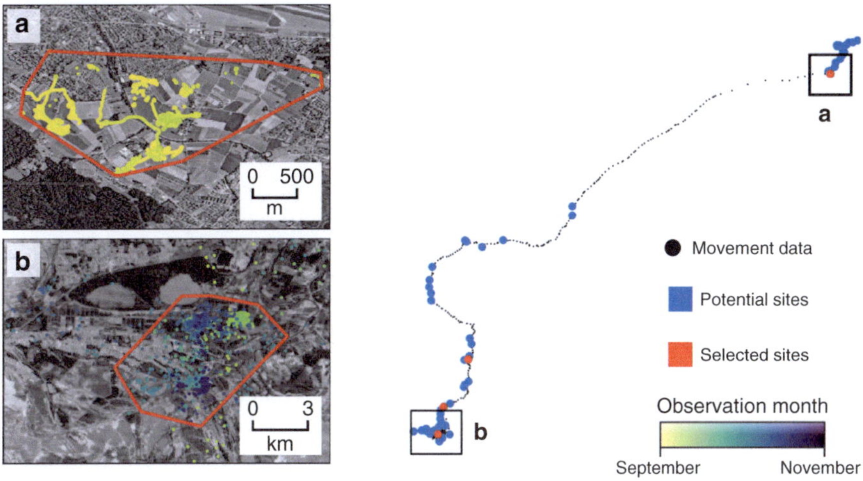

Fig. 4 Study site selection. Example of a White Stork movement track (depicted with black dots) segmented into potential study sites (blue dots). Of these sites, those with, at minimum, GPS data for a continued duration of 120 days, a GPS recording frequency of 4.9 minutes, and a day overlap of 7.7% were selected (red dots). (**a**, **b**) Show examples of selected sites (limited by polygons with a red outline) at the beginning and end of the movement data segment. Within those sites, movement data points are colored based on their observation date, from September to November

extensive periods of time, which are ideal test sites for testing new remote sensing analyses.

Below, we demonstrate the application of the *map_sites()* function to the White Stork movement data. We chose these data given it records migratory movements, leading it to extend across two continents. Using these data, we will identify potential study sites assuming a regular global grid with a resolution of 0.01 degrees, which roughly translates into a 1-km resolution at the equator. Then, using the output of the function, we can identify potential study sites based on convergence of evidence. For exemplary purposes, we will focus on regions where the number of days spent, the mean day overlap, and the GPS recording frequency are all above their respective 95th percentiles. This implies, at minimum, a temporal sequence of 120 days, a GPS recording frequency of 4.9 minutes, and a day overlap of 7.7%. Based on this analysis, only 6 candidate study sites were preserved out of a potential pool of 470 (Fig. 4):

```
# Detection of study sites
individuals = white_stork_data$id
ts = white_stork_data$timestamp
obs.dates = strptime(ts,format="%Y-%m-%d %H:%M:%S", tz="GMT")
site_selection = map_sites(white_stork_data, individuals, obs.dates, 0.01)
# Selection of study sites based on convergence of evidence
stats = site_selection$region.stats
which((stats$nr_days > quantile(stats$nr_days, .95, na.rm=T)) &
      (stats$nr_days > quantile(stats$day_overlap_mean, .95, na.rm=T)) &
      (stats$data_frequency_mean > quantile(stats$data_frequency_mean, .95, na.rm=T)))
```

7 Exploring Potential Data of Environmental Variables and Data

rsMove provides an entry point into the combined use of movement and remote sensing data. Its main objective is to support the definition of data requirements, both in terms of spatial and temporal resolutions, while alerting for potential technical challenges (e.g., due to cloud cover). However, *rsMove* does not support the concrete selection of environmental predictors, for which exploratory analyses are required.

map_sites() helps narrow our geographical focus, easing this effort substantially. Then, for each of the selected test sites, we can use other existing R packages to describe the environmental characteristics of those sites. For instance, the R package *moveVis* (Schwalb-Willmann et al. 2020) can animate movement datasets and create dynamic plots of movement interactions with land cover and other variables, informing on behavioral responses to spatial environmental gradients and temporal changes. Then, *RStoolbox* (Leutner and Horning 2015) helps analyze remote sensing data using established workflows, and *ecochange* (Lara et al. 2022) can be used to map ecological changes triggering behavioral responses in animals. Such packages can be supported by those of the likes of *OpenEOcubes* (Pondi et al. 2024), which provides access to several remote sensing data cubes. Alternative, if users can confidently code remote sensing applications, *rgee* (Aybar et al. 2020) provides an interface to Google Earth Engine, enabling the handling and processing of big remote sensing data.

The aforementioned R packages are mere examples of a software ecosystem that is constantly growing. As the use of remote sensing in R becomes faster and more popular, new packages will surely be developed, broadening the horizons of movement ecology.

8 Conclusions

Linking remote sensing and movement ecology data is essential for our understanding of animal behavior, making it equally crucial for guiding biodiversity conservation. However, linking these data is challenging due to spatial and temporal mismatches between them. *rsMove* enables its users to better perceive these challenges. This can potentially help users save time and resources by guiding a clearer definition of data requirements prior to the mobilization of large volumes of remote sensing data, or preceding their combination with movement data. This process forms a communication channel between movement ecologists and remote sensing experts, helping define concrete environmental data requirements, and guiding the development of customized remote sensing data products, and data integration workflows.

References

Abrahms B, Aikens EO, Armstrong JB, Deacy WW, Kauffman MJ, Merkle JA (2021) Emerging perspectives on resource tracking and animal movement ecology. Trends Ecol Evol 36:308–320. https://doi.org/10.1016/j.tree.2020.10.018

Alzate A, Onstein RE (2022) Understanding the relationship between dispersal and range size. Ecol Lett 25:2303–2323. https://doi.org/10.1111/ele.14089

Aybar C, Bautista L, Yali R, Barja A, Wu Q (2020) rgee: an R package for interacting with Google earth engine. J Open Source Softw 5:2272. https://doi.org/10.21105/joss.02272

Balouch S, Driscoll DA, Naseer A, Rais M, Doherty TS (2022) Impacts of land cover on reptile movement and habitat use in farming landscapes. Anim Conserv 25:837–848. https://doi.org/10.1111/acv.12789

Bradarić M, Kranstauber B, Bouten W, van Gasteren H, Baranes JS (2024) Drivers of flight altitude during nocturnal bird migration over the North Sea and implications for offshore wind energy. Conserv Sci Pract 6:e13114. https://doi.org/10.1111/csp2.13114

Brodie S, Lédée EJI, Heupel MR, Babcock RC, Campbell HA, Gledhill DC, Hoenner X, Huveneers C, Jaine FRA, Simpfendorfer CA, Taylor MD, Udyawer V, Harcourt RG (2018) Continental-scale animal tracking reveals functional movement classes across marine taxa. Sci Rep 8:3717. https://doi.org/10.1038/s41598-018-21988-5

Brum-Bastos V, Long J, Church K, Robson G, de Paula R, Demšar U (2020) Multi-source data fusion of optical satellite imagery to characterize habitat selection from wildlife tracking data. Ecol Inform 60:101149. https://doi.org/10.1016/j.ecoinf.2020.101149

Chetcuti J, Kunin WE, Bullock JM (2022) Species' movement influence responses to habitat fragmentation. Divers Distrib 28:2215–2228. https://doi.org/10.1111/ddi.13623

Claverie M, Ju J, Masek JG, Dungan JL, Vermote EF, Roger J-C, Skakun SV, Justice C (2018) The Harmonized Landsat and Sentinel-2 surface reflectance data set. Remote Sens Environ 219:145–161. https://doi.org/10.1016/j.rse.2018.09.002

Doherty TS, Hays GC, Driscoll DA (2021) Human disturbance causes widespread disruption of animal movement. Nat Ecol Evol 5:513–519. https://doi.org/10.1038/s41559-020-01380-1

Dougherty ER, Seidel DP, Blackburn JK, Turner WC, Getz WM (2022) A framework for integrating inferred movement behavior into disease risk models. Mov Ecol 10:31. https://doi.org/10.1186/s40462-022-00331-8

Duporge I, Finerty GE, Ihwagi F, Lee S, Wathika J, Wu Z, Macdonald DW, Wang T (2022) A satellite perspective on the movement decisions of African elephants in relation to nomadic pastoralists. Remote Sens Ecol Conserv 8:841–854. https://doi.org/10.1002/rse2.285

Fiedler W, Flack A, Schäfle W, Keeves B, Quetting M, Eid B, Schmid H, Wikelski M (2021) LifeTrack White Stork SW Germany (2013–2019). https:/\penalty\z@/\penalty\z@doi.org/\penalty\z@Repository; https://doi.org/10.5441/001/1.ck04mn78

Fiedler W, Flack A, Schäfle W, Keeves B, Quetting M, Eid B, Schmid H, Wikelski M (2024) Data from: study "LifeTrack White stork SW Germany" (2013–2023). https://doi.org/10.5441/001/1.ck04mn78_2

Geremia C, Merkle JA, Eacker DR, Wallen RL, White PJ, Hebblewhite M, Kauffman MJ (2019) Migrating bison engineer the green wave. Proc Natl Acad Sci 116:25707–25713. https://doi.org/10.1073/pnas.1913783116

Hansen JH, Skov C, Baktoft H, Brönmark C, Chapman BB, Hulthén K, Hansson L-A, Nilsson PA, Brodersen J (2020) Ecological consequences of animal migration: prey partial migration affects predator ecology and prey communities. Ecosystems 23:292–306. https://doi.org/10.1007/s10021-019-00402-9

Howard C, Mason THE, Baillie SR, Border J, Hewson CM, Houston AI, Pearce-Higgins JW, Bauer S, Willis SG, Stephens PA (2024) Explaining and predicting animal migration under global change. Divers Distrib 30:e13797. https://doi.org/10.1111/ddi.13797

Johnson HE, Gustine DD, Golden TS, Adams LG, Parrett LS, Lenart EA, Barboza PS (2018) NDVI exhibits mixed success in predicting spatiotemporal variation in caribou summer forage quality and quantity. Ecosphere 9:e02461. https://doi.org/10.1002/ecs2.2461

Joo R, Boone ME, Clay TA, Patrick SC, Clusella-Trullas S, Basille M (2020) Navigating through the r packages for movement. J Anim Ecol 89:248–267. https://doi.org/10.1111/1365-2656.13116

Kays R (2025) LifeTrack bald eagle [WWW document]. Movebank. https:/\penalty\z@/\penalty\z@www.movebank.org/\penalty\z@cms/\penalty\z@webapp?gwt_fragment=page%3Dstudies%2Cpath%3Dstudy20873986

Kays R, Hirsch B, Caillaud D, Mares R, Alavi S, Havmøller RW, Crofoot M (2023) Multi-scale movement syndromes for comparative analyses of animal movement patterns. Mov Ecol 11:61. https://doi.org/10.1186/s40462-022-00365-y

Keim JL, DeWitt PD, Wilson SF, Fitzpatrick JJ, Jenni NS, Lele SR (2021) Managing animal movement conserves predator–prey dynamics. Front Ecol Environ 19:379–385. https://doi.org/10.1002/fee.2358

Lara W, Londoño MC, Gonzalez I, Gutierrez-Velez VH (2022) Ecochange: an R-package to derive ecosystem change indicators from freely available earth observation products. Methods Ecol Evol 13:2379–2388. https://doi.org/10.1111/2041-210X.13986

Leutner B, Horning N (2015) RStoolbox: tools for remote sensing data analysis. R Package Version

Mo Y, Xu Y, Liu Y, Xin Y, Zhu S (2023) Comparison of gap-filling methods for producing all-weather daily remotely sensed near-surface air temperature. Remote Sens Environ 296:113732. https://doi.org/10.1016/j.rse.2023.113732

Nagaraj A, Shears E, de Vaan M (2020) Improving data access democratizes and diversifies science. Proc Natl Acad Sci 117:23490–23498. https://doi.org/10.1073/pnas.2001682117

NEO (2024) Cloud Fraction (1 month – Aqua/MODIS)

Neumann W, Martinuzzi S, Estes AB, Pidgeon AM, Dettki H, Ericsson G, Radeloff VC (2015) Opportunities for the application of advanced remotely-sensed data in ecological studies of terrestrial animal movement. Mov Ecol 3. https://doi.org/10.1186/s40462-015-0036-7

Nightingale J, Gill JA, Þórisson B, Potts PM, Gunnarsson TG, Alves JA (2023) Conservation beyond boundaries: using animal movement networks in protected area assessment. Anim Conserv 26:753–765. https://doi.org/10.1111/acv.12868

Nimmo DG, Avitabile S, Banks SC, Bliege Bird R, Callister K, Clarke MF, Dickman CR, Doherty TS, Driscoll DA, Greenville AC, Haslem A, Kelly LT, Kenny SA, Lahoz-Monfort JJ, Lee C, Leonard S, Moore H, Newsome TM, Parr CL, Ritchie EG, Schneider K, Turner JM, Watson S, Westbrooke M, Wouters M, White M, Bennett AF (2019) Animal movements in fire-prone landscapes. Biol Rev 94:981–998. https://doi.org/10.1111/brv.12486

Passoni G, Coulson T, Ranc N, Corradini A, Hewison AJM, Ciuti S, Gehr B, Heurich M, Brieger F, Sandfort R, Mysterud A, Balkenhol N, Cagnacci F (2021) Roads constrain movement across behavioural processes in a partially migratory ungulate. Mov Ecol 9:57. https://doi.org/10.1186/s40462-021-00292-4

Pekarsky S, Schiffner I, Markin Y, Nathan R (2021) Using movement ecology to evaluate the effectiveness of multiple human-wildlife conflict management practices. Biol Conserv 262:109306. https://doi.org/10.1016/j.biocon.2021.109306

Plazio E, Bubová T, Vrabec V, Nowicki P (2020) Sex-biased topography effects on butterfly dispersal. Mov Ecol 8:50. https://doi.org/10.1186/s40462-020-00234-6

Pondi B, Appel M, Pebesma E (2024) OpenEOcubes: an open-source and lightweight R-based RESTful web service for analyzing earth observation data cubes. Earth Sci Inform 17:1809–1818. https://doi.org/10.1007/s12145-024-01249-y

Remelgado R, Leutner B, Safi K, Sonnenschein R, Kuebert C, Wegmann M (2018) Linking animal movement and remote sensing – mapping resource suitability from a remote sensing perspective. Remote Sens Ecol Conserv 4:211–224. https://doi.org/10.1002/rse2.70

Remelgado R, Wegmann M, Safi K (2019) rsmove—an r package to bridge remote sensing and movement ecology. Methods Ecol Evol 10:1212–1221. https://doi.org/10.1111/2041-210X.13199

Rus AI, McArthur C, Mella VSA, Crowther MS (2021) Habitat fragmentation affects movement and space use of a specialist folivore, the koala. Anim Conserv 24:26–37. https://doi.org/10.1111/acv.12596

Russell JC, Hanks EM, Modlmeier AP, Hughes DP (2017) Modeling collective animal movement through interactions in behavioral states. J Agric Biol Environ Stat 22:313–334. https://doi.org/10.1007/s13253-017-0296-3

Russo NJ, Davies AB, Blakey RV, Ordway EM, Smith TB (2023) Feedback loops between 3D vegetation structure and ecological functions of animals. Ecol Lett 26:1597–1613. https://doi.org/10.1111/ele.14272

Schlichting PE, Boughton RK, Anderson W, Wight B, VerCauteren KC, Miller RS, Lewis JS (2022) Seasonal variation in space use and territoriality in a large mammal (Sus scrofa). Sci Rep 12:4023. https://doi.org/10.1038/s41598-022-07297-y

Schwalb-Willmann J, Remelgado R, Safi K, Wegmann M (2020) moveVis: animating movement trajectories in synchronicity with static or temporally dynamic environmental data in r. Methods Ecol Evol 11:664–669. https://doi.org/10.1111/2041-210X.13374

Shamoun-Baranes J, Alves JA, Bauer S, Dokter AM, Hüppop O, Koistinen J, Leijnse H, Liechti F, van Gasteren H, Chapman JW (2014) Continental-scale radar monitoring of the aerial movements of animals. Mov Ecol 2. https://doi.org/10.1186/2051-3933-2-9

Shaw AK (2020) Causes and consequences of individual variation in animal movement. Mov Ecol 8:12. https://doi.org/10.1186/s40462-020-0197-x

Smith JB, Keiter DA, Sweeney SJ, Miller RS, Schlichting PE, Beasley JC (2023) Habitat quality influences trade-offs in animal movement along the exploration–exploitation continuum. Sci Rep 13:4814. https://doi.org/10.1038/s41598-023-31457-3

Stratmann TSM, Dejid N, Calabrese JM, Fagan WF, Fleming CH, Olson KA, Mueller T (2021) Resource selection of a nomadic ungulate in a dynamic landscape. PLoS One 16:1–25. https://doi.org/10.1371/journal.pone.0246809

Tang X, Bratley KH, Cho K, Bullock EL, Olofsson P, Woodcock CE (2023) Near real-time monitoring of tropical forest disturbance by fusion of Landsat, Sentinel-2, and Sentinel-1 data. Remote Sens Environ 294:113626. https://doi.org/10.1016/j.rse.2023.113626

Wikelski M, Kays R (2022) Movebank: archive, analysis and sharing of animal movement data. Hosted by the Max Planck Institute for Ornithology

Wu NC, Seebacher F (2022) Physiology can predict animal activity, exploration, and dispersal. Commun Biol 5:109. https://doi.org/10.1038/s42003-022-03055-y

Xu W, Gigliotti LC, Royauté R, Sawyer H, Middleton AD (2023) Fencing amplifies individual differences in movement with implications on survival for two migratory ungulates. J Anim Ecol 92:677–689. https://doi.org/10.1111/1365-2656.13879

The `rasterdiv` Package for Measuring Diversity from Space

Matteo Marcantonio, Elisa Thouverai, and Duccio Rocchini

Abstract

This book chapter provides an extensive overview of the *rasterdiv* package which provides a comprehensive suite of functions designed to calculate diversity indices from numerical matrices, including those derived from optical remote sensing imagery. It facilitates a deeper analysis of spatial heterogeneity and ecosystem complexity by translating pixel values into ecological indicators. This package serves as a valuable tool for researchers and practitioners in the fields of landscape ecology, biodiversity monitoring, and environmental management, allowing for an enhanced understanding of landscape patterns and their ecological implications through advanced quantitative measures.

M. Marcantonio (✉)
Earth & Life Institute, University of Louvain (UCLouvain), Louvain-la-Neuve, Belgium

LandLife EcoSpatial Labs, Mezzocorona, Italy

E. Thouverai
Department of Life Sciences, University of Trieste, Trieste, Italy

BIOME Lab, Department of Biological, Geological and Environmental Sciences, Alma Mater Studiorum University of Bologna, Bologna, Italy
e-mail: elisa.thouverai2@unibo.it

D. Rocchini
BIOME Lab, Department of Biological, Geological and Environmental Sciences, Alma Mater Studiorum University of Bologna, Bologna, Italy

Faculty of Environmental Sciences, Department of Applied Geoinformatics and Spatial Planning, Czech University of Life Sciences Prague, Praha, Suchdol, Czech Republic
e-mail: duccio.rocchini@unibo.it

D. Rocchini (ed.), *R Coding for Ecology*, Use R!,
https://doi.org/10.1007/978-3-031-99665-8_9

Keywords

rasterdiv · Biodiversity indices · Habitat heterogeneity · NDVI · Spectral data · Remote sensing

1 Ecological Indexes of Ecosystem Complexity

Landscape heterogeneity, indicative of land cover diversity, is a pivotal concept in both remote sensing and ecological studies (Rocchini et al. 2021b). This diversity is traditionally quantified through algorithms like Shannon's, Gini-Simpson's, and Berger-Parker's indices. These metrics, based on spectral classes (e.g., pixels with the same digital value) within remote sensing data, serve as point-based descriptors of diversity (Rocchini et al. 2010) (refer to Table 1). Renyi's entropy index unifies these indices into a single formula, offering the flexibility to transition between them by adjusting one parameter (Rocchini et al. 2021b). Additionally, Rao's quadratic entropy (referred to as Rao's index Rao 1982) introduces more complex measures. It incorporates "spectral" distances among pixel reflectance values, thereby enriching the interpretation of diversity (Marcantonio et al. 2014). Rao's index's versatility is further enhanced through parameterisation, affecting the emphasis on spectral distances (Rocchini et al. 2021a).

The *rasterdiv* R package provides an array of user-friendly functions for calculating these indices, enabling the assessment of land cover heterogeneity from satellite imagery. *rasterdiv* employs abundance-based metrics to transform spectral data from matrix-like objects (e.g., *terra*'s *SpatRaster*) into layers representing various diversity indices. This approach considers neighbouring cell values, thus offering a more nuanced local diversity estimation (see Table 1).

In this chapter, we explore the practical applications of the *rasterdiv* package using two distinct datasets, each offering different ecological insights tied to the reported examples.

The first dataset comprises three normalised difference vegetation index (NDVI) raster layers at a spatial resolution of 60 cm sourced from the USA National Agriculture Imagery Program (NAIP). These layers represent the southern area of the Berryessa Snow Mountain National Monument in California (USA), captured on the 25th, 28th, and 30th of May 2016, 2020, and 2022, respectively. This site was selected for its remarkable environmental diversity within a compact area, from riparian woodland to chaparral and grassland plant communities (Fig. 1 2012). Moreover, it has recently undergone significant ecological changes following the destructive Wragg wildfire in the summer of 2015 and LNU Complex Fires in August 2020, making it an ideal subject for studying post-disturbance recovery and landscape temporal heterogeneity (Fig. 1 2022).

The second dataset we will explore is a multi-year (2018–2021) time series of weekly NDVI data from a 20 hectares segment of the Macchiarvana forest in the Abruzzo, Lazio, and Molise National Park, Italy, derived from 3-metre spatial resolution reflectance values from 4-band PlanetScope PS2 imagery (Planet Team

Table 1 Diversity indices in *rasterdiv* with their mathematical formulas and corresponding functions and references listed in temporal order of first publication

Index name and symbol	Mathematical formula	Function in *rasterdiv*	Reference

Shannon index (**H′**)	$-\sum_{i=1}^{N} p_i \log(p_i)$	`Shannon()`
		`logHill(alpha=1)`
		`Renyi(alpha=1)`
			Boltzmann (1872); Shannon (1948)

Gini-Simpson index (**D**)	$1-\sum_{i=1}^{N} p_i^2$	`Hill(alpha=2)`$^{-1}$	Gini (1912); Simpson (1949)
Renyi's index ($^{\alpha}$**H**)	$\begin{cases} \frac{1}{1-\alpha} log \sum_{i=}^{N} p_i^{\alpha} & \text{if } 0 \le \alpha < \infty \\ -\sum_{i=1}^{N} p_i \log(p_i) & \text{if } \alpha \to 1 \end{cases}$	`Renyi()`	Rényi (1961)
Pielou index (**J′**)	$\frac{-\sum_{i=1}^{N} p_i \log(p_i)}{ln(N)}$	`Pielou()`	Pielou (1966)
Berger-Parker index (**D$_D$**)	$\max_{i=1}^{N} p_i$	`BergerParker()` `Renyi(alpha=Inf)`	Berger and Parker (1970)
Hill's numbers (^{q}**D**)	$\begin{cases} \left(\sum_{i=1}^{N} p_i^q\right)^{\frac{1}{1-q}} & \text{if } 0 < q \neq 1 \\ \exp\left(-\sum_{i=}^{N} p_i \log(p_i)\right) & \text{if } q = 1 \end{cases}$	`Hill()`	Hill (1973)
Rao index (**Q**)	$\sum_{\substack{i,j=1 \\ i \neq j}}^{N} p_i p_j d_{ij}$	`paRao(alpha=1)`	Rao (1982)
Cumulative residual entropy (**CRE**)	$-\sum_{\substack{i,j=1 \\ i \neq j}}^{N} P(X \geq x_i) log P(X \geq x_i) dx$ $dx = (x_i - x_{i-1})$	`CRE()`	Rao et al. (2004)
Parametric Rao index (^{a}**Q**)	$\left(\sum_{\substack{i,j=1 \\ i \neq j}}^{N} \frac{1}{N^2} d_{ij}^{\alpha}\right)^{\frac{1}{\alpha}}$	`paRao()`	Rocchini et al. (2021a)

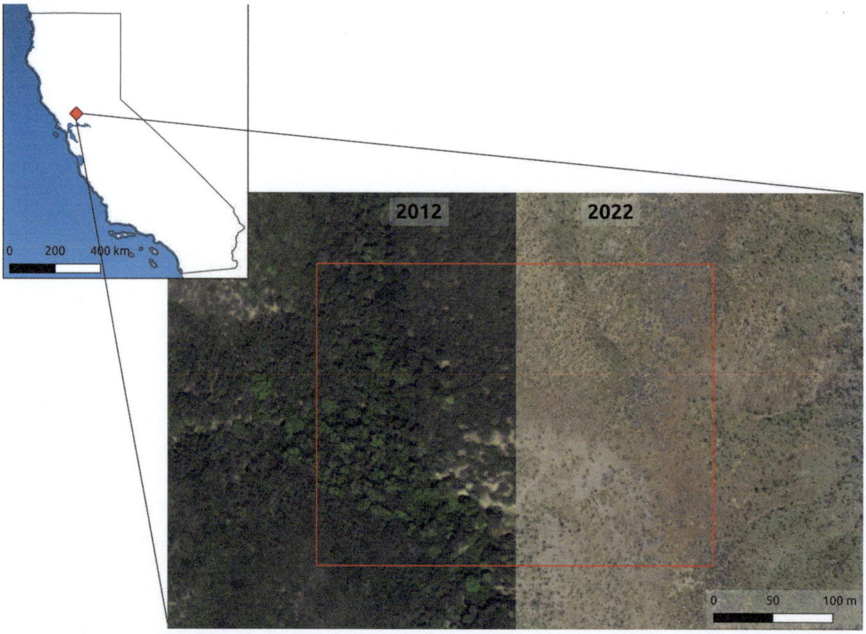

Fig. 1 A portion of the UC Davis Stebbins Cold Canyon Reserve, California, USA (lat: 38.489, lon: –122.101), pre- and post-wildfires in 2012 and 2022, respectively. The red square depicts the case study area used in the first part of this chapter

2017). This forest is predominantly an old-growth beech trees (*Fagus sylvatica*) ecosystem, known for its relatively consistent vegetative dynamics over the years (Fig. 2). The expectation here is that the year-on-year NDVI values will display a regular sinusoidal pattern, with minimal variations. Such stability in vegetative growth patterns provides an excellent opportunity to apply *rasterdiv* helical plot functions (Dorling 2020) in understanding the nuances of ecological consistency and the subtle changes that might occur in an old-growth forest environment (Thouverai et al. 2023).

Through these datasets, we aim to illustrate the diverse applications of *rasterdiv* in assessing and interpreting ecological dynamics and landscape heterogeneity, utilising NDVI as a key indicator of spatial heterogeneity in vegetative health and activity. The two datasets used in this chapter are openly available through the Open Science Framework's permanent repository, while the complete code to reproduce the example is provided on GitHub. Both resources can be accessed at the following links: GitHub repository - https://github.com/RCodingForEcology; Open Science Framework dataset - https://osf.io/pgbmq/.

The two datasets used in this chapter can be accessed through the Open Science Framework's permanent repository. For more information and to download the datasets, please visit the following link: *rasterdiv* datasets on Open Science Framework https://osf.io/pgbmq/.

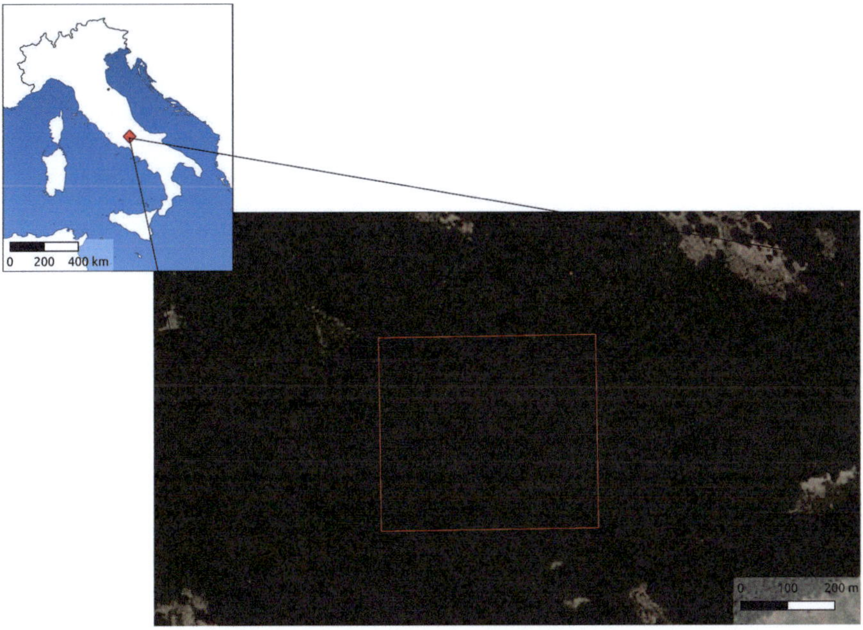

Fig. 2 A segment of the Macchiarvana old-growth beech forest located in the Central Appennins, Abruzzo, Italy (lat: 41.766, lon:13.799). The red square depicts the case study area used in the second part of this chapter

2 Classical Indexes of Diversity in the *rasterdiv* R Package

2.1 Preparing Raster Layers

To effectively apply the diversity indices provided by the *rasterdiv* package, it is imperative to begin with raster layers composed of a finite number of discrete values. Typically, these are represented as 8-bit integers ranging from 0 to 255. The rationale for this requirement lies in the fact that the indices in *rasterdiv* are designed to handle discrete data categories—such as integer counts of spectral classes or land cover types—rather than continuous data.

The indices calculate various transformations based on the relative abundance of pixel values, interpreted as distinct "types" (e.g., spectral classes) within the raster. When the raster data contains a vast range of continuous numeric values, particularly with high precision such as three decimal places, the resulting high number of unique classes coupled with an extremely low relative abundance per class can render the diversity calculations meaningless. In such cases, the diversity indices may not provide insightful information due to the dilution of meaningful class distinctions.

Essential Tips for Utilising *rasterdiv*

- **Input Data [*x*]**: Essential to all analyses, input data should be a numerical matrix or a *SpatRaster* layer with discrete cell values. Typically, an 8-bit numeric format is preferred to ensure accurate computation of diversity indices.
- **Moving Window [*window*]**: This square kernel, central to the analysis, traverses the image pixel by pixel. Its size is crucial as it determines the scope of the local neighbourhood considered for each pixel during index calculations.
- **NA Tolerance [*na.tolerance*]**: This parameter sets the proportional limit for missing data within each moving window, crucial for handling boundaries and data gaps accurately. Exceeding this limit results in an NA value for the central pixel's index, ensuring integrity in areas where the dataset is incomplete.
- **Parallel Processing [*np*]**: To enhance computational efficiency, the package allows for parallel processing. The "number of processes" parameter enables users to specify the number of concurrent processing threads, thereby optimising performance on multi-core systems.
- **Value Simplification [*simplify*]**: In dealing with high-precision floating-point numbers, simplifying pixel values by rounding them to a specified decimal place can significantly streamline index computation, balancing accuracy with processing demands.

Therefore, for the *rasterdiv* functions to yield significant and interpretable results, preprocessing steps may be necessary to categorise continuous raster data into a manageable number of discrete classes that reflect meaningful differences in the landscape or the spectral characteristics being studied. The linear transformation from floats to 8-bit (or any other range of numeric values) can be achieved using *terra::stretch()* function, by setting the output range of values between 0 and 255.

Import NDVI and transform into 8-bit categories

```
ndvi.8bit <- lapply(ndvi, terra::stretch, 0, 255)
names(ndvi.8bit) <- paste("NDVI", c(2016, 2020, 2022), sep=" ")
```

Throughout this chapter, we demonstrate the use of *rasterdiv* functions alongside *lapply*, a strategy that effectively iterates over multiple *SpatRasters*.

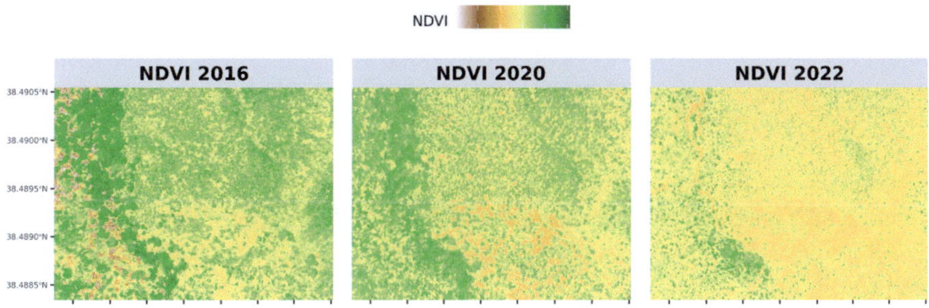

Fig. 3 NDVI values of the study areas in three different years. Higher NDVI values indicate wetter river plant communities at the bottom of the canyon, whereas lighter values chaparral and grassland communities at the top of the hill ridge

2.2 Renyi's Generalised Entropy

Renyi's entropy (Rényi 1961), a generalisation of the Shannon entropy formula denoted ${}^{\alpha}H$, offers a versatile approach to assess the prominence of different spectral classes in a landscape (Table 1). By adjusting the α parameter, one can modulate the emphasis placed on the most dominant spectral class. This adaptability is integrated into *rasterdiv* via the *alpha* option in the *Renyi()* function. The *Renyi()* function in *rasterdiv*, along with all other diversity index functions, is designed to handle large datasets by leveraging parallel computing capabilities. Users can optimise performance through the *cluster.type* option, which supports classical parallel computation backends including "SOCK", "FORK", and "MPI". Additionally, the degree of parallelism (which acts at single matrix layers) can be customised by specifying the number of processes using the *np* option. This feature enhances the processing speed, making it possible to analyse extensive spatial data.

To demonstrate the utility of Renyi's entropy index in ecological analysis, we applied it with α values of 0, 1, and 2 to three NDVI datasets (embedded in the *ndvi.8bit* multilayer *SpatRaster*) of Stebbins Cold Canyon (Fig. 3). These datasets correspond to three different ecological moments: immediately following the Wragg wildfire in 2016, just prior to the LNU wildfire in 2020, and two years post the LNU wildfire in 2022.

This application allows us to explore the heterogeneity and ecological dynamics of the landscape under varying degrees of disturbance and recovery. By adjusting the α parameter, we can shift our focus from the general diversity ($\alpha = 0$) to the more nuanced aspects of species dominance (higher α values), thereby offering a comprehensive view of the landscape's ecological resilience and response to wildfires over time.

Renyi's function with varying α

```
ndvi.Renyi <- lapply(ndvi.8bit, Renyi,
                     window=5,
```

```
alpha=c(0,1,2),
na.tolerance=0.1,
np=parallel::detectCores()-1)
```

The *ndvi.Renyi* list, generated from the application of the Renyi formula across different NDVI datasets, reflects variations in α values. Each sublist within *ndvi.Renyi* is a lower level list corresponding to a specific moving window value, and then each element in the window sublist is a *SpatRaster* (or matrix) object corresponding to each α value:

Output of functions in *rasterdiv*:

```
├── window size 5
│   ├── alpha 0
│   ├── alpha 1
│   └── alpha 2
├── window size 5
│   ├── alpha 0
│   ├── alpha 1
│   └── alpha 2
└── ...
```

Such hierarchical organisation is common for the outputs of all the diversity index functions provided in *rasterdiv*. This structure facilitates conversion into a multilayer SpatRaster by using *lapply* in conjunction with *terra::rast()*, enabling visualisation through *ggplot2* and *tidyterra* (Fig. 4).

In *rasterdiv*, the *window* parameter in diversity index functions specifies the size in pixels of the square kernel (often referred to as the "moving window") utilised to compute the relative abundances of spectral classes (see Info Box 2.1). It is crucial that *window* is an odd integer, as it represents a square kernel's dimensions. While varying α offers insights into the dominance of spectral classes, altering the window size can reveal how local landscape heterogeneity influences diversity metrics. Varying the window size in *Renyi()* and all other diversity function contained in *rasterdiv* can be achieved using a vector of odd integers as follows:

Renyi's function with varying moving windows

```
ndvi.Renyi <- lapply(ndvi.8bit, Renyi,
                 window=c(3,7,9),
                 alpha=1,
                 na.tolerance=0.1,
                 np=parallel::detectCores()-1
                 )
```

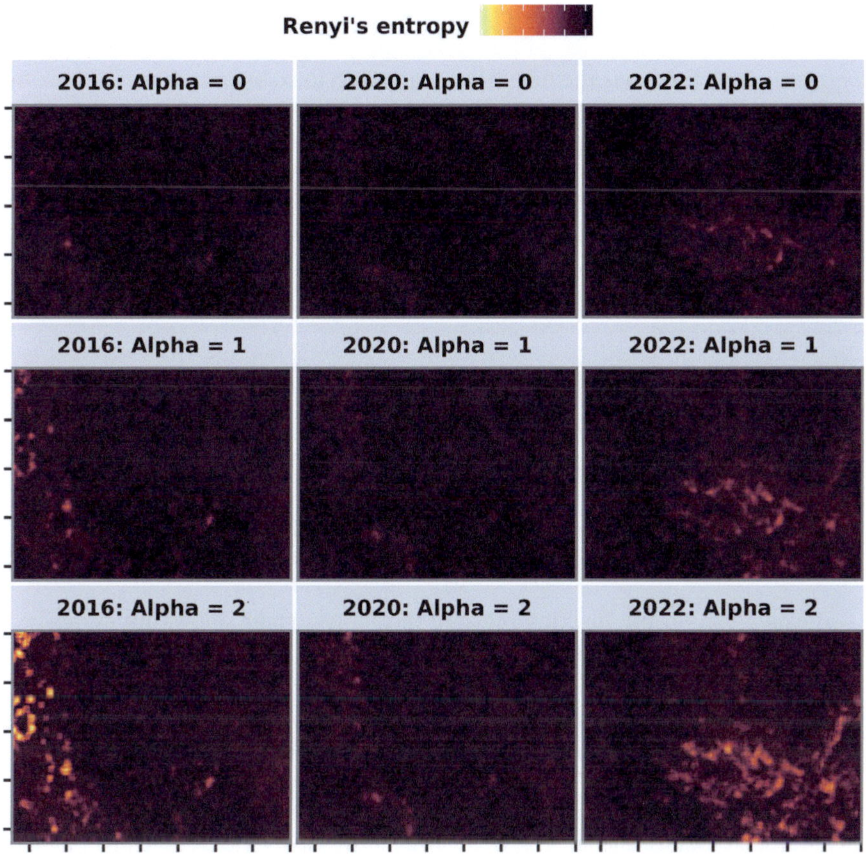

Fig. 4 Overview of Renyi's index with varying α values (rows) for NDVI in different years (columns). Lighter colours show lower Renyi's index values which generally indicate more homogenous habitats

This approach enables ecologists and remote sensing experts to tailor the analysis to specific spatial resolutions, providing a flexible tool for investigating ecological patterns and processes at multiple scales.

2.3 Other Classical Index of Diversity

Hill's numbers, denoted as ${}^qD$ and also known as true diversity indices, are a versatile set of measures in biodiversity studies (Table 1). These indices can be effectively computed using the *Hill()* function in the *rasterdiv* package by specifying various α values. This functionality parallels that of the *Renyi()* function, offering a broad spectrum of diversity measurements based on different α parameters. Beyond these generalised formulas, *rasterdiv* also includes functions dedicated to calculating specific point diversity indices (e.g., fixed snapshots within the broader, continuous spectrum of Rényi entropies). For instance, the *Shannon()* and *BergerParker()* functions provide a direct approach to computing Renyi's index for α values of 1 and inf, respectively (Table 1).

Additionally, *rasterdiv* offers the *Pielou()* function (J'), designed to calculate Pielou's evenness index (Table 1). This index represents Shannon's entropy normalised by its maximum possible value, thus ranging between 0 and 1. Pielou's evenness index is particularly valuable for comparative analyses across different ecological scenarios or temporal scales, as it provides a standardised measure of species distribution uniformity within a given area.

Pielou's index

```
ndvi.Pielou <- lapply(ndvi.8bit, Pielou,
                 window=c(3,7,9),
                 na.tolerance=0.1,
                 np=parallel::detectCores()-1)
```

Pielou() and all the function reported in this paragraph allows for parameters relevant for their calculation as in the *Renyi()* function.

3 Incorporating Distance into Diversity Assessment: Rao's Quadratic Entropy

In the realm of landscape heterogeneity analysis, diversity indexes are pivotal tools. They often involve various transformations of the relative abundances of pixel "types" (e.g., spectral classes) within a defined spatial context, a moving window in *rasterdiv*. Among these indexes, Rao's quadratic entropy stands out for its ability to integrate "distances" between spectral classes into the diversity index calculation. This more nuanced measure considers not just the presence or abundance of spectral classes but also how different they are from each other in a given characteristic, adding a layer of ecological or spectral interpretation to the analysis (Fig. 5).

In the *rasterdiv* package, this index is accessible through the *paRao()* function. This function implements a generalised version of Rao's formula, as detailed in Rocchini et al. (2021a)'s work. Much like the Renyi's and Hill's formulas, Rao's parametrised quadratic entropy formula allows for the adjustment of the index's emphasis through the *alpha* parameter. In this context, the *alpha* parameter plays a critical role—the higher its value, the greater the emphasis on the distances between spectral classes. When α is set to 1, the function reverts to computing the classic Rao's index, which considers only the distances without additional weighting, thus providing a baseline measure of diversity that reflects the heterogeneity in pixel classes based on their spectral distances. The *paRao()* accommodates a variety of distance metrics to suit diverse analytical needs. Supported distance measures include commonly used options such as “euclidean”, “manhattan”, “canberra”, “minkowski”, and “mahalanobis”. Beyond these standard choices, *paRao()* extends its functionality to include all distance calculations available in the *proxy::dist()* function as well as user-defined distance functions.

To compute Rao’s index using an α value of 1, along with the Euclidean distance between pixel values, for the Stebbins NDVI set of imagery, we employ the following approach:

Rao’s quadratic entropy

```
ndvi.paRao <- lapply(ndvi.8bit, paRao,
                     window=17,
                     alpha=1,
                     distance="euclidean",
                     na.tolerance=0.1,
                     np=parallel::detectCores()-1
                     )
```

paRao() offers the alternative to calculate Rao’s index using defined areas instead of fixed moving windows. This functionality is accessible by supplying a polygonal vector layer through the *area* option, alongside the corresponding polygon identifier via the *field* option. In our upcoming example, we will demonstrate the computation of Rao’s index within specific areas of Stebbins Canyon, delineated by isopleths (contours of equal altitude), for NDVI observations spanning the three distinct years:

Area-based Rao’s quadratic entropy

```
ndvi.paRaoArea <- lapply(ndvi.8bit, paRao,
       area=isopleths,
```

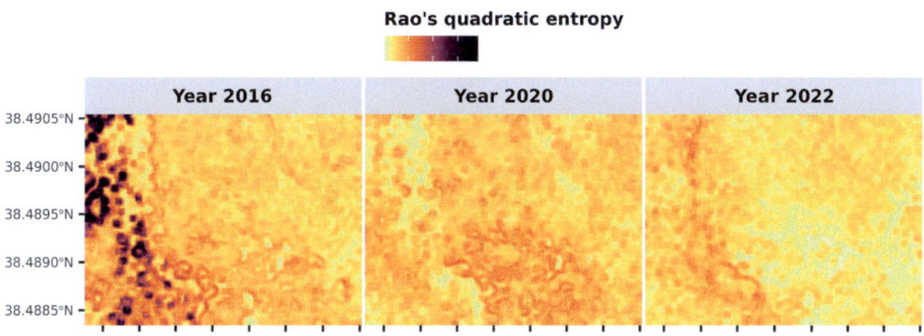

Fig. 5 Overview of parametric Rao’s index for NDVI in different years. Darker values show areas where pixel distances and thus spectral heterogeneity are higher. In 2016, areas with portion having freshly burned vegetation are highlighted due to the strong pixel distance between green healthy and burned vegetation patches

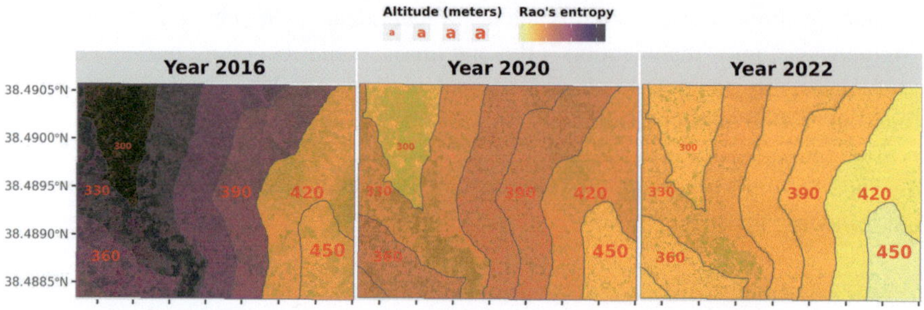

Fig. 6 Overview of parametric Rao's index calculated using areas defined by isopleths for NDVI in different years. Darker values show areas where pixel distances and thus spectral heterogeneity is higher

```
        field="altitude",
        alpha=1,
        distance="euclidean",
        na.tolerance=0.1
        )
```

It is evident how Rao's index varied considerably spatially in the 2016 imagery, which was taken less than two months after the Wragg wildfire (Fig. 6). In the following years, the landscape underwent a strong homogenisation due to a decrease in Rao's entropy index for formerly spectrally diverse habitats located at the bottom of the valley.

3.1 Multidimensional Rao's Quadratic Entropy

The extended version of the parametric Rao index implemented in *rasterdiv* facilitates the incorporation of multiple raster layers or numerical matrices into the diversity index computation. This approach calculates the distance matrix across the *n* raster layers within a multidimensional framework using the distance measures previously discussed. To utilise this methodology, it is essential to specify the option *method="multidimension"* and to provide a minimum of two single-layer *spatRasters* either as a list or a vector or as a multilayer *spatRaster*. In scenarios where a multilayer *spatRaster* is provided, the *paRao()* function will automatically incorporate all layers in the index calculation.

In this example, we demonstrate the multidimensional application of *paRao()* by using it on the red and near-infrared bands (supplied as a multilayer spatRaster) from a sub-selection of the 2016 Stebbins Cold Canyon imagery. Thus, here we focus on the single spectral bands that are crucial for assessing vegetation dynamics, rather than on their combination to derive NDVI (Fig. 7).

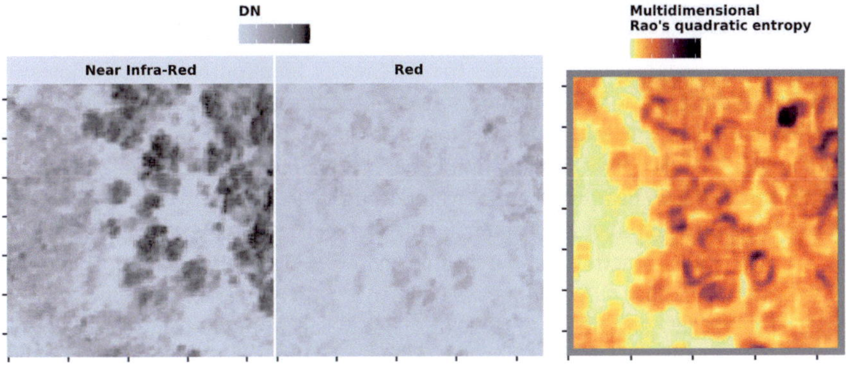

Fig. 7 Overview of parametric Rao's index calculated using areas defined by isopleths for NDVI in different years. Darker values show areas where pixel distances and thus spectral heterogeneity is higher

Multidimension Rao's quadratic entropy

```
bands.MpaRao <- paRao(
                  x=select(RIR.bands,1:2),
                  window=9,
                  alpha=1,
                  np=parallel::detectCores()-1,
                  na.tolerance=0.1,
                  method="multidimension"
                  )
```

As for the other indexes in *rasterdiv*, multidimension *paRao()* allows for multiple moving window (*window*), α values (*alpha*), distance metrics (*dist m*), and simplification of the numerical matrices (*simplify*). Moreover, the option *rescale* can be used in multidimensional *paRao()* to centre and scale the values of each layer provided for calculation.

Recently, a new metric to account for phenological changes has been added to the multidimensional framework of rasterdiv. Specifically, the *paRao()* function now allows the use of Time-Weighted Dynamic Time Warping as a distance measure (dist_m = "twdtw"). This advance enables the incorporation of phenological differences among plant communities and provides a useful tool to distinguish between natural habitats, which follow seasonal phenological patterns, and artificial land-cover types. A full description of the methodology is available in Shayle et al. (1948).

3.2 Rao's Accumulation Function

The *rasterdiv* package also includes the *AUCRao()* function, which calculates the area under the curve (AUC) for the Rao's index across a range of alpha values. This is especially valuable in exploratory data analysis where the choice of alpha for the Rao's index may not be immediately apparent. By assessing the AUC, researchers can gain insights into the overall trend of diversity across different scales of emphasis on spectral classes abundance and distance (Thouverai et al. 2022).

The *AUCRao()* function is based on the *paRao()* function, meaning it supports all the arguments that *paRao()* does. This provides flexibility in handling different types of input data and distance measures. For instance, in the following example, the *AUCRao()* function is applied to integrate the Rao's index over alpha values from 1 to 5, allowing us to compare the cumulative effect of these alpha parameters on the diversity index outcome.

Area under the curve of Rao's parametric function

```
ndvi.paRaoAUC <- lapply(ndvi.8bit, RaoAUC,
                                   alphas=1:5,
                                   window=17,
                                   dist_m="euclidean",
                                   na.tolerance=0.1,
                                   np=parallel::detectCores()-1)
```

4 Helical Plots to Visualise Time-Series Variability

Remote sensing data play a crucial role in monitoring and visualising temporal trends in various ecological and environmental contexts and spatial scales. These datasets, rich in spatial and temporal information, are pivotal in discerning subtle yet significant changes occurring over time. Traditional visualisation methods, while offering simplicity and clarity, often compress or overlook nuanced details that are vital for a comprehensive understanding of ecological dynamics (Cooksey 2020).

In contexts where the complex interplay of various factors is common, more sophisticated visualisation techniques are required. One innovative approach is the use of "Helical Graphs" (H-graphs), developed by statistician Danny Dorling and illustrator Kirsten McClure. These graphs offer a unique visualisation method by intertwining the magnitude and directional trends of a metric with its temporal progression (Dorling 2020). H-graphs stand out for their ability to convey complex data narratives, blending quantitative changes with their qualitative aspects, thus providing a more holistic view of the ecological changes observed. This approach not only enhances the interpretability of remote sensing data but also opens new avenues for communicating complex environmental transformations in a more intuitive and insightful manner (Thouverai et al. 2023).

In the *rasterdiv* package, helical graphs (H-graphs) are implemented via two distinct functions: *heliPrep()* and *heliPlot()*. The *heliPrep()* function is designed to calculate the rate of change between time-series values. It then applies a two-sided moving average linear filter to both the original values and their calculated rate of change. This function accepts three arguments: *values* for the time series of numeric values, *dates* for the corresponding chronological data, and *window_size* for determining the size of the linear filter.

The *heliPlot()* function is then responsible for visualising the dataframe generated by *heliPrep()*. It does so by leveraging B-splines within the *ggplot2* framework. This function, supplied with the data argument, enables the separation of input data into distinct helical lines when the *group* argument is utilised. Additionally, *heliPlot()* is versatile, allowing the incorporation of various *ggplot()* arguments using the + syntax from *ggplot2*, thus offering extensive customisation and flexibility in data visualisation.

To show how to build H-plot with *rasterdiv*, we will start from a dataframe presenting a multi-year time series of remotely sensed weekly NDVI values from a patch of old-growth beech forest from central Italy. This plant community represents a relatively stable vegetation dynamics typical of broadleaved mountain forests. The dataframe *NDVI.ts* has three columns, *ndvi* is the average weekly NDVI, and *dates* is the date corresponding to the first day of each weekly NDVI value.

Helical graph preparation

```
ndvi.smooth <-  heliPrep(
                   dates=NDVI.ts$date,
                   values=NDVI.ts$ndvi,
                   filterWidth=3)
```

The output of *heliPrep()*, *ndvi.smooth*, is a dataframe with three columns, the averaged values, their rate of change, and the corresponding date. We then use *heliPlot()* to visualise the H-plot:

Helical graph

```
ndvi.Hgraph <- heliPlot(
            data=ndvi.smooth,
            group="year",
            sizeRange = c(0.05, 3),
            ylab="NDVI weekly average",
            xlab="NDVI [rate of weekly change]",
            dateInterval=24,
            dateFormat="%b",
            dateFont=2)

# We can use ggplot2 syntax to further improve a rasterdiv H-graph
```

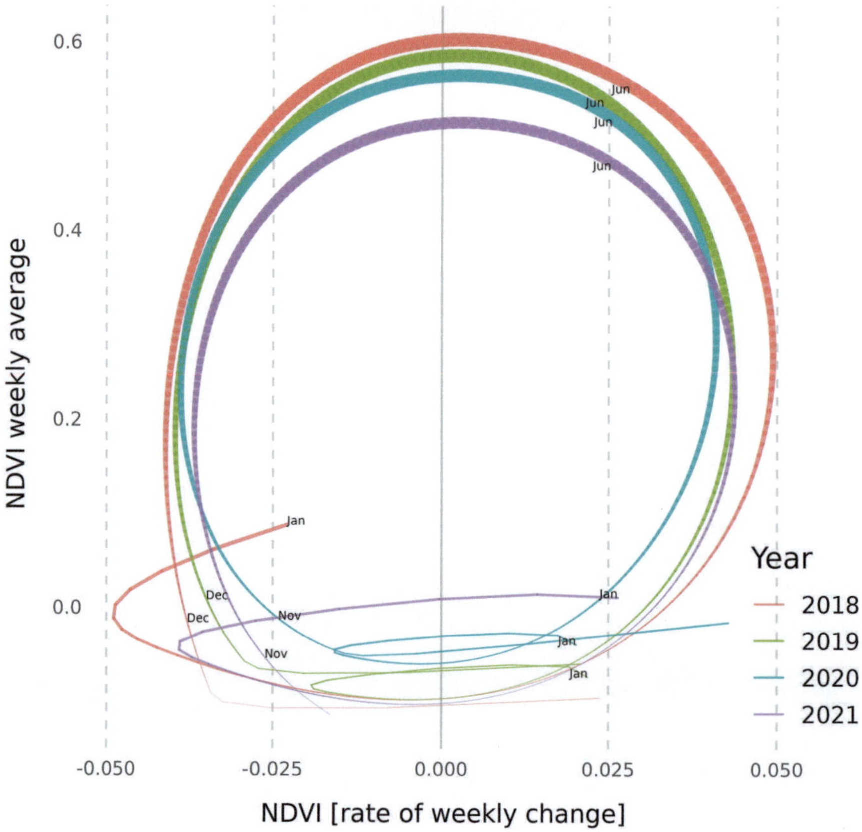

Fig. 8 Helical plots of a multi-year weekly NDVI times series from an old-growth beech forest in central Italy

```
ndvi.Hgraph + theme(legend.position = c(1,0.15)) +
guides(color=guide_legend(title="Year"))
```

The H-plot provides a multidimensional view of vegetation dynamics, capturing both the magnitude of NDVI values over time and the rate at which these values change, thereby offering a comprehensive picture of vegetative health and seasonal changes in the beech forest (Fig. 8). Lines circling upwards towards the "June" marker denote increasing NDVI values, which typically correspond to growing biomass during the spring and early summer months. The closeness of these lines suggests that the annual vegetative patterns are consistent, with similar growth and senescence rates each year. This is expected for old-growth forests, which tend to have stable yearly cycles. The horizontal axis shows how quickly the NDVI values rise or fall. A steeper slope on this plot would indicate a more rapid change, such as during the onset of spring or the decline in autumn. The flatter parts of the curves suggest periods of little change, likely corresponding to the peak of summer or the dormancy of winter. If the different coloured lines show some divergence, this could indicate inter-annual variability in maximum NDVI values or in the timing of the vegetative peak. For instance, if one year's line peaks later than others, it may suggest delayed spring growth, potentially due to climatic conditions. 2021 experienced comparatively lower levels of vegetation greenness for the beech forest patch, potentially as a consequence of an unusually cold and wet spring, which then gave way to a heatwave during the months of July and August. These climatic fluctuations may have stressed the vegetation, as evidenced by the NDVI's reduced values. Conversely, the year 2018 is characterised by notably higher NDVI values in the spring, aligning with the fact that 2018 ranks as the second warmest year on record for Italy (Istituto di Scienza dell'atmosfera e del clima 2018). The warmer conditions likely promoted an earlier and more vigorous onset of the growing season, leading to heightened vegetative growth and greenness.

The utility of H-graphs extends beyond the visualisation of simple NDVI time series; they can effectively represent the temporal evolution of various diversity indices calculated through the *rasterdiv* package. By capturing the dynamic nature of ecological data, H-graphs offer a powerful tool for illustrating trends and detecting anomalies in time-series data of ecological diversity.

Acknowledgments We thank the U.S. Department of Agriculture, Farm Production and Conservation, Geospatial Enterprise Operations, as well as PlanetData (Education and Research Standard access plan, ID: 336490) for providing the source imagery data used to develop the example presented in this manuscript.

References

Berger WH, Parker FL (1970) Diversity of planktonic foraminifera in deep-sea sediments. Science 168(3937):1345–1347

Boltzmann L (1872) Weitere studien über das wärmegleichgewicht unter gasmolekälen. S. K. Akad. Wiss. Wein 66

Cooksey RW (2020) Descriptive statistics for summarising data. In: Illustrating Statistical Procedures: Finding Meaning in Quantitative Data, Cooksey RWedn. Springer, Singapore, pp 61–139. https://doi.org/10.1007/978-981-15-2537-7_5

Dorling D (2020) Slowdown: The End of the Great Acceleration–and Why It's Good for the Planet, the Economy, and Our Lives. Yale University Press, New Haven

Gini C (1912) Gini, Corrado. 1912. Variabilità e mutabilità: contributo allo studio delle distribuzioni e delle relazioni statistiche. [Fasc. I.]. Tipogr. di P. Cuppini

Hill MO (1973) Diversity and evenness: a unifying notation and its consequences. Ecology 54(2):427–32. https://doi.org/10.2307/1934352

Istituto di Scienza dell'atmosfera e del clima (2018) 2018 Warmest Year for Italy since 1800. Accessed November 24, 2023. https://www.isac.cnr.it/en/news/2018-warmest-year-italy-1800

Marcantonio M, Rocchini D, Ottaviani G (2014) Impact of alien species on dune systems: a multifaceted approach. Biodivers Conserv 23:2645–2668. https://doi.org/10.1007/s10531-014-0742-2

Pielou EC (1966) The measurement of diversity in different types of biological collections. J Theor Biol 13:131–44

Planet Team (2017) Planet Application Program Interface: In Space for Life on Earth. San Francisco. https://api.planet.com

Radhakrishna RC (1982) Diversity and dissimilarity coefficients: a unified approach. Theor Popul Biol 21(1):24–43. https://doi.org/10.1016/0040-5809(82)90004-1

Rao M, Chen Y, Vemuri BC, Wang F (2004) Cumulative residual entropy: a new measure of information. IEEE Trans Inf Theory 50(6):1220–28

Rényi A (1961) On measures of entropy and information. In: Proceedings of the Fourth Berkeley Symposium on Mathematical Statistics and Probability, Volume 1: Contributions to the Theory of Statistics, vol 4, no 1. University of California Press, Oakland, pp 547–62

Rocchini D, Balkenhol N, Carter GA, Foody GM, Gillespie TW, He KS, Kark S et al (2010) Remotely sensed spectral heterogeneity as a proxy of species diversity: recent advances and open challenges. Ecol Inf Spec Issue Adv Ecol Remote Sens Under Global Change 5(5):318–29. https://doi.org/10.1016/j.ecoinf.2010.06.001

Rocchini D, Marcantonio M, Da Re D, Bacaro G, Feoli E, Foody GM, Furrer R et al (2021a) From zero to infinity: minimum to maximum diversity of the planet by spatio-parametric Rao's quadratic entropy. Global Ecol Biogeography 30(5):1153–62. https://doi.org/10.1111/geb.13270

Rocchini D, Thouverai E, Marcantonio M, Iannacito M, Da Re D, Torresani M, Bacaro G et al (2021b) Rasterdiv–an information theory tailored R package for measuring ecosystem heterogeneity from space: to the origin and back. Methods Ecol Evol 12(6):1093–1102. https://doi.org/10.1111/2041-210X.13583

Shannon CE (1948) A mathematical theory of communication. Bell Syst Tech J 27:379-423, 623–656

Shayle, ES, Matteo M, Rocco L, Chiara R, Saverio V et al (2024) Phenological diversity trends with remote sensing datacubes. BioHackrXiv. https://doi.org/10.37044/osf.io/pqfws.

Simpson EH (1949) Measurement of diversity. Nature 163:688

Thouverai E, Marcantonio M, Lenoir J, Galfré M, Marchetto E, Bacaro G, GattiRC et al (2022) Integrals of life: tracking ecosystem spatial heterogeneity from space through the area under the curve of the parametric Rao's Q index. Ecol Compl 52(December):101029. https://doi.org/10.1016/j.ecocom.2023.101029

Thouverai E, Marcantonio M, Cosma E, Bottegoni F, Roberto Gatti C, Conti L, Musciano MD et al (2023) Helical graphs to visualize the NDVI temporal variation of forest vegetation in an open source space. Ecol Inf 74(May):101956. https://doi.org/10.1016/j.ecoinf.2022.101956

The `landscapemetrics` and `motif` Packages for Measuring Landscape Patterns and Processes

Jakub Nowosad and Maximilian H. K. Hesselbarth

Abstract

This book chapter emphasizes the significance of categorical raster data in ecological studies, specifically land use or land cover (LULC) data, and highlights the pivotal role of landscape metrics and pattern-based spatial analysis in comprehending environmental patterns and their dynamics. It explores the usage of R packages, particularly **landscapemetrics** and **motif**, for quantifying and analyzing landscape patterns using LULC data from three distinct European regions. It showcases the computation, visualization, and comparison of landscape metrics, while also addressing additional features such as patch value extraction, subregion sampling, and moving window computation. Furthermore, the chapter delves into the intricacies of pattern-based spatial analysis, explaining how spatial signatures are computed and how the **motif** package facilitates comparisons and clustering of landscape patterns. The chapter concludes by discussing the potential of customization and expansion of the presented tools.

Keywords

Categorical raster data · Land use · Land cover · Landscape metrics · Pattern-based spatial analysis · Spatial ecology

J. Nowosad (✉)
Institute of Geoecology and Geoinformation, Adam Mickiewicz University, Poznań, Poland

M. H. K. Hesselbarth
Statistics Austria, Directorate Spatial Statistics, Energy and Environment, Vienna, Austria

International Institute for Applied Systems Analysis, Biodiversity, Ecology, and Conservation Group, Laxenburg, Austria

D. Rocchini (ed.), *R Coding for Ecology*, Use R!,
https://doi.org/10.1007/978-3-031-99665-8_10

1 Introduction

Categorical raster data, such as ones representing land use or land cover (LULC), is often used in ecological studies (Fassnacht et al., 2006; Wulder et al., 2018; Chandra Pandey et al., 2021). This data is typically derived using remote sensing data, for example, satellite images in combination with statistical learning (Talukdar et al., 2020; Wang et al., 2022; Wang and Mountrakis, 2023). LULC describes the spatial distribution of different anthropogenic uses of land or the natural land cover in a landscape. Thus, it provides information about human activities (usage) and natural features (physical material at the surface) in a given area within a specific time frame (Fisher et al., 2005). This information can then be used to better understand the processes and changes taking place in the landscape, such as urbanization (Fu and Weng, 2016), deforestation (Floreano and De Moraes, 2021), or the spread of invasive species (Manzoor et al., 2021).

From a computational perspective, LULC data can be conceptualized as a collection of cells organized in a regular grid, with each cell assigned to a specific category (With, 2019). This framework enables us to analyze the data in terms of its two fundamental components: the composition, referring to the number of cells for each category, and the configuration, pertaining to the spatial arrangement of cells within each category (Riitters, 2019). These components jointly create spatial patterns of categorical raster data, or landscape patterns, in short, which are commonly used to characterize the structure of landscapes. Furthermore, additional characteristics of a categorical raster, for example, the diversity of categories, connectivity, or patch shape, can also be considered (Gustafson, 1998; Uuemaa et al., 2013).

The aim of this chapter is to provide an overview of the methods and tools that allow to quantify and analyze landscape patterns. We will focus on the methods that are mainly implemented in two R packages: **landscapemetrics** (Hesselbarth et al., 2019) and **motif** (Nowosad, 2021). The **landscapemetrics** package allows to compute a collection of landscape metrics, which quantify the composition and configuration of categorical raster data. The package also provides a set of additional functions, making it possible to visualize the results of the calculations, extract values of patches, sample metrics within subregions, or apply a moving window computation. Thus, its main purpose is to provide a set of tools to describe landscape patterns. The **motif** package, on the other hand, is focused on the analysis of these patterns. It allows to compare landscape patterns between different times, search for areas with resembling landscape patterns, or cluster areas with similar landscape patterns.

In this chapter, we will consistently utilize a set of example data. These data consist of spatial raster representing the LULC of three distinct regions in Europe: the Centre-Val de Loire region in France, the Noord-Brabant region in the Netherlands, and the Norra Mellansverige region in Sweden. We obtained the data from the Copernicus mission 2018 (European Environment Agency, 2023), available at https://land.copernicus.eu/), and cropped them to the respective regions.

Additionally, all original 45 LULC classes were reclassified into five general classes (i.e., "urban," "agriculture," "vegetation," "marshes," and "water") to simplify all examples. To reproduce the results presented in this chapter, you can download the corresponding data and code from the GitHub repository at https://github.com/Nowosad/landscapemetrics_motif_2024.

1.1 Landscape Metrics

Landscape metrics allow to quantify the composition (i.e., amount) of and configuration (i.e., arrangement) of spatial landscape characteristics within the context of categorical raster data (Gustafson, 1998, 2019; Uuemaa et al., 2013). The advantages of the landscape metrics approach include their easy application, interpretation, and communication, especially in landscapes with clear distinctions between classes (Lausch et al., 2015). Thus, they are a common tool in landscape and spatial ecology (Kupfer, 2012; Lausch et al., 2015; Frazier and Kedron, 2017). Even though the first conceptual developments date back to the 1980s (Turner, 1989; Gustafson, 1998, 2019), the application of landscape metrics gained popularity with the release of the FRAGSTATS software (McGarigal et al., 2012) in 1995, which allows to calculate an extensive collection of metrics (Kupfer, 2012; Gustafson, 2019).

While FRAGSTATS pioneered the field and made landscape metrics available to many scientists using a graphical user interface, more recently, the **landscapemetrics** R package (Hesselbarth et al., 2019) allows to calculate a large collection of metrics using the R programming environment, which is popular among ecologists (Lai et al., 2019; Hesselbarth et al., 2021). The **landscapemetrics** package is based on the **terra** R package (Hijmans, 2021), but it additionally supports raster objects from the **stars** (Pebesma, 2019) and **raster** (Hijmans, 2019) packages. Furthermore, some functionality also supports vector data (e.g., sample points) using the **sf** package (Pebesma, 2018)

Following (McGarigal et al., 2012), landscape metrics can be classified depending on the characteristics they describe, namely, area and edge, shape, core area, contrast, aggregation, and diversity metrics. Furthermore, landscape metrics can be calculated on patch-, class-, and landscape-scale. While patch-level metrics describe each patch, i.e., contiguous cells belonging to the same class as identified by a connected component labeling algorithm, class-level metrics describe all patches belonging to the same class, and finally, landscape-level metrics describe the entire landscape, Fig. 1. In this chapter, we will demonstrate how to calculate all these types of landscape metrics.

In order to use all functionality of the **landscapemetrics** package and to pre- and post-process data, we need to load several packages related to spatial analysis and data wrangling. This includes the **terra** and **sf** packages for methods related to predominantly raster and vector data, respectively. Furthermore, the **dplyr** (Wickham, 2019) and **tidyr** (Wickham et al., 2023) packages provide a grammar for data manipulation and cleaning, while **ggplot2** (Wickham, 2016) is a data

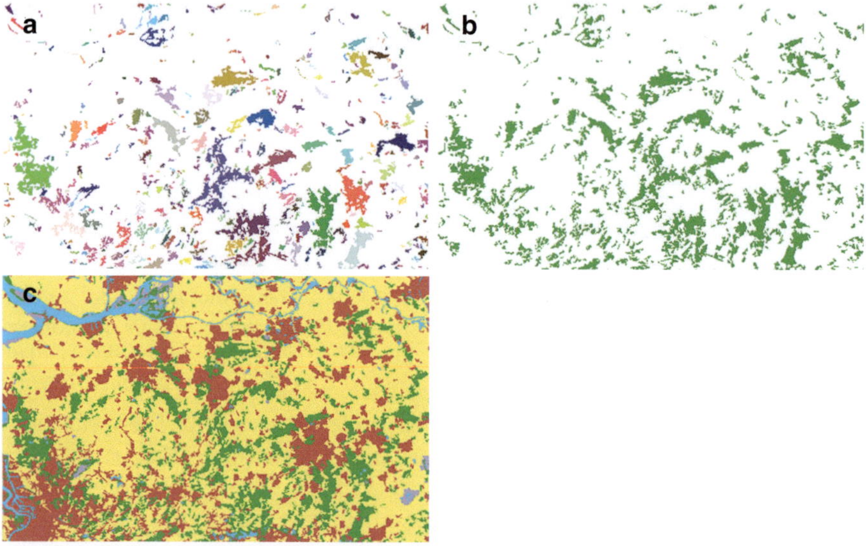

Fig. 1 (**a**) Patch level: patches of an exemplary class indicated by color, i.e., all connected cells belonging to the same LULC class. (**b**) Class level: exemplary class indicated by color, i.e., all patches belonging to the same LULC class. (**c**) Landscape level: exemplary landscape including all LULC classes

visualization package. Additionally, we create a color scheme for all maps we are going to create in later examples.

```
library(landscapemetrics)
library(terra)
library(sf)
library(dplyr)
library(tidyr)
library(ggplot2)
color_scale <- c(urban="#C86058", agriculture="#FCE569",
                 vegetation="#44A321", marshes="#A3A6FF",
                 water="#00CFFD", nodata="#666666")
```

Next, we import the three raster objects using the **terra** package representing the LULC data of the three exemplary regions as `SpatRaster` objects.

```
france <- rast("data/raster_france.tif")
netherlands <- rast("data/raster_netherlands.tif")
sweden <- rast("data/raster_sweden.tif")
```

Before calculating metrics, we need to check if the input raster fulfills certain requirements of the **landscapemetrics** package. For this, we use the `check_landscape()` function. The function checks if the coordinate reference system (CRS) is projected (i.e., using Cartesian coordinates on a planar surface), the units of the map, and the number of classes. If some of the checks fail, we are still able to calculate all metrics; however, especially metrics relying on distances might not be correct or hard to interpret, e.g., due to the decimal degree units of most geographic CRS (on the contrary, metric distance units of most projected CRS). In this case, the function will return a corresponding warning.

```
check_landscape(france)
```

```
  layer       crs units   class n_classes OK
1     1 projected     m integer         5  v
```

The input raster, `france`, has a projected CRS with units in meters and five discrete classes in total. Because the raster does not violate any of the checks, the function returns an OK check mark.

1.1.1 Deriving Landscape Metrics

All functions to calculate metrics start with the prefix `lsm_`, followed by an abbreviation for the level (either `p` = patch, `c` = class, and `l` = landscape), and finally, an abbreviation of the metric. We can get an overview of all metrics that are currently provided by the **landscapemetrics** package using the `list_lsm()` function. Additionally, we can use the function to display only a certain subset of metrics based on type or level.

```
list_lsm()
```

```
# A tibble: 133 x 5
  metric name                          type                  level function_name
  <chr>  <chr>                         <chr>                 <chr> <chr>
1 area   patch area                    area and edge metric patch lsm_p_area
2 cai    core area index               core area metric     patch lsm_p_cai
3 circle related circumscribing circle shape metric         patch lsm_p_circle
# i 130 more rows
```

```
list_lsm(type="area and edge metric", level="class")
```

```
# A tibble: 11 x 5
  metric  name       type                 level function_name
  <chr>   <chr>      <chr>                <chr> <chr>
1 area_cv patch area area and edge metric class lsm_c_area_cv
2 area_mn patch area area and edge metric class lsm_c_area_mn
3 area_sd patch area area and edge metric class lsm_c_area_sd
# i 8 more rows
```

To calculate a singular metric on patch-, class-, or landscape-level, we simply use the function name and provide the input raster.

```
df_p_shape <- lsm_p_shape(france)
df_c_area <- lsm_c_area_mn(france)
df_l_lpi <- lsm_l_lpi(france)
```

Thus, to calculate the shape index of each patch, the mean patch area of each class *i*, or the largest patch index of the landscape, we use the three functions shown above. The returned data frames always include the same columns regardless of the specified metric.

```
df_l_lpi
```

```
# A tibble: 1 x 6
  layer level     class    id metric value
  <int> <chr>     <int> <int> <chr>  <dbl>
1     1 landscape    NA    NA lpi     65.1
```

These are an identifier of the raster layer, the level of the calculated metric, the class identifier, the patch identifier, the name of the metric, and finally, the metric value. If any of the columns are not applicable (e.g., the patch identifier on the landscape level), the column will contain `NA` values. This allows us to combine different data frames or apply the same post-processing workflow, even for different metrics and/or levels. For example, we can use the `rbind()` function to combine all three results into a single data frame and use functions from the **dplyr** package to calculate the minimum and maximum value for each metric.

```
result_combined <- rbind(df_p_shape, df_c_area, df_l_lpi)
result_range <- group_by(result_combined, metric) |>
  summarise(min=min(value), max=max(value))
result_range
```

```
# A tibble: 3 x 3
  metric    min    max
  <chr>   <dbl>  <dbl>
1 area_mn  64.4 3049.
2 lpi      65.1   65.1
3 shape     1     84.5
```

If we want to derive more than a few metrics, we can also use the `calculate_lsm()` function, which allows to calculate several metrics with only one function call. However, as pointed out by other authors (Gustafson, 2019), "metric fishing" is an issue, i.e., computing as many metrics as possible and searching for any signal. Thus, we strongly advise not to calculate all metrics, but rather select metrics based on research questions and hypotheses.

The calculate_lsm() function has many options for how to specify metrics, e.g., a vector of function names or the types and levels of metrics that can be specified.

```
df_area <- calculate_lsm(france, what=c("lsm_p_area",
                                        "lsm_c_area_mn",
                                        "lsm_l_ta"))
df_aggr_lsm <- calculate_lsm(france, level="landscape",
                                     type="aggregation metric")
```

For many metrics, we can specify further arguments, such as the direction of the connected labeling algorithm, which cells are considered core or edge cells, or the potential number of maximum classes. To see what arguments are available, we can have a look at the help page of a function, e.g., ?lsm_p_core, but all arguments of a specific metric function can also be used with calculate_lsm(). For example, we can change the directions of the connected labeling algorithm and calculate the number of patches on the class level. Comparing the results using a workflow including the **dplyr** package, we can see that the number of patches is larger when using "rook's" (4-neighborhood) instead of "queen's" rule (8-neighborhood; the default).

```
df_np_queen <- lsm_c_np(france, directions=8)
df_np_rook <- lsm_c_np(france, directions=4)
df_comparison <- full_join(x=df_np_queen, y=df_np_rook,
                           by=c("layer", "level", "class", "id",
                                "metric"),
                           suffix=c(".queen", ".rook")) |>
  mutate(value.diff=abs(value.queen - value.rook))
df_comparison
```

```
# A tibble: 5 x 8
  layer level class    id metric value.queen value.rook value.diff
  <int> <chr> <int> <int> <chr>        <dbl>      <dbl>      <dbl>
1     1 class     1    NA np            3231       4589       1358
2     1 class     2    NA np            1541       3592       2051
3     1 class     3    NA np            5512       9242       3730
# i 2 more rows
```

We can also use the calculate_lsm() function (or any other lsm_ function) to calculate metrics of several raster layers or objects simultaneously. For this, we need to use a SpatRaster with several layers or a list of SpatRaster's as input. We are using the later example to calculate the perimeter and area of each patch for two regions in Europe. For this example, we only want to keep the vegetation class, i.e., class 3. Additionally, because some of the patches within each landscape are disproportional large, we are going to use the 95% quantiles of each region and metric to remove some larger values. Next, we are relabeling the unique layer identification column using the country names. Last, we are reshaping the data frame from a long to a wider format to produce a **ggplot2** figure in the final step, Fig. 2.

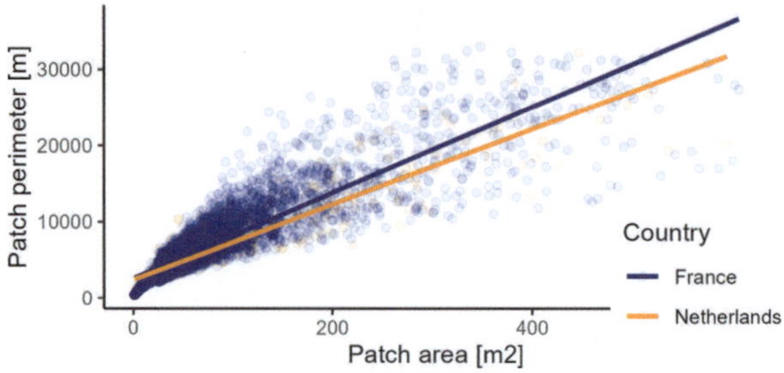

Fig. 2 Relationship between the patch area and patch perimeter in the Centre-Val de Loire region in France (blue) and Noord-Brabant region in the Netherlands (orange)

```
df_perim_core <- list(nl=netherlands, fr=france) |>
  calculate_lsm(what=c("lsm_p_perim", "lsm_p_area")) |>
  filter(class %in% 3) |>
  mutate(layer=case_when(layer==1~"Netherlands",
                         layer==2~"France")) |>
  pivot_wider(names_from=metric, values_from=value) |>
  filter(area<=quantile(area, probs=0.95) &
         perim<=quantile(perim, probs=0.95))
ggplot(data=df_perim_core, aes(x=area, y=perim, color=layer)) +
  geom_point(alpha=0.1) +
  geom_smooth(se=FALSE, method="lm", formula="y ~ x") +
  scale_color_manual(name="Country",
                     values=c(Netherlands="#F79400",
                              France="#001E96")) +
  labs(x="Patch area [m2]", y="Patch perimeter [m]") +
  theme_classic() + theme(legend.position = c(0.9, 0.1))
```

This example also demonstrates how easily the **landscapemetrics** package can be incorporated into larger workflows. As expected, there is a strong relationship between the patch area and perimeter. Logically, as the patch area increases, also the perimeter increases, and this trend is comparable between the two regions.

1.1.2 Visualization of Landscape Metrics

We can also use built-in functions to visualize the landscapes or calculated metrics. All visualization functions start with the `show_` prefix and return a list storing **ggplot2** objects. This is done to be type-stable (i.e., always returning the same data type) regardless of the number of raster objects/layers or selected metrics. First, we can visualize all patches, i.e., all connected cells as specified by the `directions` argument of the connected components labeling algorithm. We can show the patches of the entire landscape (using the argument `class="global"`) or only patches of specific classes (e.g., using `class=3`). Here, we show all patches of the vegetation class in the French region, Fig. 3.

```
show_patches(france, class=3)
```

Fig. 3 Patches of the natural vegetation class in the Centre-Val de Loire region in France

Furthermore, showcasing a more advanced approach, we have the capability to fill each cell of a patch accordingly to a specific landscape metric. For instance, in the following example, the fill color of each patch in the vegetation class corresponds to its area. Since we utilize a list to store two landscapes, the resulting list contains two elements, each representing a landscape and storing the associated ggplot2 objects. It becomes apparent that the landscape in France is dominated by a single large patch situated at the center of the region, Fig. 4.

```
list_gg_area <- list(fr=france, nl=netherlands) |>
  show_lsm(class=3, what="lsm_p_area")
plot_grid(plotlist=list_gg_area, ncol=2, labels=c("NL", "FR"))
```

Correlation between several metrics can be an issue (Cushman et al., 2008; Schindler et al., 2008; Nowosad and Stepinski, 2018), especially in combination with the previously discussed practice of "metric fishing." Thus, we can use the show_correlation() function to check correlations between metrics that were previously calculated using either calculate_lsm() or singular metric functions, Fig. 5.

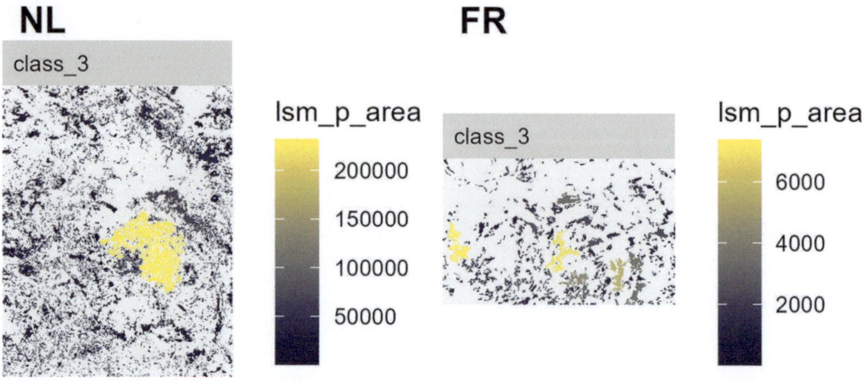

Fig. 4 Visualization of the patch area for the natural vegetation LULC class in the Centre-Val de Loire region in France (left) and the Noord-Brabant region in the Netherlands (right)

Fig. 5 Correlation matrix of all class-level metrics for the Noord-Brabant region in the Netherlands

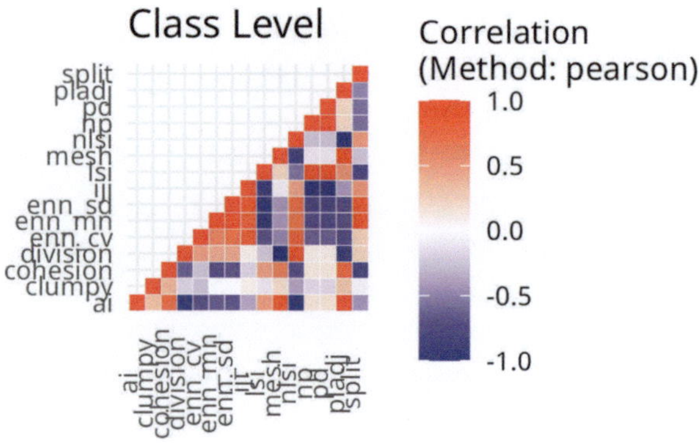


```
class_metrics <- calculate_lsm(netherlands, level="class",
                               type="aggregation metric")
show_correlation(class_metrics)
```

1.1.3 Additional Features

The **landscapemetrics** also provides some more advanced functionality. This includes extracting values of patches, sampling of metrics within subregions, or applying a moving window approach. In terms of metrics selection or arguments passed on to the metric functions, all these functions behave similarly to the `calculate_lsm()` function.

In order to demonstrate the extraction and sampling of metrics, we first need to create sample points. For this, we are going to randomly create ten points within a region, Fig. 6.

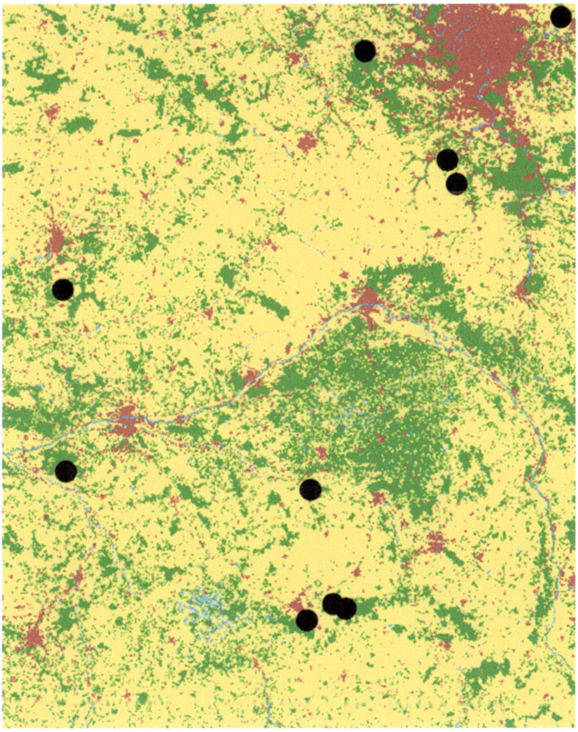

Fig. 6 Location of exemplary sample points used to extract and sample landscape metrics in the Centre-Val de Loire region in France

```
samplepoints <- spatSample(france, size=10, as.points=TRUE) |>
  st_as_sf()
```

Now, we can extract patch-level metrics of all patches that contain a sample point. For this, we use the extract_lsm() function. The resulting data frame has 20 rows, one for each metric and sample point. It is worth noting that compared to previous result data frames, there is an additional column named extract_id, which facilitates matching the metric results with the respective sample points.

```
df_extract <- extract_lsm(france, samplepoints,
                          what=c("lsm_p_area", "lsm_p_perim"))
df_extract
```

```
# A tibble: 20 x 7
  layer level class    id metric  value extract_id
  <int> <chr> <int> <int> <chr>   <dbl>      <int>
1     1 patch     3  5207 area    15202          1
2     1 patch     3  5207 perim  297200          1
3     1 patch     3  7899 area     7652          2
# i 17 more rows
```

Similarly, we can also sample metrics within a buffer around each sample point. For this, we need to additionally provide a size of the sampling buffer and its shape (either a square or a circle), but it is also possible to directly use polygon objects as sampling buffers.

```
df_sample <- sample_lsm(france, samplepoints,
                        what=c("lsm_c_pland", "lsm_l_ta"),
                        size=10000, shape="circle")
filter(df_sample, percentage_inside>=75)
```

```
# A tibble: 47 x 8
  layer level class    id metric value plot_id percentage_inside
  <int> <chr> <int> <int> <chr>  <dbl>   <int>             <dbl>
1     1 class     1    NA pland   3.54       1              101.
2     1 class     2    NA pland  48.6        1              101.
3     1 class     3    NA pland  46.8        1              101.
# i 44 more rows
```

The resulting data frame has two additional columns. First, similar to the metrics extraction, an additional column identifies the sampling buffer. Second, an additional column stores information about the actual clipped sampling buffer area. Theoretically, this value should be equal to 100%. However, it includes all of the raster cells whose centroids are inside the selector polygon (Lovelace et al., 2019), and thus the actual sampling buffer area might be slightly larger or smaller than the specified area resulting in a deviation from the theoretical value. Furthermore, sample points at the edge of the landscape might include a smaller area than specified by the size argument. In these cases, a warning is returned, and these sample plots can potentially be excluded from further analyses.

Similar to the previously used `show_lsm()` function, patch-level metrics can be returned as a raster object in which each cell stores the metric value of the patch it belongs to. Thus, each cell in the raster object stores its corresponding metrics value. For this, the function `spatialize_lsm()` can be used. However, in order to be type-stable, this function always returns a nested list for each layer and metric (even if only one layer and/or metric is present). The first level of the list corresponds to the number of layers, while the second level of the list corresponds to the number of selected metrics. So we need to use some indexing in order to get only one raster, e.g., the fractal dimension index raster, which we now can use for further analysis, such as calculating kernel density estimates (results not shown).

```
list_shape <- spatialize_lsm(netherlands,
                             what=c("lsm_p_shape", "lsm_p_frac"))
class(list_shape)
str(list_shape)
values(list_shape$layer_1$lsm_p_frac, mat=FALSE) |> density()
```

Last, **landscapemetrics** provides a moving window approach, i.e., metrics can be calculated for the local neighborhood of each focal cell. For this, the `window_lsm()` function can be used. Similar to the `focal()` function of the **terra** package, a local neighborhood must be specific using a matrix object. As with the previously discussed functions, metrics can be selected using the type of metric or specific metrics using the `what` argument. However, currently only landscape-level metrics are implemented. Depending on the size of the landscape and the local neighborhood matrix, this can be computationally demanding with a long run time.

```
mat_window <- matrix(1, nrow=501, ncol=501)
window_lsm(netherlands, window=mat_window, what="lsm_l_pr")
```

1.1.4 Utility Functions

Many of the functions internally used by the **landscapemetrics** package might also be useful for users during pre- or post-processing of raster data or to develop new methods to quantify landscape characteristics. All of these functions start with the `get_` suffix and usually return a (nested) list to be type-stable for different inputs.

One of the most fundamental functions of the package is the `get_patches()` function, which returns the patches of each class using the connected component labeling algorithm. Yet, the result is a nested list in which the first level refers to the number of raster layers/objects and the second level stores all patches separated by class *i*. All other cells not belonging to the current class *i* are set to `NA`. In the following example, we first get all patches of the region in French separated by class. Next, we use all wetland and water-related classes and combine them into one `SpatRaster` object as layers. Last, we use raster algebra to sum all cells which returns a singular layer in which all cells are labeled by their cell identifier and all other cells are `NA`. This results in a single raster layer combining all patches of one of the two water-related classes.

```
list_patches <- get_patches(france)
ras_water <- list_patches$layer_1[4:5] |>
  rast() |>
  sum(na.rm=TRUE)
ras_water
```

```
class       : SpatRaster
dimensions  : 2860, 2333, 1  (nrow, ncol, nlyr)
resolution  : 100, 100  (x, y)
extent      : 3568800, 3802100, 2619700, 2905700  (xmin, xmax, ymin, ymax)
coord. ref. : ETRS89-extended / LAEA Europe (EPSG:3035)
source(s)   : memory
name        :   sum
min value   : 10285
max value   : 10844
```

The `get_boundaries()` function returns a raster in which all boundary/edge cells of a patch are labeled 1 and all core cells are labeled 0. We define an edge cell as a cell that shares a neighboring cell with a different LULC value than itself. Usually, we apply the function to previously connected component labeled landscapes and receive a list with one element for each layer/object. This list can be used for further analysis, such as counting the number of edge and core cells.

```
ras_boundaries_urban <- get_boundaries(list_patches$layer_1$class_1)
lapply(ras_boundaries_urban, freq, wide=TRUE)
```

```
[[1]]
  layer      0      1
1     1 304595 164282
```

We can use the two functions `get_unique_values()` and `get_adjacencies()` to receive a vector with all unique class values and their adjacency matrix, respectively. Because both return a list in order to be type-stable, we can directly index the first element if we only provide one raster layer/object. Of course, the dimensions of the adjacency matrix must be identical to the length of the unique class values given.

```
vec_unique <- get_unique_values(sweden)[[1]]
mat_adjacencies <- get_adjacencies(sweden)[[1]]
dim(mat_adjacencies)
```

```
[1] 5 5
```

```
length(vec_unique)
```

```
[1] 5
```

Relatedly, we can use the `get_nearestneighbour()` function on a previously connected component labeled landscape, which returns a data frame with the minimum Euclidean distances to the nearest patch of the same class. In the following example, we calculate the minimum distances between all patches of class 1, i.e., urban areas. There are 1,175 patches in total and the distance, as well as the identifier of the nearest neighbor, is included. We see that the distance between patches 626 and 650 of the urban class is the largest in the LULC data of Sweden (results are not shown).

```
sweden_cl1 <- get_patches(sweden, class=1)
df_neighbour <- sweden_cl1$layer_1$class_1 |>
  get_nearestneighbour(return_id=TRUE) |>
  arrange(-dist)
```

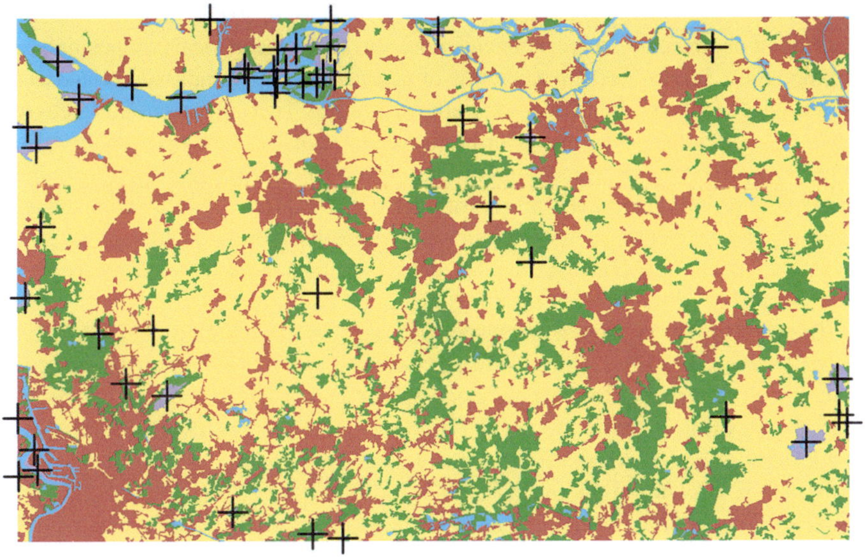

Fig. 7 Patch centroids of all marsh patches in the Noord-Brabant region in the Netherlands

Last, there are two utility functions related to the shape of the patches. First, the function `get_circumscribingcircle()` returns a data frame with the diameter of the smallest circumscribing circle around each patch and the x- and y-coordinate of it. Similar to this, the `get_centroids()` function returns a data frame with the coordinates of the centroids of each patch. For this example, we will filter the data frame to include only the centroids of the marshes class and create a map plotting the patches and all the centroids, Fig. 7.

```
df_circle <- get_circumscribingcircle(netherlands)
df_centroids <- get_centroids(netherlands) |>
  dplyr::filter(class==4)
as.data.frame(netherlands, xy=TRUE) |>
  ggplot(aes(x=x, y=y)) +
  geom_raster(aes(fill=as.factor(cover))) +
  geom_point(data=df_centroids, aes(x=x, y=y), shape=3, size=2.5) +
  scale_fill_manual(values=color_scale) +
  coord_equal() + theme_void() + theme(legend.position="none")
```

1.2 Pattern-Based Spatial Analysis

Pattern-based spatial analysis is a set of methods allowing for performing various tasks on landscape patterns, such as comparing landscape patterns or searching for areas with similar landscape patterns to the query one (Wickham and Norton,

1994; Jasiewicz et al., 2015; Netzel et al., 2018; Nowosad, 2021). Instead of treating each cell independently, this approach focuses on considering local composition and configuration. Firstly, each area, a group of adjacent cells encompassing a landscape pattern, is described numerically as spatial signatures. Next, the similarity between spatial signatures for two areas can be measured using various dissimilarity measures. Low dissimilarity values suggest that the two areas have similar composition and configuration. These ideas allow, for example, to compare spatial pattern change between an area in time, search for similarities between areas of interest, or group areas with similar patterns together. Spatial signatures may also be used directly as new variables, for example, in machine learning models.

The pattern-based spatial analysis methods are available in the R package **motif** (Nowosad, 2021), which allows to describe spatial patterns of one or more categorical raster data for any defined regular and irregular regions. It accepts spatial raster objects from the **terra** (Hijmans, 2021) and **stars** (Pebesma, 2019) R packages as inputs. All **motif** functions use the `lsp_` (*local spatial pattern*) prefix consistently, allowing users to find the desired tool quickly. Most functions in the package are based on a computationally fast C++ code, and the software is designed to work on larger-than-RAM raster datasets.

We need to start by attaching the **motif** package for pattern-based spatial analysis and two main packages for handling spatial data, **terra** and **sf** (Pebesma, 2018).

```
library(motif)
library(terra)
library(sf)
```

Our main dataset for the following examples will be the spatial raster representing land cover classes in the Centre-Val de Loire region in France for the year 2018.

```
france <- rast("data/raster_france.tif")
plot(france) # result not shown
```

1.2.1 Spatial Signatures

The backbone of pattern-based spatial analysis is the concept of spatial signatures. A spatial signature is a short numerical descriptor of landscape patterns featured in a particular area and represented by a categorical raster. Landscape patterns can be described in various ways, and thus many possible spatial signatures exist. That being said, they usually express the composition and/or configuration (arrangement) of categorical raster cells (Table 1). It is also worth noting that spatial signatures can be calculated for a large area or many smaller subareas divided by regular or irregular zones (*windows*).

Spatial signatures are derived using the `lsp_signature()` function. We need to provide two arguments to calculate a single spatial signature: our input categorical raster data (`x`) and the spatial signature type (`type`). The most basic spatial signature is `"composition"`: It is a share of cells of each category in a given area.

Table 1 Spatial signatures available in the motif package

Type	Name	Description
Composition	Composition	A representation of the share of cells for each category within a local landscape. The length of the composition vector corresponds to the number of unique categories in a raster. Input: one categorical raster
Cove	Co-occurrence vector	A count of all of the pairs of the adjacent cells for each category in a local landscape. By default, its length equals $(n * n - n)/2 + n$, where n is a number of unique categories. Input: one categorical raster
Wecove	Weighted co-occurrence vector	A modification of a co-occurrence vector, in which each adjacency contributes to the output based on the values from the weight raster. By default, its length equals $(n * n - n)/2 + n$, where n is a number of unique categories. Input: one categorical and one numerical raster
Incove	Integrated co-occurrence vector	A count of all of the pairs of the adjacent cells for each category in a local landscape for all of the input rasters. Input: two or more categorical rasters
-	A user-defined function	Any user-defined function that can summarize categorical spatial raster data

```
france_comp <- lsp_signature(france, type="composition")
france_comp
```

```
# A tibble: 1 x 3
     id na_prop signature
* <int>   <dbl> <list>
1     1       0 <dbl [1 x 5]>
```

The `lsp_signature()` function always returns a data frame with three columns: a unique identifier (`id`), the proportion of cells in an area with missing values (`na_prop`), and the calculated signature (`signature`). This last column is a list containing a single signature per area of interest. We can see the obtained spatial signature by accessing the first element of the `signature` list.

```
france_comp$signature[[1]]
```

```
              1         2         3            4         5
[1,] 0.07027133 0.7041287 0.2184272 0.0004055524 0.0067673
```

It shows that ~70% of the area is covered by agriculture (2), ~22% by vegetation (3), and ~7% by the urban class (1).

While the "composition" signature is straightforward to understand, it does not encompass the spatial arrangement of the categories: It does not tell if a given category represents one large continuous area or if it is distributed as many small patches. Thus, more complex and informative signatures should be used in most cases. For instance, when dealing with a single categorical raster, a co-occurrence vector ("cove") serves as an appropriate choice for such a signature.

The co-occurrence vector is a spatial signature that encapsulates both the composition and configuration of raster classes in a given area (Haralick et al., 1973; Jasiewicz et al., 2015; Nowosad and Stepinski, 2021). This is achieved by counting not how many times each class occurs but how often a pixel with a given class is adjacent to another pixel of some class. Thus, the result is a long vector (15 elements in this example) that counts how many times pairs of pixels with given classes are adjacent, e.g., urban cells are adjacent to other urban cells, and how many times urban cells are adjacent to, for example, vegetation cells. In general, you can think of spatial signatures as a way to compress information: Our original raster has 6,672,380 values, which we now compressed into a vector of only 15 elements.

```
france_cove <- lsp_signature(france, type="cove")
france_cove
```

```
# A tibble: 1 x 3
     id na_prop signature
* <int>   <dbl> <list>
1     1       0 <dbl [1 x 15]>
```

As shown above, spatial signatures may be used as multi-value descriptors of landscape patterns of a given area. This allows us to calculate spatial signatures of two or more areas and then compare them using a dissimilarity measure.

By default, the `window` argument is set to `NULL`, and **motif** performs calculations for the entire area. However, the package also offers the flexibility to compute numerous spatial signatures for a single large area by dividing it into multiple subareas referred to as "local landscapes." This partitioning is achieved by specifying either a numeric value or an `sf` polygon data as the `window` argument.

When a numeric value is provided, the input raster is subdivided into square-shaped subareas, with each side length equal to the specified numeric value, measured in cells. In the following example, we demonstrate the calculation of co-occurrence vectors for square subareas measuring 50 by 50 cells, which is equivalent to 5 by 5 km in this example.

```
france_cove2 <- lsp_signature(france, type="cove", window=50)
france_cove2
```

```
# A tibble: 2,726 x 3
     id na_prop signature
* <int>   <dbl> <list>
1     1       0 <dbl [1 x 15]>
2     2       0 <dbl [1 x 15]>
3     3       0 <dbl [1 x 15]>
# i 2,723 more rows
```

Now, notice that the output here is still a data frame with three columns, but instead of having only one row, we now have 2,726 rows—one row per subarea. We may visualize these subareas by creating a new `sf` object with `lsp_add_sf()` and then using the `plot()` function (Fig. 8).

```
france_cove2_sf <- lsp_add_sf(france_cove2)
plot(france, ext=ext(france_cove2_sf))
plot(st_geometry(france_cove2_sf), add=TRUE)
```

The `window` argument can also accept an `sf` polygon, allowing for the independent derivation of spatial signatures within each polygon-defined area. In the provided example, we begin by randomly selecting 100 sample points within the study area. For each point, we create a buffer of 300 m, resulting in a corresponding buffer polygon. By setting this buffer polygon as the `window` argument, we can obtain a spatial signature for the 300 m buffer surrounding each input point.

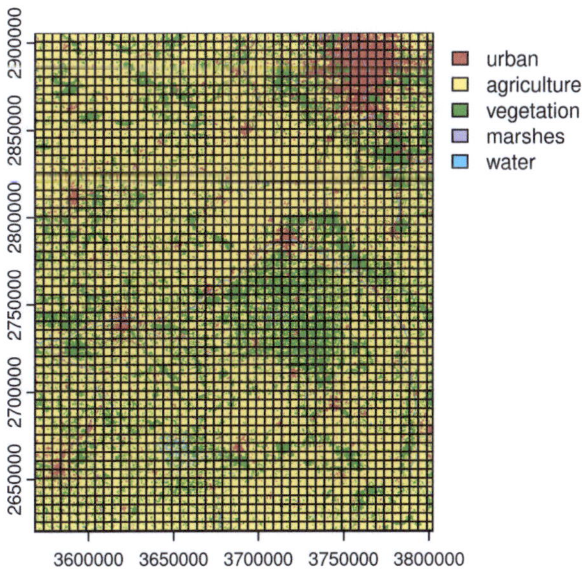

Fig. 8 Land cover of the study area overlaid with a grid with each subarea of 50 by 50 input raster cells

```
sample_points <- read_sf("data/france_sample_points.gpkg")
sample_polys <- st_buffer(sample_points, dist=300)
france_cove3 <- lsp_signature(france, type="cove",
↪  window=sample_polys)
france_cove3 # result not shown
```

1.2.2 Landscape Patterns' Comparison

Landscape patterns change through time, and we are able to evaluate and measure such a change. Here, we will use a second dataset, land cover classes for the Centre-Val de Loire region in France for the year 2000.

```
france2000 <- rast("data/raster_france2000.tif")
```

Now that we have two datasets, for the years 2000 and 2018, we can proceed with their comparison. In pattern-based spatial analysis, the comparison involves calculating the dissimilarity between the spatial signatures derived from both rasters. This approach is complementary to the pixel-based comparison that is based on comparing all the values for each dataset, as it allows us to evaluate the change in the landscape pattern.

To perform this comparison, we utilize the `lsp_compare()` function and provide our two datasets, `france2000` and `france`, along with specifying the type of spatial signature (`type`) and the chosen dissimilarity measure (`dist_fun`). Additionally, we have the option to obtain the results in one of several output classes. For instance, we can choose to use the `SpatRaster` class from the **terra** package by setting `output="terra"`.

```
lc_change1 <- lsp_compare(france2000, france,
                          type="cove", dist_fun="jensen-shannon",
                          output="terra")
lc_change1["dist"]
```

```
class       : SpatRaster
dimensions  : 1, 1, 1  (nrow, ncol, nlyr)
resolution  : 233300, 286000  (x, y)
extent      : 3568800, 3802100, 2619700, 2905700  (xmin, xmax, ymin, ymax)
coord. ref. : ETRS89-extended / LAEA Europe (EPSG:3035)
source(s)   : memory
name        :         dist
min value   : 0.0002682711
max value   : 0.0002682711
```

The above example compared the change in landscape patterns using the co-occurrence vector signature (`"cove"`) and the Jensen-Shannon distance (`"jensen-shannon"`) for the whole area. However, we can change the spatial scale of our analysis and look at the spatial pattern changes also for smaller subareas.

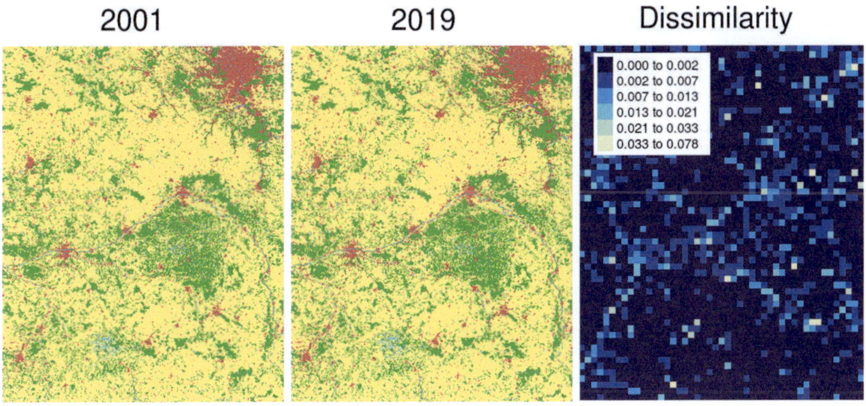

Fig. 9 (Left) land cover data for year 2000; (center) land cover data for year 2018; and (right) change in land cover landscape patterns between 2000 and 2018 for 50 by 50 cells areas (Jensen-Shannon distance)

This can be done by adding a value to the `window` parameter. In the example below, we set the `window` parameter to 50, meaning each subarea will consist of 50 by 50 cells from the input rasters, Fig. 9.

```
lc_change <- lsp_compare(france2000, france,
                         type="cove", dist_fun="jensen-shannon",
                         output="terra", window=50)
```

The highest values in the right panel of Fig. 9 represent areas with the most prominent change of land cover landscape patterns. They mainly relate to the change of herbaceous areas into shrubs. We can also find examples of the areas with the largest changes by subsetting only the local landscapes with dissimilarity above some threshold. Then, the `lsp_extract()` function can be used to extract a local landscape with a given `id`.

```
lc_change_df <- as.data.frame(lc_change)
subset(lc_change_df, dist>0.05)
```

```
       id na_prop_x na_prop_y       dist
1764 1764         0         0 0.05966520
2107 2107         0         0 0.07848007
```

```
compare_1 <- lsp_extract(c(france2000, france), window=50, id=2107)
plot(compare_1, legend = FALSE)
```

Figure 10 shows a local landscape with the largest change of land cover landscape patterns between 2000 and 2018. This location is an example of an area with a large loss of natural vegetation (to agriculture) and a small expansion of urban areas.

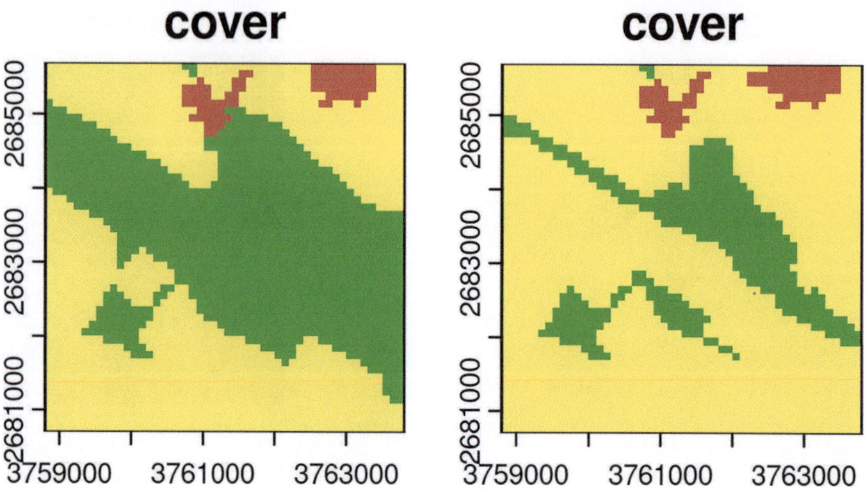

Fig. 10 An example of a local landscape of 5 by 5 km with the largest change of land cover landscape patterns between 2000 and 2018

1.2.3 Landscape Patterns' Search

Another example of pattern-based spatial analysis is searching for areas with similar landscape patterns to a query one. Pattern-based search is a comparison of a spatial pattern of a given local landscape to many subareas of a larger raster.

Here, we start by reading polygon data with our study area and cropping the `france` raster to its borders. Thus, the `france_study_area` object is a raster storing LULC data for the study area.

```
study_area <- read_sf("data/france_study_area.gpkg")
france_study_area <- crop(france, study_area, mask=TRUE)
```

Search can be performed with the `lsp_search()` function, while providing two raster datasets (an area of interest and a larger area), the signature type, distance measure, and the window size.

```
nlcd_search <- lsp_search(france_study_area, france,
                          type="cove", dist_fun="jensen-shannon",
                          output="terra", window=50)
```

```
# result not shown
plot(nlcd_search[["dist"]])
```

Its output is a **terra** object with three raster layers: a unique `id`, the proportion of cells in an area with missing values (`na_prop`), and the calculated dissimilarity

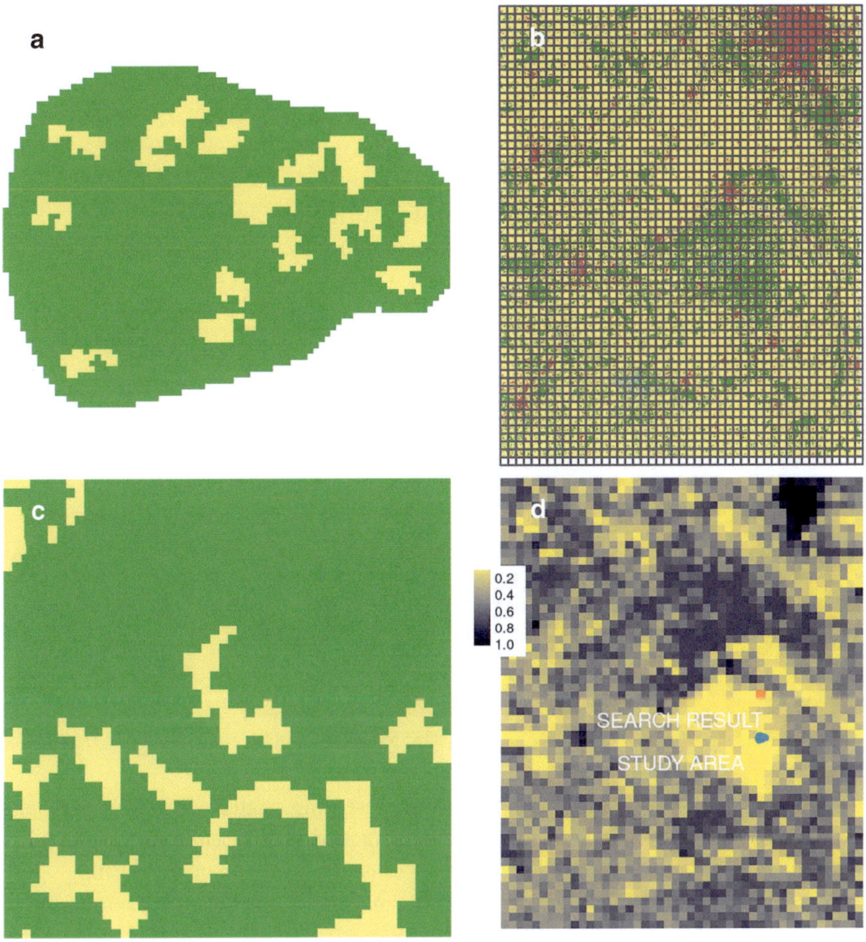

Fig. 11 (**a**) Area of interest, (**b**) search area divided into a set of regular local landscapes, (**c**) a local landscape with the most similar land cover spatial pattern to the area of interest, and (**d**) dissimilarity between land cover landscape patterns of the area of interest and the search area

(dist) (Fig. 11d). The areas with the smallest dissimilarity values are the ones that have the most similar spatial pattern to the area of interest.

```
nlcd_search_df <- as.data.frame(nlcd_search)
subset(nlcd_search_df, dist<0.001)
```

```
       id na_prop         dist
1303 1303       0 0.0003939899
1677 1677       0 0.0007520309
1867 1867       0 0.0007956902
```

To extract a selected local landscape, the `lsp_extract()` function can also be used here (Fig. 11c).

```
search_1 <- lsp_extract(france, window=50, id=1303)
```

1.2.4 Landscape Patterns' Regionalization and Clustering

Spatial structures derived with `lsp_signature()` may also be converted to spatial objects of `sf`, `stars`, and `terra` classes. The `lsp_add_terra()` function, for example, takes a spatial signature object and converts it into a **terra** raster object.

```
france_cove2 <- lsp_signature(france, type="cove", window=10)
france_cove_terra <- lsp_add_terra(france_cove2, metadata=FALSE)
france_cove_terra
```

```
class       : SpatRaster
dimensions  : 286, 234, 15  (nrow, ncol, nlyr)
resolution  : 1000, 1000  (x, y)
extent      : 3568800, 3802800, 2619700, 2905700  (xmin, xmax, ymin, ymax)
coord. ref. : ETRS89-extended / LAEA Europe (EPSG:3035)
source(s)   : memory
names       : X1,        X2, X3,        X4,        X5, X6, ...
min values  :  0, 0.0000000,  0, 0.0000000, 0.0000000,  0, ...
max values  :  1, 0.3191489,  1, 0.3055556, 0.3944444,  1, ...
```

This new object has several raster layers representing spatial signature elements. This opens new possibilities for using the information of landscape patterns in other workflows related to spatial regionalization, clustering, and machine learning.

A regionalization example below uses the **supercells** package to group adjacent local areas with similar landscape patterns. We can apply the **supercells** algorithm (Nowosad and Stepinski, 2022), which expects four essential arguments: input raster data, `k`: a number of regions desired by the user, `compactness`: a compactness value, where larger values cause clusters to be more compact, and `dist_fun`: a distance function.

```
library(supercells)
france_sc <- supercells(france_cove_terra, k=200, compactness=0.6,
                        dist_fun="jensen-shannon", metadata=FALSE)
```

The output here is a spatial vector with a set of polygons, where each polygon represents a larger area with a certain level of homogeneity of its landscape patterns (Fig. 12a).

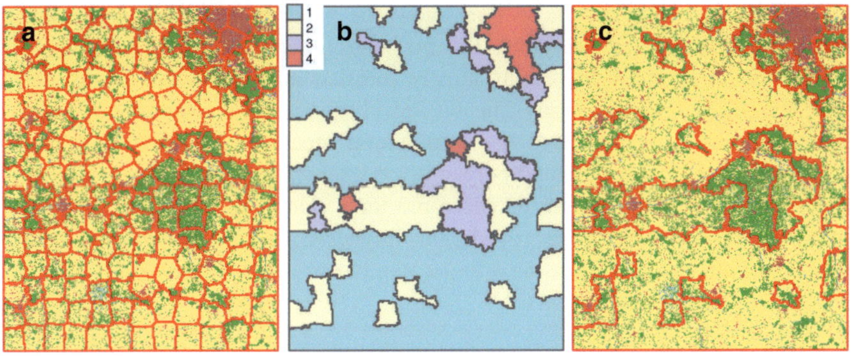

Fig. 12 (**a**) Study area divided into supercells of homogeneous land cover landscape patterns, (**b**) supercells grouped into four clusters using the hierarchical clustering method, and (**c**) borders of the polygons derived with the hierarchical clustering method

```
plot(france) # result not shown
plot(st_geometry(france_sc), add=TRUE, border="red")
```

You may notice that while the land cover landscape patterns are homogeneous internally, there are several examples of similar adjacent polygons. In cases like this, we could merge such polygons, for example, using a clustering method, such as hierarchical clustering or k-means. Firstly, we need to drop the geometry column from the `france_sc` object, as it is not needed for the clustering, and then calculate the value distances between the landscape patterns of each supercell. Next, we can apply the hierarchical clustering method with the `hclust()` function. Based on the dendrogram visualization (`plot(hc)`), we can select the number of clusters: Here we choose four clusters. Finally, we can assign a cluster number to each supercell using the `cutree()` function.

```
france_sc_df <- st_drop_geometry(france_sc)
france_sc_dist <- philentropy::distance(france_sc_df,
                                         method="jensen-shannon",
                                         as.dist.obj=TRUE)
hc <- hclust(france_sc_dist)
france_sc$k <- cutree(hc, 4)
```

The hierarchical clustering method assigns a cluster number to each, previously derived supercell. Thus, to obtain a separate polygon for each cluster, we need to perform a post-processing step: `aggregate()` the clusters to dissolve the borders between clusters.

```
france_sc2 <- aggregate(france_sc, by=list(france_sc$k), FUN=mean)
```

The dissolved areas of clusters are shown on panel b of Fig. 12, while panel c shows the final created polygons. The first cluster represents mostly agricultural areas, the third cluster represents mostly forested areas, and the fourth cluster represents mostly urban areas. The second cluster is a bit more complex, as it contains a mosaic of forested and agricultural areas. The result seems mostly satisfactory, where most created polygons are internally homogeneous and visibly distinct from their neighbors. At the same time, there are a few areas in which homogeneity needs to be improved. This suggests that the result could still be improved, for example, by using a larger number of initial supercells or changing the number of expected clusters.

1.3 Conclusions

In conclusion, this chapter has highlighted the significance of categorical raster data, specifically land use or land cover (LULC) data, in ecological studies. We have explored the role of landscape metrics and pattern-based spatial analysis in quantifying and analyzing landscape patterns, using the R packages **landscapemetrics** and **motif**.

Landscape metrics, computed using the **landscapemetrics** package, have proven to be valuable tools in spatial ecology and landscape ecology. They allow us to quantify the composition and configuration of spatial landscape characteristics, enabling a deeper understanding of environmental processes and changes. By calculating metrics at various levels and visualizing them, researchers can gain insights into the patterns and structures present in the landscape. However, it is crucial to select metrics purposefully, aligned with research questions and hypotheses, to avoid the pitfall of "metric fishing" (Gustafson, 2019).

Moreover, pattern-based spatial analysis, facilitated by the **motif** package, offers a robust framework for analyzing and comparing landscape patterns. By considering local groups of adjacent cells as landscape patterns and calculating spatial signatures, we can capture both composition and configuration information. This approach allows us to compare landscape patterns over time, identify areas with similar patterns, and group similar areas together. The **motif** package's ability to handle large raster datasets is particularly advantageous.

The chapter has provided practical examples and code snippets demonstrating the calculation of landscape metrics and spatial signatures, as well as their visualization. We have shown how to use these techniques to analyze landscape patterns, explore changes over time, and extract local landscapes with significant pattern variations. By combining the power of landscape metrics and pattern-based spatial analysis, researchers can gain a comprehensive understanding of landscape dynamics, composition, and configuration. These techniques facilitate the investigation of the pattern-process link that is a central framework of many ecological fields, but specifically landscape ecology (Turner, 1989). For example, the pattern-process link using LULC data was studied for species distributions (Marshall et al., 2018), (functional) connectivity (Vogt et al., 2009), or genetic population structure (Borthwick et al., 2020). However, patterns and processes are often in an interacting relationship influencing each other, and many drivers and processes act simultaneously—such as climate, disturbances, succession, biotic competition and dispersal, or human land use (Turner, 2005). The pattern-process link is a complex and challenging topic, and the analysis of landscape patterns is only one part of the puzzle. Nevertheless, landscape metrics and pattern-based spatial analysis are powerful tools to investigate the pattern-process link and to gain a better understanding of the landscape.

Both packages, being open source, offer extensive possibilities for customization and expansion. This enables researchers to tailor them to their specific inquiries and requirements. The **landscapemetrics** package, for instance, allows for the integration of new metrics through the utilization of diverse utility functions, thereby enhancing its capabilities. Similarly, the **motif** package can be expanded by incorporating user-defined spatial signatures, which can be used to measure similarities between different properties of landscape patterns. Consequently, contributions to both packages are highly encouraged and welcomed. You can actively participate in the development of these packages at https://github.com/r-spatialecology/landscapemetrics and https://github.com/Nowosad/motif.

Acknowledgments This work was supported by the Initiative of Excellence—Research University project at Adam Mickiewicz University, Poznan [grant number 107/07/POB1/0002]. We also thank the developers and contributors of the **landscapemetrics** and **motif** packages for their work and support.

References

Borthwick R, de Flamingh A, Hesselbarth MHK et al (2020) Alternative quantifications of landscape complementation to model gene flow in banded longhorn beetles [typocerus v. Velutinus (olivier)]. Front Genet 11:307. https://doi.org/10.3389/fgene.2020.00307

Chandra Pandey P, Koutsias N, Petropoulos GP et al (2021) Land use/land cover in view of earth observation: Data sources, input dimensions, and classifiers—a review of the state of the art. Geocarto Int 36:957–988. https://doi.org/10.1080/10106049.2019.1629647

Cushman SA, McGarigal K, Neel MC (2008) Parsimony in landscape metrics: strength, universality, and consistency. Ecol Indicat 8:691–703. https://doi.org/10.1016/j.ecolind.2007.12.002

European Environment Agency (EEA) (2023) Copernicus land monitoring service. CORINE land cover

Fassnacht KS, Cohen WB, Spies TA (2006) Key issues in making and using satellite-based maps in ecology: a primer. Forest Ecol Manag 222:167–181. https://doi.org/10.1016/j.foreco.2005.09.026

Fisher PF, Comber AJ, Wadsworth R (2005) Land use and land cover: Contradiction or complement. In: Fisher PF, Unwin DJ (eds) Re-Presenting GIS. Wiley, Hoboken

Floreano IX, De Moraes LAF (2021) Land use/land cover (LULC) analysis (2009–2019) with google earth engine and 2030 prediction using Markov-CA in the Rondônia state, Brazil. Environ Monit Assess 193:239. https://doi.org/10.1007/s10661-021-09016-y

Frazier AE, Kedron P (2017) Landscape metrics: past progress and future directions. Current Landscape Ecol Rep 2:63–72. https://doi.org/10.1007/s40823-017-0026-0

Fu P, Weng Q (2016) A time series analysis of urbanization induced land use and land cover change and its impact on land surface temperature with Landsat imagery. Remote Sens Environ 175:205–214. https://doi.org/10.1016/j.rse.2015.12.040

Gustafson EJ (1998) Quantifying landscape spatial pattern: what is the state of the art? Ecosystems 1:143–156. https://doi.org/10.1007/s100219900011

Gustafson EJ (2019) How has the state-of-the-art for quantification of landscape pattern advanced in the twenty-first century? Landscape Ecol 34:1–8. https://doi.org/10.1007/s10980-018-0709-x

Haralick RM, Shanmugam K, Dinstein I (1973) Textural features for image classification. IEEE Trans Syst Man Cybern SMC-3:610–621. https://doi.org/bdqvtn

Hesselbarth MHK, Sciaini M, With KA et al (2019) Landscapemetrics: an open-source R tool to calculate landscape metrics. Ecography 42:1648–1657. https://doi.org/10.1111/ecog.04617

Hesselbarth MHK, Nowosad J, Signer J, Graham LJ (2021) Open-source tools in R for landscape ecology. Current Landscape Ecol Rep 6:97–111. https://doi.org/10.1007/s40823-021-00067-y

Hijmans RJ (2019) Raster: Geographic data analysis and modeling. R package version 2.9-5. https://cran.r-project.org/package=raster

Hijmans RJ (2021) Terra: Spatial data analysis. R package version 1.0-10. https://cran.r-project.org/package=terra

Jasiewicz J, Netzel P, Stepinski T (2015) GeoPAT: a toolbox for pattern-based information retrieval from large geospatial databases. Comput Geosci 80:62–73. https://doi.org/f7fz7r

Kupfer JA (2012) Landscape ecology and biogeography: rethinking landscape metrics in a post-FRAGSTATS landscape. Progress Phys Geography 36:400–420. https://doi.org/10.1177/0309133312439594

Lai J, Lortie CJ, Muenchen RA et al (2019) Evaluating the popularity of R in ecology. Ecosphere 10:e02567. https://doi.org/10.1002/ecs2.2567

Lausch A, Blaschke T, Haase D et al (2015) Understanding and quantifying landscape structure - a review on relevant process characteristics, data models and landscape metrics. Ecol Model 295:31–41. https://doi.org/10.1016/j.ecolmodel.2014.08.018

Lovelace R, Nowosad J, Muenchow J (2019) Geocomputation with R. Chapman and Hall/CRC Press, Boca Raton

Manzoor SA, Griffiths G, Lukac M (2021) Land use and climate change interaction triggers contrasting trajectories of biological invasion. Ecol Indicat 120:106936. https://doi.org/10.1016/j.ecolind.2020.106936

Marshall L, Biesmeijer JC, Rasmont P et al (2018) The interplay of climate and land use change affects the distribution of EU bumblebees. Global Change Biol 24:101–116. https://doi.org/10.1111/gcb.13867

McGarigal K, Cushman SA, Ene E (2012) FRAGSTATS v4: Spatial pattern analysis program for categorical and continuous maps. Computer software program produced by the authors at the University of Massachusetts, Amherst. http://www.umass.edu/landeco/research/fragstats/fragstats.html

Netzel P, Nowosad J, Jasiewicz J et al (2018) GeoPAT 2: User's manual. https://doi.org/10.5281/zenodo.1185407

Nowosad J (2021) Motif: an open-source R tool for pattern-based spatial analysis. Landscape Ecol 36:29–43. https://doi.org/ghfsnh

Nowosad J, Stepinski TF (2018) Global inventory of landscape patterns and latent variables of landscape spatial configuration. Ecol Indicat 89:159–167. https://doi.org/10.1016/j.ecolind.2018.02.007

Nowosad J, Stepinski TF (2021) Pattern-based identification and mapping of landscape types using multi-thematic data. Int J Geograph Inf Sci 35:1634–1649. https://doi.org/gk437s

Nowosad J, Stepinski TF (2022) Extended SLIC superpixels algorithm for applications to non-imagery geospatial rasters. Int J Appl Earth Observ Geoinfor 112:102935. https://doi.org/10.1016/j.jag.2022.102935

Pebesma EJ (2018) sf: simple features for R. https://cran.r-project.org/package=sf

Pebesma EJ (2019) Stars: Scalable, spatiotemporal tidy arrays for R. R package version 0.3-1. https://cran.r-project.org/package=stars

Riitters K (2019) Pattern metrics for a transdisciplinary landscape ecology. Landscape Ecol 34:2057–2063. https://doi.org/gkb48v

Schindler S, Poirazidis K, Wrbka T (2008) Towards a core set of landscape metrics for biodiversity assessments: a case study from Dadia National Park, Greece. Ecol Indicat 8:502–514. https://doi.org/10.1016/j.ecolind.2007.06.001

Talukdar S, Singha P, Mahato S et al (2020) Land-use land-cover classification by machine learning classifiers for satellite observations—A review. Remote Sens 12:1135. https://doi.org/10.3390/rs12071135

Turner MG (1989) Landscape ecology: the effect of pattern on process. Ann Rev Ecol Syst 20:171–197

Turner MG (2005) Landscape ecology: what is the state of the science? Ann Rev Ecol Evol Syst 36:319–344. https://doi.org/10.1146/annurev.ecolsys.36.102003.152614

Uuemaa E, Mander Ü, Marja R (2013) Trends in the use of landscape spatial metrics as landscape indicators: a review. Ecol Indicat 28:100–106. https://doi.org/10.1016/j.ecolind.2012.07.018

Vogt P, Ferrari JR, Lookingbill TR et al (2009) Mapping functional connectivity. Ecol Indicat 9:64–71. https://doi.org/10.1016/j.ecolind.2008.01.011

Wang Z, Mountrakis G (2023) Accuracy assessment of eleven medium resolution global and regional land cover land use products: a case study over the conterminous United States. Remote Sens 15:3186. https://doi.org/10.3390/rs15123186

Wang J, Bretz M, Dewan MAA, Delavar MA (2022) Machine learning in modelling land-use and land cover-change (LULCC): Current status, challenges and prospects. Sci Total Environ 822:153559. https://doi.org/10.1016/j.scitotenv.2022.153559

Wickham H (2016) ggplot2: Elegant Graphics for Data Analysis. Springer, New York

Wickham H, François R, Henry L, Müller K (2019) Dplyr: A grammar of data manipulation. R package version 0.8.1. https://cran.r-project.org/package=dply

Wickham JD, Norton DJ (1994) Mapping and analyzing landscape patterns. Landscape Ecol 9:7–23. https://doi.org/fwz5bm

Wickham H, Vaughan D, Girlich M (2023) Tidyr: Tidy messy data. R package version 1.3.0 https://cran.r-project.org/package=tidyr

With KA (2019) Essentials of Landscape Ecology, 1st edn. Oxford University Press, Oxford

Wulder MA, Coops NC, Roy DP et al (2018) Land cover 2.0. Int J Remote Sens 39:4254–4284. https://doi.org/10.1080/01431161.2018.1452075

The CAST Package for Training and Assessment of Spatial Prediction Models

Hanna Meyer, Marvin Ludwig, Carles Milà, Jan Linnenbrink, and Fabian Schumacher

Abstract

One key task in environmental science is to map environmental variables continuously in space or even in space and time. Machine learning algorithms are frequently used to learn from local field observations to make spatial predictions by estimating the value of the variable of interest in places where it has not been measured. However, the application of machine learning strategies for spatial mapping involves additional challenges compared to "nonspatial" prediction tasks that often originate from spatial autocorrelation and from training data that are not independent and identically distributed.

In the past few years, we developed a number of methods to support the application of machine learning for spatial data which involves the development of suitable cross-validation strategies for performance assessment and model selection, spatial feature selection, and methods to assess the area of applicability of the trained models. The objective of the CAST package is to support the application of machine learning strategies for predictive mapping by implementing such methods and making them available for easy integration into modeling workflows.

Here we introduce the CAST package and its core functionalities. At the case study of mapping plant species richness, we will go through the different steps

H. Meyer (✉) · M. Ludwig · J. Linnenbrink · F. Schumacher
Institute of Landscape Ecology, University of Münster, Münster, Germany
e-mail: hanna.meyer@uni-muenster.de; marvin.ludwig@uni-muenster.de; jan.linnenbrink@uni-muenster.de; fabian.schumacher@uni-muenster.de

C. Milà
ISGlobal, Barcelona Institute for Global Health, Barcelona, Spain

Universitat Pompeu Fabra, Barcelona, Spain
e-mail: carles.mila@isglobal.org

D. Rocchini (ed.), *R Coding for Ecology*, Use R!,
https://doi.org/10.1007/978-3-031-99665-8_11

of the modeling workflow and show how CAST can be used to support more reliable spatial predictions.

Keywords

Area of applicability · Cross-validation · Machine learning · Predictive mapping · Spatial modeling

1 Introduction

One key task in environmental science is to map environmental variables continuously in space or even in space and time as a baseline to inform decision-making in various applications, such as agriculture, land-use planning, or natural resource management, or to study ecological research questions based on spatial patterns. However, most ecological variables are only available as point data, e.g., from field surveys. Modeling approaches are hence required to move from local field observations to continuous maps of ecological variables by estimating the value of the variable of interest in places where it has not been measured. In recent years, machine learning methods have become a popular tool to learn patterns in nonlinear and complex systems. These methods have been applied to map various ecological variables, even on a global scale, such as soil properties (Hengl et al. 2017; Lembrechts et al. 2022), plant traits (Moreno-Martínez et al. 2018), or occurrence and abundance of plant (Sabatini et al. 2022; Cai et al. 2023) and animal species (e.g., soil nematodes van den Hoogen et al. 2019).

These studies all follow a very similar workflow (Fig. 1). First, reference data of the response variable are collected, both for model training and for model validation. These are often derived from field surveys (e.g., soil sampling, plot-based vegetation surveys, and measurements from weather stations), where often data from various studies are compiled in larger databases such as sPlotOpen (Sabatini et al. 2021) for vegetation surveys or TRY (Kattge et al. 2020) for plant

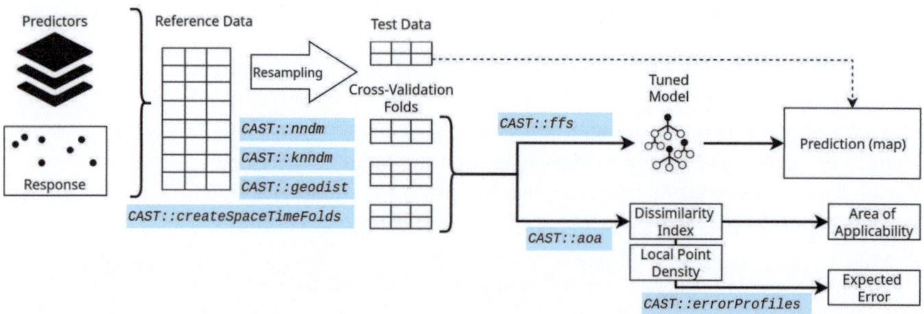

Fig. 1 A very simple workflow for a spatial prediction mapping workflow, indicating which function in CAST can be used in different steps to support the spatial prediction

trait data. Second, a set of potential predictor variables (or "features") that are assumed to be drivers of the response variable is compiled. These variables are available for the entire area of interest and often originate from remote sensing data sources such as spectral channels from optical sensors. These might also be variables describing the climate (e.g., WorldClim, Fick and Hijmans 2017) or terrain properties. Third, a supervised model is then trained to learn the relationships between the response variable and the corresponding predictor values. Among the many available algorithms, random forest is one the most popular machine learning algorithms for spatial modeling tasks. The model training process typically involves several steps including hyperparameter tuning and feature selection, where typically cross-validation is used to optimize the parameters or to select the variables used in the model. Fourth, the trained model is then applied to each pixel of the predictor set to generate spatial maps of the environmental response variable of interest. Finally, the assessment of the map quality, as well as uncertainty assessment, is performed as an essential part of the modeling process to inform the producer and user of the reliability of the generated product.

R is a very popular software environment and programming language for spatial predictive modeling and all its involved steps. There are a large number of machine learning algorithms implemented in various packages, which can also be accessed via wrapper packages such as caret (Kuhn 2008), tidymodels (Kuhn and Wickham 2020), or mlr3 (Lang et al. 2019). All of these three wrapper packages simplify and standardize the machine learning workflow in R, providing users with a consistent interface for model training, prediction, and evaluation. However, the application of machine learning strategies for spatial mapping involves additional challenges compared to other "nonspatial" prediction tasks that often originate from spatial autocorrelation and from training data that are not independent and identically distributed (i.i.d.) (see, e.g., Meyer and Pebesma 2022). In the past few years, we developed a number of methods to support the application of machine learning for spatial data which involves the development of suitable cross-validation strategies for performance assessment and model selection (Milà et al. 2022; Linnenbrink et al. 2024), spatial feature selection (Meyer et al. 2018, 2019), and methods to assess the area of applicability of the trained models (Meyer and Pebesma 2021).

The purpose of the CAST package is to support the application of machine learning strategies for predictive spatial mapping by implementing such methods and making them available for easy integration into modeling workflows. CAST stands for "Caret Applications for Spatio-Temporal Models," which indicates that the package was initially (in 2018) developed as a wrapper around caret (Kuhn 2008). However, with more users shifting toward mlr3 or tidymodels, we intend to make the functionality available beyond the caret package by making it usable with other wrapper packages as well.

With this chapter, we want to show how the CAST package can be used to close important gaps in the machine learning-based spatial mapping workflow. For each of the main CAST functionalities (Fig. 1), we will first explain the concept and idea behind the methods and then show its practical application with an example dataset of plant species richness in South America.

2 Functionality of the CAST Package

Alongside the implementation in CAST, all methods are published in scientific papers. Currently, the main features of CAST are:

- The cross-validation strategy "Nearest Neighbor Distance Matching" (NNDM) cross-validation (Milà et al. 2022) and its k-fold variant kNNDM cross-validation (Linnenbrink et al. 2024), which are designed to overcome limitations of existing cross-validation strategies to provide estimates of map accuracy. The package further implements additional support for user-defined spatial or spatiotemporal assignment of cross-validation folds as explained in Meyer et al. (2018). See `CAST::nndm`, `CAST::knndm`, and `CAST::createSpaceTimeFolds`.
- Visualization methods as used in Milà et al. (2022), Meyer and Pebesma (2022), Ludwig et al. (2023) to inspect the representativeness of the cross-validation folds for a spatial prediction task. See `CAST::geodist`.
- Methods for spatial feature selection that are designed to select predictor variables that are most suitable for spatial mapping and minimize the risk of overfitting (Meyer et al. 2018, 2019). Applications for large-scale spatial prediction tasks are shown in Ludwig et al. (2023). See `CAST::ffs`.
- A method for the assessment of the area of applicability (Meyer and Pebesma 2022) that is defined as the area where the model was enabled to learn about relationships and where the estimated cross-validation performance may hold. See `CAST::aoa`.
- A method to estimate prediction uncertainties based on distances and data point densities in the predictor space (Meyer and Pebesma 2021; Schumacher et al. 2024). See `CAST::errorProfiles`.

The package is available on CRAN https://CRAN.R-project.org/package=CAST; the developer version is on GitHub https://github.com/HannaMeyer/CAST. Here we used version 1.0.0. The package is supported by several vignettes that explain the functionality of each of its components. They can be found on the documentation page of the package: https://hannameyer.github.io/CAST/.

3 Example Data and Modeling Task

We assume that the reader of this chapter is already familiar with basic principles of applied predictive modeling (if not, we recommend Kuhn and Johnson 2013; James et al. 2013, as a starting point). For demonstration of the CAST functionalities, we will go through a typical spatial prediction task, which aims here at producing a spatially continuous map of plant species richness for South America. As reference data, we use plant species richness data from plot-based vegetation surveys that are compiled in the sPlotOpen database described in Sabatini et al. (2021) (datasets used: Pauchard et al. 2013; Peyre et al. 2015; Lopez-Gonzalez et al. 2011; Vibrans

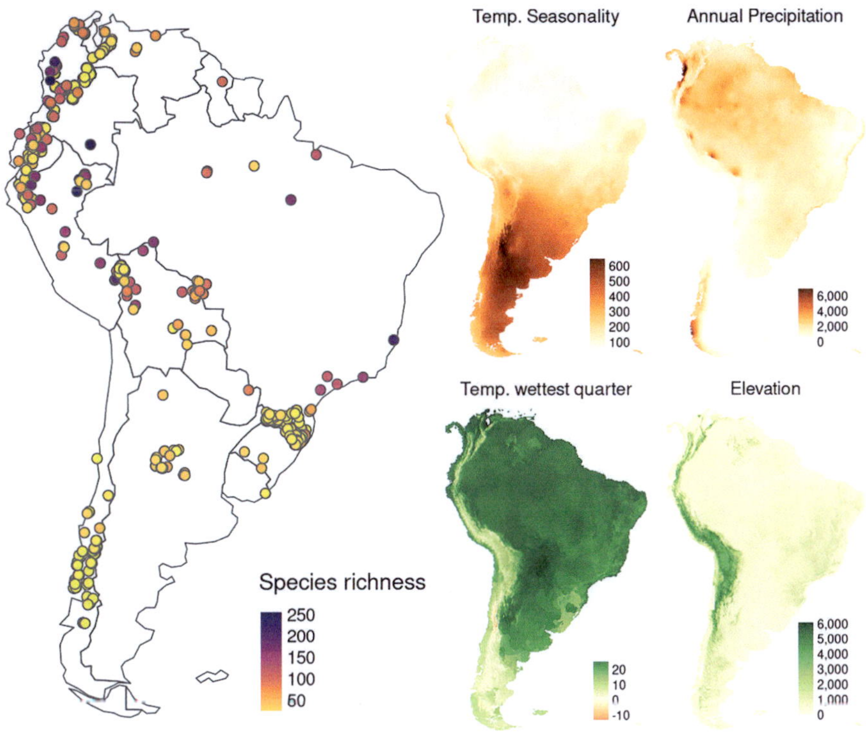

Fig. 2 Location of the reference data from sPlotOpen and example predictor variables (an excerpt of *predictors_raster*) for the desired prediction domain

et al. 2020). We will use WorldClim climatic variables (Fick and Hijmans 2017) and elevation (Jarvis et al. 2008) as predictors, assuming that they are relevant drivers of species richness. The data are visualized in Fig. 2. Random forest is used here as machine learning algorithm. Please note that this example is built here for technical demonstration purposes and should not be regarded as a reference for mapping plant species richness.

The code and data used for the example described in this chapter is available at https://github.com/RCodingForEcology. Therein, the predictor variables for entire South America are available as multiband GeoTIFF (here referred to as *predictors_raster*). The reference data from sPlotOpen are available as sf object as part of the CAST package (called here *training_data*) and contain the predictor information as attributes (called here *predictor_names*), as well as the species richness information (called here *response_name*). For handling of raster data, we will rely on the terra package (Hijmans 2023). For the vector data, sf will be used (Pebesma 2018; Pebesma and Bivand 2023).

3.1 A First Simple Prediction Model

As a first example we will see how we can train a "default" model without consideration of the challenges of spatial prediction that will be discussed next. When using the caret package, we can use the `train` function, which takes the predictors and response from the training data, as well as the information of the algorithm to be used as arguments (here random forest with 100 trees from the ranger package, Wright and Ziegler 2017). Default random cross-validation is used for hyperparameter tuning and assessment. Once the model is trained, we can use it to make spatial predictions (see the code below).

```
# train model:
rfmodel_rcv <- caret::train(x = training_data[,predictor_names],
                            y = training_data[,response_name],
                            method = "ranger",
                            trControl = trainControl(method = "cv"))
# get validation statistics and predict:
CAST::global_validation(rfmodel_rcv)
prediction <- terra::predict(predictors_raster, rfmodel_rcv, na.rm = TRUE)
```

In this first model we see that, technically, we can easily make predictions for the entire study area as shown in Fig. 3 (and might even go beyond that). However, these predictions also raise many questions, e.g., how much should we trust them? How could we improve the model? Can we really apply the model trained on so few and clustered reference locations to make predictions for entire South America? What are the uncertainties in doing so? In the next sections, we will see how we can use the CAST package to deal with these concerns.

4 Cross-Validation to Estimate the Map Accuracy

A map as produced by the workflow described above is only valuable when we have an accuracy estimate that measures how much (or little) we can trust it. For an unbiased estimation of the map accuracy, a probability sample of the entire prediction area is required (e.g., Stehman 1999; Wadoux et al. 2021). The problem is that this is rarely available in ecology (see, e.g., Meyer and Pebesma 2022) because reference data are often clustered in space, while some areas are left completely unsampled. This is problematic not only for the map evaluation but also for the modeling process itself since an estimation of the map accuracy is used during model tuning, i.e., to identify the best set of hyperparameters and/or predictor variables. To account for the absence of an external probability sample used for testing the final predictions, the available reference data are usually split into three datasets: training data, validation data (used during model tuning), and test data (used to evaluate the final predictions). Data resampling into training and validation sets (so-called folds) is usually done several times in a cross-validation procedure. Test data, on the other hand, are usually selected once beforehand and ideally according to the same resampling rules that are applied for defining the

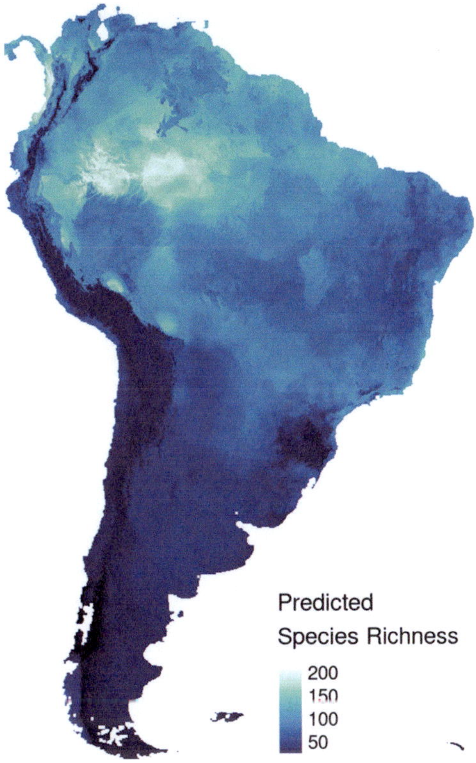

Fig. 3 A first prediction of plant species richness in South America

cross-validation folds. Here, we will focus on cross-validation; however, the same concepts and functions can be used for a prior split into training and test data.

Whether the (cross-)validation can give a reliable estimate of the map accuracy depends on the strategy we use to split the data. It has been reported several times that the commonly applied random cross-validation is not suitable for spatial data. It assumes data are independent and results in overoptimistic estimates when sampling data are clustered (Wadoux et al. 2021; Milà et al. 2022). Spatial cross-validation strategies that try to achieve independence between test and train data (e.g., Valavi et al. 2019; Roberts et al. 2017; Brenning 2012), however, may have shortcomings as well, especially if training data are randomly or regularly distributed (Wadoux et al. 2021; Milà et al. 2022). To address this issue, in Meyer and Pebesma (2022), Milà et al. (2022) we argue that the cross-validation strategy used for map accuracy estimation should be prediction-oriented; that is, it should aim to produce similar predictive conditions to those found when using the model to predict a new area. In the case of spatial prediction models, the "predictive conditions" can refer to the geographical Nearest Neighbor Distances (NNDs) between prediction and training points. Figure 4 shows the challenge in our example: Since our training samples are clustered within the prediction area, we see that NND between training points is generally much shorter than NND between prediction and training points. The

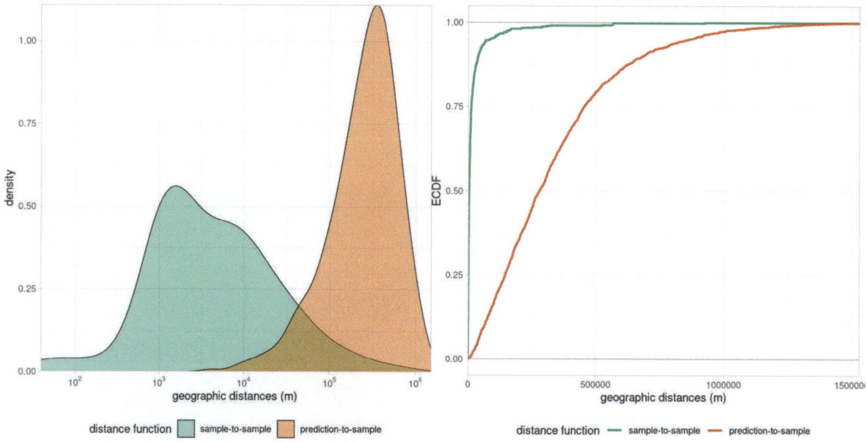

Fig. 4 Nearest neighbor distance distribution represented as density plot (left) and empirical cumulative distribution function plot (right). Both show that prediction requires an application of the model far beyond the clustered reference data. The aim of the cross-validation strategies implemented in CAST is to produce similar prediction-to-sample distances based on the available training data

latter is the situation the model faces during prediction (i.e., the model needs to be applied to areas that are far away from a reference data point, and thus we should try to resemble it during cross-validation to get a meaningful estimate of the map accuracy). Note that `geodist` can be calculated in both geographical space and feature space. Visualization allows for density plots (default, see Fig. 4, left) or empirical cumulative distribution function plots (see Fig. 4, right).

```
# compare distances in geographic space:
geo_distance <- CAST::geodist(training_data,
                              modeldomain=predictors_raster)
plot(geo_distance)
plot(geo_distance, stat="ecdf")
```

With the objective of providing a prediction-oriented cross-validation strategy suitable for spatial prediction tasks, we proposed the Nearest Neighbor Distance Matching (NNDM) algorithm (Milà et al. 2022), a Leave-One-Out (LOO) cross-validation method that aims to create a distribution of NND between test and train data during cross-validation that matches the one found during prediction. Briefly, NNDM uses spatial point pattern concepts to exclude additional samples when performing a LOO cross-validation so that the Empirical Cumulative Distribution Function (ECDF, see Fig. 4 right) of NND between test and train points during cross-validation matches the ECDF of NND between prediction and train points (see Milà et al. 2022, for more information and simulations evaluating its performance). Nonetheless, as a LOO method, it is computationally intensive and cannot be used with large datasets. Therefore, in Linnenbrink et al. (2024), we further suggest a k-fold variant of this approach: the kNNDM algorithm.

kNNDM uses a clustering approach in which the reference samples are clustered in a varying number of groups and then merged into the desired final number of k folds. Among all the candidate groupings, the one that minimizes the Wasserstein distance (absolute value integral between the two curves) between the ECDF of NND between test and train points during cross-validation and the ECDF of NND between prediction and train points, i.e., the one that offers the best match, is chosen.

Going back to our example, if we just use a random k-fold cross-validation, the distribution of NND between cross-validation folds closely matches the distribution of NND between train points (Fig. 5 top). Thus, when using random k-fold cross-validation, we are not testing the ability of the model to make predictions for all South America, but rather evaluating how good the model can reproduce the training data. We now use the kNNDM algorithm to assign the sample points to cross-validation folds in a way that resembles the prediction conditions. The resulting cross-validation split assigns the reference points to spatially distinct folds, thus extending nearest neighbor distances. This results in a better match between the NND during cross-validation and the NND between prediction and training points (Fig. 5).

```
# define cross-validation folds with knndm:
knndm_folds <- CAST::knndm(tpoints = plots,
                           modeldomain = predictors_raster, k = 5)
```

We can now use the resulting cross-validation folds to estimate the map accuracy and find the best hyperparameters during model tuning. The R-code is mostly the same as the one to train the default random forest model; we just need to specify the cross-validation folds as the index parameter in caret's `trainControl` function:

```
# indicate the use kNNDM cross-validation folds:
tr_control <- caret::trainControl(method = "cv",
                                  index = knndm_folds$indx_train,
                                  savePredictions = TRUE)
# tune and train model:
rfmodel_knndmcv <- caret::train(x = training_data[,predictor_names],
                                y = training_data[,response_name],
                                method = "ranger",
                                trControl = tr_control)
# get validation statistics and predict:
CAST::global_validation(rfmodel_knndmcv)
prediction <- terra::predict(rfmodel_knndmcv, predictors_raster)
```

Note that we also set the parameter `savePredictions = TRUE`. This allows us to stack the different cross-validation predictions afterward to calculate the validation statistics using the function `CAST::global_validation`. This is relevant because kNNDM matches the distribution of NND based on all the folds simultaneously and can produce folds with a different number of samples, and thus the traditional approach of computing validation statistics in each fold and then averaging them would not produce the desired results (especially for validation statistics such as R^2). If we now compare the results of the random and the kNNDM cross-validation, we see that the random k-fold cross-validation results in higher

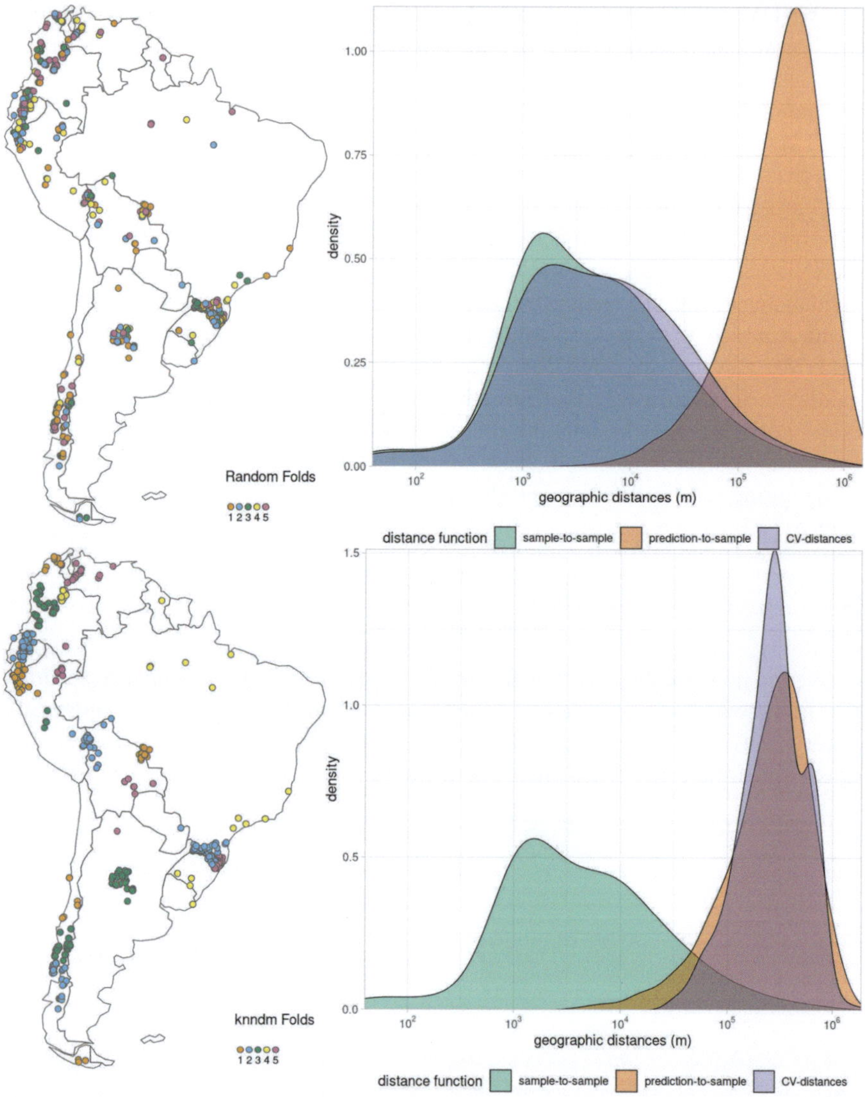

Fig. 5 Comparison of cross-validation methods: random folds and their corresponding nearest neighbor distance distribution (top), as well as kNNDM folds and their nearest neighbor distance distribution (bottom)

values for the validation metrics (Table 1). This is due to the fact that we tested the performance of the model holding out points that are very close to the training points, making it easy for the model to predict their values, which probably leads to overoptimism in the estimated map accuracy (Milà et al. 2022; Wadoux et al. 2021). The kNNDM cross-validation results in worse validation metrics, because the

Table 1 Validation metrics resulting from a random k-fold cross-validation and kNNDM cross-validation (based on global validation). The table shows results for the full model and for the model that was simplified using forward feature selection based on kNNDM cross-validation. For comparison, the random cross-validation performance of the simplified model (* simplified via kNNDM) is shown. A higher Root Mean Square Error (RMSE) and a smaller R^2 indicate a worse map accuracy

Model	Cross-validation	RMSE	R^2	Predictors
Full	Random k-fold	24.16	0.71	11
Full	kNNDM	33.34	0.47	11
Simplified	kNNDM	31.97	0.52	5
Simplified	Random k-fold	24.31	0.71	5*

holdout points are further away from the training points and, similar to what we will find when using the model for all South America, are thus harder to predict. Since the latter better reflects the actual predictive conditions for which the model is going to be used (Fig. 4), kNNDM cross-validation will likely offer a more realistic map accuracy estimate (see Milà et al. 2022; Linnenbrink et al. 2024, for simulations).

Note that here we only showed k(NNDM) in geographic space. CAST, however, also supports k(NNDM) in the predictor space (see argument space = "feature"). Hence, data are split in a way that each fold resembles the feature space distances observed when predicting the final map from the training data. Besides the implementations of NNDM and kNNDM, CAST also offers additional spatiotemporal cross-validation strategies (`CAST::CreateSpacetimeFolds`, see Meyer et al. (2018) for an overview and examples of its use). Note that, in general, the cross-validation strategy will not necessarily change the predictions since the model used to create the final map generally uses all samples. Therefore, if the tuning parameters (e.g., mtry in a random forest) and features stay the same, both models will result in the same prediction of species richness in South America.

5 Spatial Model Tuning and Feature Selection

A suitable cross-validation strategy may not only support a more reliable estimation of the map accuracy (in the absence of suitable test data) but can also improve predictions if used in the process of model selection, i.e., when choosing the predictors (feature selection) and hyperparameters (hyperparameter tuning) that yield the best performance. By accounting for a suitable spatial cross-validation strategy, the aim is to select the hyperparameters/variables that lead to the best spatial prediction for a given task. The relevance of spatial hyperparameter tuning has been shown by Schratz et al. (2019). The process is rather straightforward: Different sets of hyperparameters are tested and compared via a suitable cross-validation strategy that adequately reflects the predictive conditions of the model (e.g., kNNDM).

Using a suitable cross-validation strategy is also of high relevance during feature selection. This is especially the case if the difference between a spatial and a random cross-validation performance is large, as this may indicate significant spatial overfitting: The model may well predict the training data but is not able to make spatial predictions beyond them. This difference is possibly due to the "clever Hans effect," a phenomenon whereby a model appears to have high predictive performance but is, in reality, relying on non-predictive features (in line with the horse "Clever Hans" that had been famous for being capable of mathematics but in reality was only reacting to unintentional signals of the owner). In the context of spatial prediction models, the Clever Hans effect can occur when the model learns to predict the outcome based on spurious correlations that exist in the data rather than true underlying relationships (e.g., Lapuschkin et al. 2019). This can lead to inaccurate predictions when the model is applied to new areas that have not been covered by the training data. To avoid the Clever Hans effect, domain knowledge regarding which are the relevant predictors for a given outcome is key to identify a suitable set of candidate features. This prior selection can be complemented in a semiautomatic way by conducting a spatial feature selection, where we explicitly test which variables are suitable for predictive mapping and which variables are not because they are either irrelevant or act as Clever Hans predictors.

During feature selection, we usually try to avoid testing all sets of potential predictor variable combinations for computational reasons; therefore, different feature selection strategies such as recursive feature elimination and others have been implemented in caret, as well as in the other wrapper packages. Here, it is of relevance that the selection of variables is based on an appropriate cross-validation strategy that reflects the predictive conditions of the model; for example, recursive feature elimination as implemented in caret is based on variable importance ranking from bootstrapping, a resampling method that, similar to k-fold random cross-validation, does not consider the geographical space in its default implementation. Hence, variables that lead to overfitting are likely to be top ranked and hence selected. In CAST, we implemented a forward feature selection that first tests, for each combination of two predictor variables, which features lead to the best model, defined as the model leading to the lowest cross-validation error (i.e., kNNDM cross-validation error). Then, the number of variables is sequentially increased by choosing the predictor leading to the lowest error until none of the remaining variables can further improve the performance. The R-code to run such a forward feature selection (`CAST::ffs`) is, again, very similar to the `caret::train` function we used before.

```
# feature selection and final model training:
# Note that a suitable cross-validation is required as trControl
rfmodel_ffs <- CAST::ffs(training_data[,predictors],
                         training_data[,response],
                         method = "ranger",
                         importance = "permutation",
                         trControl = tr_cntrl)
# get validation statistics and predict:
CAST::global_validation(rfmodel_ffs)
```

```
prediction <- terra::predict(rfmodel_ffs, predictors_raster)
plot(rfmodel_ffs, plotType = "selected")
```

As we can see in Table 1, the model that used only a subset of the predictors (see Fig. 6 for the selected predictors) led to a higher spatial mapping performance compared to the full model.

6 Assessment of the Area of Applicability

So far, we have developed a machine learning workflow that is designed for spatial predictions by implementing a suitable cross-validation strategy for model optimization and accuracy assessment. We have also seen how, technically, predictions are possible for the entire South America by simply using the fitted model for the entire continent. However, we have not yet tested whether we should have done so. While machine learning models such as random forests can fit nonlinear relationships between predictors and response well, they are not suitable for extrapolation in the predictor space. Hence, to arrive at predictions for the entire South America, the predictor values (and their combinations) found in the prediction area need to be represented in the training data to avoid extrapolation. This, however, is rarely the case in predictive mapping applications in the field of ecology. For instance, in our example, considering the diversity of climatic conditions in the South American continent, it is unlikely that our comparably few reference samples cover all these distinct environments.

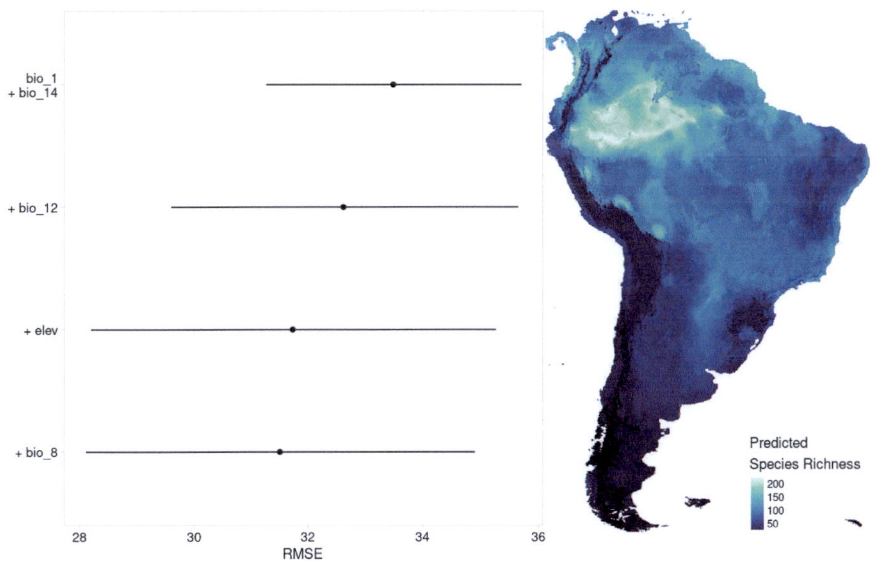

Fig. 6 Selected variables that decrease the RMSE estimated with kNNDM cross-validation (left) and predicted species richness using the simplified model (right)

To avoid predictions to such "unknown" areas, we recently suggested the area of applicability, defined as the area for which the model was enabled to learn relationships, and where we assume that the estimated cross-validation performance holds (see Meyer and Pebesma 2021, for details). The delineation of the area of applicability is based on a dissimilarity index (DI) that measures, for each prediction location, the distance to the nearest training data point in the multidimensional predictor space. The predictor space is previously normalized and (optionally) weighted by the relevance of the predictors within the trained model. Currently, Euclidean and Mahalanobis distances are implemented as distance measures.

The area of applicability is then derived by applying a threshold to the computed values of the dissimilarity index in order to identify pixels that are too dissimilar from the reference samples. In Meyer and Pebesma (2021) we suggested that this threshold should be derived from cross-validation. If a prediction location is more dissimilar than what has been observed during cross-validation, it is considered to be outside the area of applicability. For those locations, the estimated cross-validation performance will not be considered as valid, since these dissimilar conditions have not been explored during the cross-validation.

The dissimilarity index is based on the distance to the nearest training data point in the multidimensional predictor data space; thus local training data point densities (LPDs) are not considered during this process. Due to this, the index does not discriminate between areas where few, or even one isolated training data point is the nearest, and a prediction location that is densely covered by training data points. As we assume that training data point densities can be highly decisive for prediction uncertainty, we suggest additionally calculating and communicating the local data point density for a prediction location Schumacher et al. (2024). The local data point density reflects the number of training data points that are within the threshold of the area of applicability.

The dissimilarity index, the local training data point density, and the area of applicability can be derived via the `CAST::aoa` function, and the result is an object of class "AOA." However, it is also possible to first calculate the dissimilarity index of the training data (which is the baseline to derive the threshold and hence the area of applicability) as a previous step, which is especially useful if the prediction area is large and the AOA function is applied to several tiles of the area in parallel. In this case, `CAST::trainDI` is calculated first, and the result is then passed to the `CAST::aoa` function. Here, we will show the first way, i.e., directly using the `CAST::aoa` function, but a vignette on the parallelization is included in the package.

```
# estimate the area of applicability (AOA):
AOA <- CAST::aoa(predictors, rfmodel_ffs, LPD=TRUE)
plot(AOA)
# visualize the DI, LPD and the AOA:
plot(AOA$DI)
plot(AOA$LPD)
plot(AOA$AOA)
```

As we can see, some areas show high dissimilarity index values (Fig. 8, left) because they have very different values for the predictor variables used in the

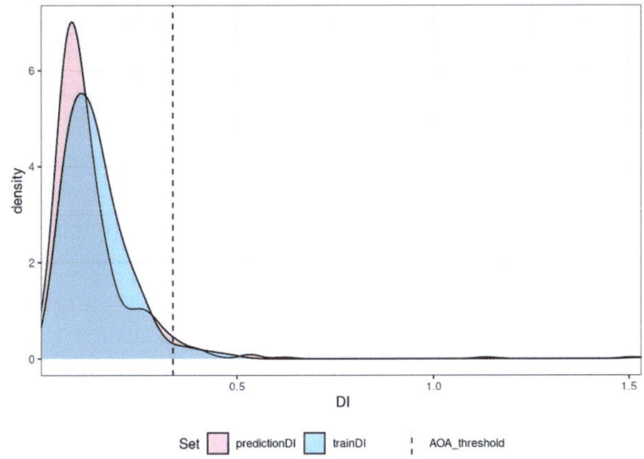

Fig. 7 Output of `plot(AOA)`, dissimilarity index of training and prediction data with AOA threshold

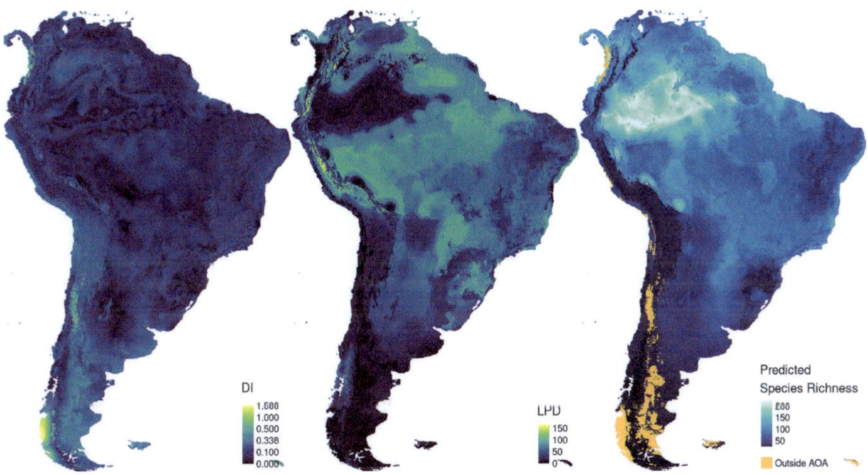

Fig. 8 Dissimilarity index (DI; left), local point density (LPD; middle), and predicted species richness based on the simplified model, but only for the area of applicability (AOA; right)

model. Based on the dissimilarity index of the training data and the model fitted in the previous section using kNNDM cross-validation, the threshold was derived (Fig. 7), which was used to mask the area for which the model is not considered to be applicable (Fig. 8 right).

7 Pixel-Wise Performance Estimation

The dissimilarity index is not only the basis to delineate the area of applicability but can also be used to estimate the prediction performance on a pixel level based on the relationship between the dissimilarity index and prediction errors derived via cross-validation (described in more detail in Meyer and Pebesma 2021). This is done using

CAST::errorProfiles. To be able to test the model performance for very low and high values of the dissimilarity index, the model may be refitted and validated with multiple cross-validations where the cross-validation folds are defined by clusters in the predictor space, ranging from three clusters to N clusters (equivalent to leave-one-out cross-validation). In this way, a large range of dissimilarity index values is created during the different cross-validations. Again, the errors on a pixel level should only be estimated within the area of applicability. Note that multiple cross-validations may change the area of applicability: Since multiple cross-validations will produce a large range of dissimilarities (including extrapolation situations), the threshold will likely become larger because a validation was possible for areas with higher dissimilarity. In this case, the area of applicability has to be updated accordingly by applying the new threshold to the dissimilarity index.

```
# calculate model:
performancemodel <- CAST::errorProfiles(rfmodel_ffs, AOA)
plot(performancemodel)
# apply model:
expected_RMSE <- terra::predict(AOA$DI, performancemodel)
# mask predictions with the AOA:
expected_RMSE = terra::mask(expected_RMSE, AOA$AOA, maskvalues = 0)
```

The resulting object contains the model representing the relationship between the dissimilarity index and the performance metric. It can be used to predict the error (or performance) on a pixel level. The results for our example are shown in Fig. 9, left: The performance (per default the same that was already used during

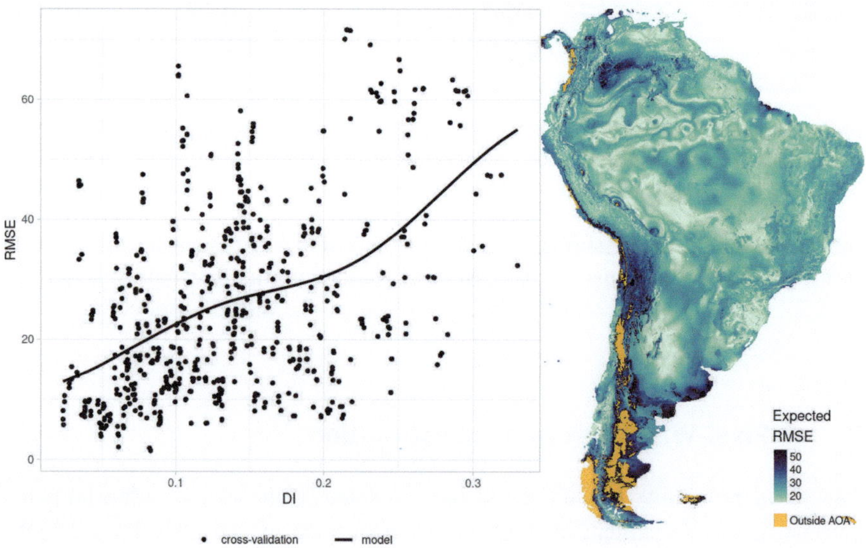

Fig. 9 The modeled relationship between the dissimilarity index (DI) and cross-validation RMSE (left) and the resulting map of the expected error (right)

model tuning; here RMSE) is calculated for moving windows of dissimilarity index values based on the cross-validated training data. A shape-constrained additive model is then used to model the relationship between the performance metric and the dissimilarity values (see Meyer and Pebesma 2021, for details). The model is, in a second step, applied to each prediction location (i.e., pixel) to map the expected performance (Fig. 9, right). Analogous to the workflow described above, it is also possible to use the local data point density for a pixel-wise performance estimation. We have, however, to keep in mind that the estimated performance is based here on the dissimilarity (or local data point density) in predictor space only. Hence, other sources of uncertainty are not reflected and may be accounted for by other methods (e.g., quantile regression forest).

8 Conclusion and Outlook

Here, we presented the R package CAST to support the process of spatial model training and evaluation. We first introduced NNDM and kNNDM cross-validation to provide more realistic estimates of map accuracy and showed how it can be used in a feature selection procedure to reduce the risk of spatial overfitting and eventually improve the spatial model performance. Next, the area of applicability was presented with the aim of limiting predictions to the areas where the model was enabled to learn about relationships. Finally, we showed how the dissimilarity index based on distances in the predictor space, or the local training point density, can be used to quantify uncertainties that originate from different predictor properties than those of the training data.

The CAST package is still under active development, and we intend to extend the functionalities. Future developments will include further model inspection tools and a stronger support for integration into mlr3 and tidymodels workflows. The code to reproduce the results and figures from this chapter is available at https://github.com/RcodingForEcology, and the tutorial along with additional CAST tutorials can be found at https://hannameyer.github.io/CAST/.

References

Brenning A (2012). Spatial cross-validation and bootstrap for the assessment of prediction rules in remote sensing: the R package sperrorest. In: 2012 IEEE International Geoscience and Remote Sensing Symposium, pp 5372–5375. https://doi.org/10.1109/IGARSS.2012.6352393

Cai L, Kreft H, Taylor A, Denelle P, Schrader J, Essl F, et al (2023) Global models and predictions of plant diversity based on advanced machine learning techniques. New Phytolog 237(4):1432–1445. Retrieved from https://nph.onlinelibrary.wiley.com/doi/abs/10.1111/nph.18533. https://doi.org/10.1111/nph.18533

Fick SE, Hijmans RJ (2017) WorldClim 2: new 1-km spatial resolution climate surfaces for global land areas. Int J Climatol 37(12):4302–4315. Retrieved from https://rmets.onlinelibrary.wiley.com/doi/abs/10.1002/joc.5086. https://doi.org/10.1002/joc.5086

Hengl T, Mendes de Jesus J, Heuvelink GBM, Ruiperez Gonzalez M, Kilibarda M, Blagotić A et al (2017) SoilGrids250m: global gridded soil information based on machine learning. PLOS ONE 12(2):e0169748. https://doi.org/10.1371/journal.pone.0169748

Hijmans RJ (2023) terra: Spatial data analysis [Computer Software manual]. https://rspatial.org/. (R package version 1.7-34)

James G, Witten D, Hastie T, Tibshirani R (2013) An Introduction to Statistical Learning: With Applications in R, 1st ed. Springer, New York

Jarvis A, Reuter H, Nelson A, Guevara E (2008) Hole-filled seamless SRTM data V4. International Centre for Tropical Agriculture (CIAT). https://srtm.csi.cgiar.org

Kattge J, Bönisch, G., Díaz, S., Lavorel S, Prentice IC, Leadley P, et al (2020) Try plant trait database – enhanced coverage and open access. Global Change Biol 26(1):119–188. Retrieved from https://onlinelibrary.wiley.com/doi/abs/10.1111/gcb.14904. https://doi.org/10.1111/gcb.14904

Kuhn M (2008) Building predictive models in r using the caret package. J Stat Softw 28(5):1–26. Retrieved from https://www.jstatsoft.org/index.php/jss/article/view/v028i05. https://doi.org/10.18637/jss.v028.i05

Kuhn M, Johnson K (2013) Applied Predictive Modeling, 1st ed. Springer, New York

Kuhn M, Wickham H (2020) Tidymodels: a collection of packages for modeling and machine learning using tidyverse principles. [Computer Software manual]. Retrieved from https://www.tidymodels.org

Lang M, Binder M, Richter J, Schratz P, Pfisterer F, Coors S et al (2019) mlr3: a modern object-oriented machine learning framework in R. J Open Source Softw 4:1903. Retrieved from https://joss.theoj.org/papers/10.21105/joss.01903. https://doi.org/10.21105/joss.01903

Lapuschkin S, Wäldchen S, Binder A, Montavon G, Samek W, Müller K-R (2019) Unmasking Clever Hans predictors and assessing what machines really learn. Nat Commun 10(1):1096. Retrieved from https://doi.org/10.1038/s41467-019-08987-4

Lembrechts JJ, Hoogen J van den, Aalto, J., Ashcroft MB, De Frenne P, Kemppinen J et al (2022) Global maps of soil temperature. Global Change Biol 28(9):3110–3144. Retrieved from https://onlinelibrary.wiley.com/doi/abs/10.1111/gcb.16060. https://doi.org/10.1111/gcb.16060

Linnenbrink J, Milà C, Ludwig M, Meyer H (2024) kNNDM CV: k-fold nearest-neighbour distance matching cross-validation for map accuracy estimation. Geosci Model Dev 17, 5897–5912. https://doi.org/10.5194/gmd-17-5897-2024

Lopez-Gonzalez G, Lewis SL, Burkitt M, Phillips OL (2011) ForestPlots.net: a web application and research tool to manage and analyse tropical forest plot data: ForestPlots.net. J Vegetat Sci 22(4):610–613. https://doi.org/10.1111/j.1654-1103.2011.01312.x

Ludwig M, Moreno-Martinez A, Hölzel N, Pebesma E, Meyer H (2023) Assessing and improving the transferability of current global spatial prediction models. Global Ecol Biogeography 32(3):356–368. https://doi.org/10.1111/geb.13635

Meyer H, Pebesma E (2021) Predicting into unknown space? estimating the area of applicability of spatial prediction models. Methods Ecol Evol 12(9):1620–1633. Retrieved from https://besjournals.onlinelibrary.wiley.com/doi/abs/10.1111/2041-210X.13650. https://doi.org/10.1111/2041-210X.13650

Meyer H, Pebesma E (2022) Machine learning-based global maps of ecological variables and the challenge of assessing them. Nat Commun 13(1):2208. Retrieved from https://doi.org/10.1038/s41467-022-29838-9

Meyer H, Reudenbach C, Hengl T, Katurji M, Nauss T (2018) Improving performance of spatio-temporal machine learning models using forward feature selection and target-oriented validation. Environ Modell Softw 101:1–9. Retrieved from https://www.sciencedirect.com/science/article/pii/S1364815217310976. https://doi.org/10.1016/j.envsoft.2017.12.001

Meyer H, Reudenbach C, Wöllauer S, Nauss T (2019) Importance of spatial predictor variable selection in machine learning applications – moving from data reproduction to spatial prediction. Ecol Model 411:108815. Retrieved from https://www.sciencedirect.com/science/article/pii/S0304380019303230. https://doi.org/10.1016/j.ecolmodel.2019.108815

Milà C, Mateu J, Pebesma E, Meyer H (2022) Nearest neighbour distance matching leave-one-out cross-validation for map validation. Methods Ecol Evol 13(6):1304-1316. Retrieved from https://besjournals.onlinelibrary.wiley.com/doi/abs/10.1111/2041-210X.13851. https://doi.org/10.1111/2041-210X.13851

Moreno-Martínez A, Camps-Valls G, Kattge J, Robinson N, Reichstein M, Bodegom PV et al (2018) A methodology to derive global maps of leaf traits using remote sensing and climate data. Remote Sens Environ 218:69–88. Retrieved from https://www.sciencedirect.com/science/article/pii/S0034425718304176

Pauchard A, Fuentes N, Jiménez A, Bustamante R, Marticorena A (2013) Alien plants homogenise protected areas: evidence from the landscape and regional scales in South Central Chile. In: Foxcroft LC, Pyšek P, Richardson DM, Genovesi P (eds) Plant Invasions in Protected Areas. Springer Netherlands, Dordrecht, pp 191–208. https://doi.org/10.1007/978-94-007-7750-7_10

Pebesma E (2018) Simple Features for R: standardized support for spatial vector data. R J 10(1):439–446. Retrieved from https://doi.org/10.32614/RJ-2018-009

Pebesma E, Bivand R (2023) Spatial Data Science: With Applications in R. Chapman and Hall/CRC, Boca Raton. Retrieved from https://r-spatial.org/book/

Peyre G, Balslev H, Martí D, Sklenář P, Ramsay P, Lozano P et al (2015) VegPáramo, a flora and vegetation database for the Andean páramo. Phytocoenologia 45(1):195–201. https://doi.org/10.1127/phyto/2015/0045

Roberts DR, Bahn V, Ciuti S, Boyce MS, Elith J, Guillera-Arroita G et al (2017) Cross-validation strategies for data with temporal, spatial, hierarchical, or phylogenetic structure. Ecography 40(8):913-929. https://onlinelibrary.wiley.com/doi/pdf/10.1111/ecog.02881. https://doi.org/10.1111/ecog.02881

Sabatini FM, Lenoir J, Hattab T, Arnst EA, Chytrý M, Dengler J et al (2021) sPlotOpen – an environmentally balanced, open-access, global dataset of vegetation plots. Global Ecol Biogeography 30(9):1740–1764. Retrieved from https://onlinelibrary.wiley.com/doi/abs/10.1111/geb.13346. https://doi.org/10.1111/geb.13346

Sabatini FM, Jiménez-Alfaro B, Jandt U, Chytrý M, Field R, Kessler M et al (2022) Global patterns of vascular plant alpha diversity. Nat Commun 13(1):4683. https://doi.org/10.1038/s41467-022-32063-z

Schratz P, Muenchow J, Iturritxa E, Richter J, Brenning A (2019) Hyperparameter tuning and performance assessment of statistical and machine-learning algorithms using spatial data. Ecol Model 406:109–120. Retrieved from https://www.sciencedirect.com/science/article/pii/S0304380019302145. https://doi.org/10.1016/j.ecolmodel.2019.06.002

Schumacher FL, Knoth C, Ludwig M, Meyer H et al (2024) Estimation of local training data point densities to support the assessment of spatial prediction uncertainty. EGUsphere [preprint]. https://doi.org/10.5194/egusphere-2024-2730.

Stehman SV (1999) Basic probability sampling designs for thematic map accuracy assessment. Int J Remote Sens 20(12):2423–2441. Retrieved from https://doi.org/10.1080/014311699212100

Valavi R, Elith J, Lahoz-Monfort JJ, Guillera-Arroita G (2019) blockCV: an R package for generating spatially or environmentally separated folds for k-fold cross-validation of species distribution models. Methods Ecol Evol 10(2):225–232. https://doi.org/10.1111/2041-210X.13107

van den Hoogen J, Geisen S, Routh D, Ferris H, Traunspurger W, Wardle DA et al (2019) Soil nematode abundance and functional group composition at a global scale. Nature 572(7768):194–198. https://doi.org/10.1038/s41586-019-1418-6

Vibrans AC, Gasper ALD, Moser P, Oliveira LZ, Lingner DV, Sevegnani L (2020) Insights from a large-scale inventory in the southern Brazilian Atlantic Forest. Scientia Agricola 77(1):e20180036. https://doi.org/10.1590/1678-992x-2018-0036

Wadoux AM-C, Heuvelink GB, de Bruin S, Brus DJ (2021) Spatial cross-validation is not the right way to evaluate map accuracy. Ecol Model 457:109692. Retrieved from https://www.sciencedirect.com/science/article/pii/S0304380021002489. https://doi.org/10.1016/j.ecolmodel.2021.109692

Wright MN, Ziegler A (2017) **Ranger**: a Fast Implementation of Random Forests for High Dimensional Data in *C++* and *R*. J Stat Softw 77(1):1–17. https://doi.org/10.18637/jss.v077.i01

The `imageRy` Package for Teaching Remote Sensing in a Straightforward Way

Duccio Rocchini, Ludovico Chieffallo, Giovanni Andrea Nocera, Giacomo Panza, Michele Torresani, Diletta Santovito, and Elisa Thouverai

Abstract

Effectively conveying coding concepts requires an approach that considers the diverse learning styles and levels of expertise among students. One of the most challenging tasks is related to a proper import and export of data. This is due to (i) proper download of data and storage in local folders and (ii) proper import starting from local folders. In this book chapter, we will describe the `imageRy` R package, which provides a specific set of data for making remote sensing analysis

D. Rocchini (✉)
BIOME Lab, Department of Biological, Geological and Environmental Sciences, Alma Mater Studiorum University of Bologna, Bologna, Italy

Faculty of Environmental Sciences, Department of Applied Geoinformatics and Spatial Planning, Czech University of Life Sciences Prague, Praha, Suchdol, Czech Republic
e-mail: duccio.rocchini@unibo.it

L. Chieffallo · G. A. Nocera · G. Panza · D. Santovito
BIOME Lab, Department of Biological, Geological and Environmental Sciences, Alma Mater Studiorum University of Bologna, Bologna, Italy
e-mail: ludovico.chieffallo2@unibo.it; giovanni.nocera2@studio.unibo.it; giacomo.panza2@studio.unibo.it; diletta.santovito@studio.unibo.it

M. Torresani
Free University of Bolzano/Bozen, Faculty of Agricultural, Environmental and Food Sciences, Bolzano/Bozen, Italy

Competence Centre for Plant Health, Piazza Universitá 5, Bolzano, Italy
e-mail: Michele.Torresani@unibz.it

E. Thouverai
Department of Life Sciences, University of Trieste, Trieste, Italy

BIOME Lab, Department of Biological, Geological and Environmental Sciences, Alma Mater Studiorum University of Bologna, Bologna, Italy
e-mail: elisa.thouverai2@unibo.it

D. Rocchini (ed.), *R Coding for Ecology*, Use R!,
https://doi.org/10.1007/978-3-031-99665-8_12

together with a simplified version of some fundamental functions for a better didactical experience.

Keywords

Coding · Learning · Remote sensing · Spatial data

1 Introduction

Teaching while coding is a matter of flow. Effectively conveying coding concepts requires an approach that considers the diverse learning styles and levels of expertise among students. A good strategy would be breaking down complex coding procedures into digestible steps, providing clear explanations, and offering hands-on exercises to reinforce learning. In particular, when dealing with ecological data, it is also important to provide a proper understanding of the meaning of the code and the theoretical concepts behind it. This process is especially hard when practitioners and students are making use of their own computers. In fact, in such cases, issues might arise related to operating systems instead of software being adopted interrupting the flow of the lecture.

One of the most challenging tasks is related to a proper import and export of data. This is due to (i) proper download of data and storage in local folders and (ii) proper import starting from local folders.

To solve this issue some of the packages in R provide for data directly inside single packages (e.g., the `sdm` package, Naimi and Araújo 2016, the `vegan` package, Oksanen et al. 2022, and the `overlap` package, Ridout and Linkie 2009).

Another problem is related to the complexity of the functions provided by R packages: Novice students, still navigating the basics, may find it overwhelming to comprehend and utilize functions with numerous arguments. This can potentially impede their ability to grasp foundational concepts, hindering their overall learning experience. Furthermore, practitioners face the challenge of simplifying these complex functions when communicating with students or conveying essential concepts. The abundance of arguments can obscure the core principles being taught, making it difficult for both practitioners and students to focus on the fundamental aspects of the coding or analytical process.

In this context, the `imageRy` package provides a specific set of data for making remote sensing analysis together with a simplified version of some fundamental functions for a better didactical experience.

2 Packages Needed and Package Installation

The main dependency of the `imageRy` package is the `terra` package from which it is deriving several functions for data import and visualization.

Furthermore, in order to install `imageRy`, the package `devtools` is needed to access GitHub in which `imageRy` is stored. Hence, the installation is simply done by

```
devtools::install_github("ducciorocchini/imageRy")
```

3 Data Description

The data contained in the `imageRy` package are fully described at https://github.com/ducciorocchini/imageRy/blob/main/data_description.md. They include both raw datasets such as Sentinel-2 data and preprocessed images coming out from the NASA Earth Observatory website (https://earthobservatory.nasa.gov/).

4 Main Functions of the imageRy Package

A rapid guide of the main functions of the imageRy package is available at https://htmlpreview.github.io/?https://github.com/ducciorocchini/imageRy/blob/main/imageRy_rapid_manual.html.

We developed specific functions to deal with three main aspects: (i) data import (Sect. 4.1), (ii) data visualization (Sect. 4.2), and (iii) (Sect. 4.3). All the functions of the package contain the suffix `im.`. In this manner, users are always aware of the functions linked to our package.

4.1 Functions Related to Data Import

Data are continuously being updated in GitHub with potentially new imagery to be used. To check the data available a complete list can be gathered by the `im.list()` function.

Once the list is produced, images can be loaded by the `im.import` function. As an example, among the different images, an image of the Mato Grosso area coming from the Earth Observatory (NASA) gathered in 1992 can be loaded by

```
mato1992 <- im.import("matogrosso_l5_1992219_lrg.jpg")
```

Typing the name of the created object `b2` will return all the information of the image, including the class (`SpatRaster`, being based on the `terra` package), the dimensions of the image, its resolution, the extent, the coordinate system, the source, the name, and value ranges. An automatic plot of the image is also provided.

4.2 Functions Related to Data Visualization

Multispectral images can be viewed in `imageRy` through the use of RGB spaces, through the `im.plotRGB()` function as follows:

```
im.plotRGB(mato1992, r=2, g=1, b=3)
```

The function will lead to a plot in an RGB space like that shown in Fig. 1.

Fig. 1 The Mato Grosso area is represented in an RGB color scheme, with NIR on top of the green component. The image has been gathered by the Thematic Mapper sensor on NASA Landsat 5 satellite on August 6, 1992. It is part of the `imageRy` package and freely downloadable at the NASA Earth Observatory site (https://earthobservatory.nasa.gov/)

For a first automatic plot, the function `im.plotRGB.auto()` can be used to automatically plot the first three bands of any image, like

```
im.plotRGB.auto(mato1992)
```

As previously stated, in this image the near infrared channel is on band 1. Since in the direct plot `r=1`, vegetation that highly reflects in the NIR channel will become red (Fig. 2).

4.3 Functions Related to Data Analysis

4.3.1 Spectral Indices

Spectral indices allow condensing information in one layer based on the combination of two or more original bands in a single metric. In `imageRy`, two spectral indices are considered: the Difference Vegetation Index (DVI) and the Normalized Difference Vegetation Index (NDVI), both related to the status of vegetation health.

DVI is generally calculated as

$$\mathrm{DVI} = \lambda_{\mathrm{NIR}} - \lambda_{\mathrm{red}} \tag{1}$$

Fig. 2 The Mato Grosso area with the NIR band on top of the red component of an RGB space

where λ_{NIR} is the reflectance in the NIR channel and λ_{red} is the reflectance in the red channel. The more healthy a plant, the higher will be the reflectance in the NIR channel due to the structure of the leaf, and the lower will be the reflectance in the red channel, due to absorption for photosynthesis. The DVI function in imageRy is straightforward:

```
dvi <- im.dvi(mato1992, 1, 2)
```

The user is asked to enter the image on which the analysis should be performed and the bands corresponding to NIR and red reflectances, respectively.

A straightforward plot of the result based on the viridis palette can be performed as

```
clviridis <- colorRampPalette(viridis(7))(255) plot(dvi, col=clviridis)
```

In the Mato Grosso area, higher values of DVI clearly represent the original natural forest, while lower values indicate areas cleared for agriculture through human intervention (Fig. 3).

4.4 Image Classification

Unsupervised image classification can lead to classified maps allowing quantification of change in space and time of ecosystems.

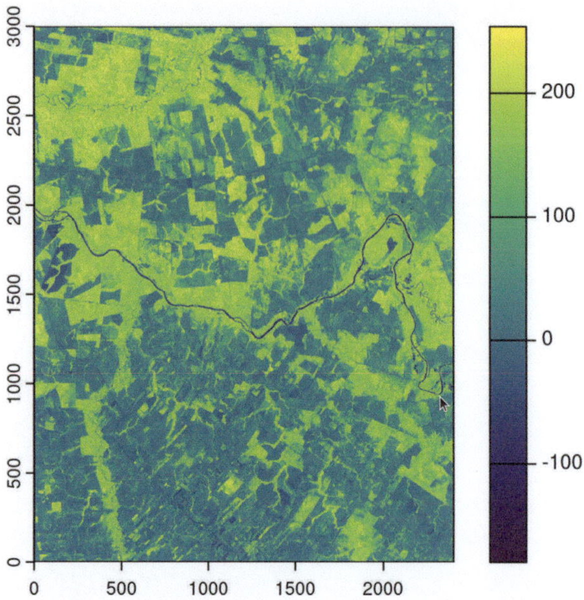

Fig. 3 The Difference Vegetation Index (DVI) calculated for the Mato Grosso area

The `im.classify()` function allows the classification of images using the k-means algorithm. The sole parameter required for this function is the number of clusters (classes) needed:

```
matoc <- im.classify(mato, num\_clusters = 3)
```

The output map is directly plotted with the individualized clusters (Fig. 4).

The percentage of occupancy of each class can then be extracted by

```
freq(matoc) * 100 / ncell(matoc)
```

which will lead to 17% for bare soil and 83% for tropical forest.

Notice that the classification process is iteration-based. The numbering of classes can change, and the final percentage values may show low variability from one calculation to the other. In this context the `set.seed()` function could help soften the issue.

This process can be repeated with various images at different times to monitor spatiotemporal changes of ecosystems.

4.5 Multivariate Analysis

In most cases, the information contained in a satellite image is replicated over several bands. In other words, there is a high multicollinearity in the reflectance among different bands, especially considering that of the visible part of the

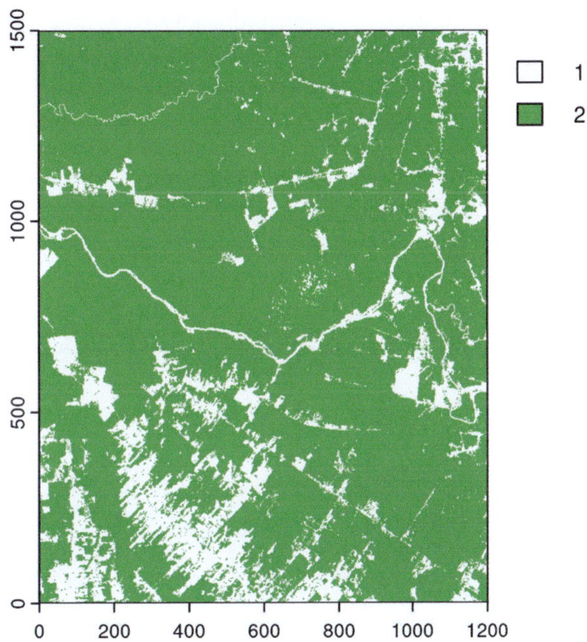

Fig. 4 The Mato Grosso image classified in two classes. Class 1 = agricultural areas, bare soil, and water and class 2 = Tropical Forest

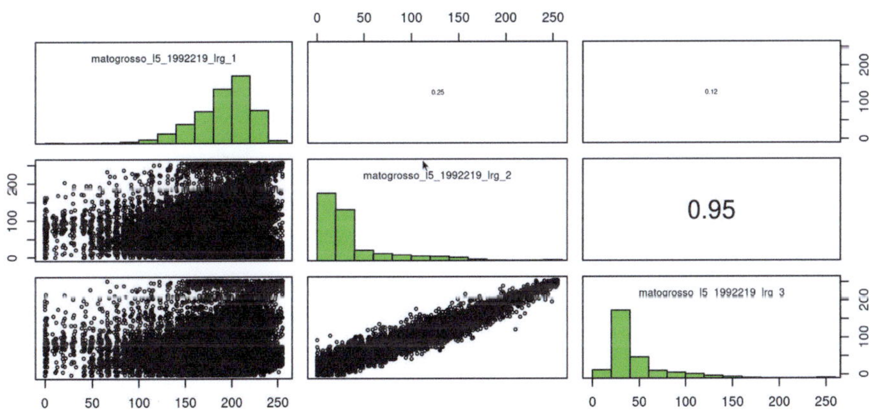

Fig. 5 The pairwise relationship between the different bands of the Mato Grosso image. Notice the high correlation between the visible bands (band 2 = red and band 3 = green)

spectrum. This can be simply tested by the function `pairs`, applied to an image, like

```
pairs(mato)
```

which will lead to a graph showing the correlation between pairs of bands (Fig. 5). As an example, in this case, bands 2 and 3, representing the reflectance in the red and green channels, respectively, showed a very high correlation in terms of reflectance values.

In order to decrease multicollinearity problems, multivariate analysis is a valuable tool for compressing the original set of data into new uncorrelated axes. In `imageRy`, multivariate analysis is implemented via Principal Components Analysis and can be performed as

```
pc <- im.pca(mato)
```

The first component alone will explain in this case 70% of the input variance and is also graphically representing most of the variability of the system (Fig. 6).

In case of analyses related to single layers, like, e.g., the calculation of variance or the calculation of entropy metrics like the Shannon index, the first principal component would then be the best candidate as a single layer for further calculations.

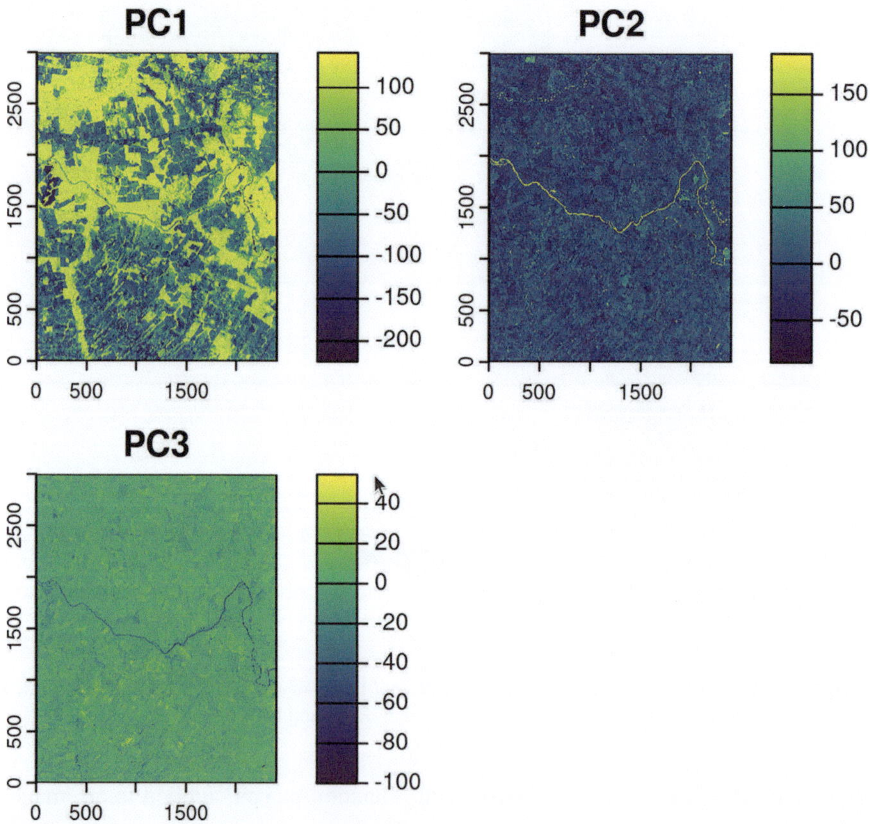

Fig. 6 Principal Components (PCs) as an output of the `im.pca()` function. The first PC represents most of the variability of the image in space, since it statistically explains—by definition—the highest amount of variance

5 Conclusion

In this book chapter, we described the `imageRy` package, which is devoted to providing functionalities for teaching remote sensing in a straightforward way.

Obviously, the functions described for remote sensing data analyses attaining to the `imageRy` package can also be applied to any image in any format, e.g., to data imported by the `rast` function in the `terra` package. We suggest training students with the images available in `imageRy` before opting for import/export of data.

The importance of open access and data sharing in general and of Earth Observation data, in particular, has been acknowledged by many authors (e.g., Michener and Jones 2012; Rocchini and Neteler 2012; Wulder et al. 2012; Turner 2013), as well as the importance of code sharing or open-source software (Stallman 1985; Ince et al. 2012; Stokstad 2011; Morin et al. 2012), mainly based on transparency and openness promotion (McNutt 2016).

An increased usage of Free and Open-Source Software in remote sensing training (i) enables students and researchers to apply professional remote sensing analysis in their daily work, without being limited by their institutes' budgets, and (i) pushes for reproducible research. In fact, reproducibility, feasible using open-source software (Stokstad 2011), is a primary pillar of science and should also be adopted in scientific training.

Remote sensing and ecology are both dynamic disciplines encompassing a number of research areas (Gillespie et al. 2008) and being able to use modern methods, and software is a crucial aspect to address the global conservation challenges which lay ahead. We believe that a package devoted to remote sensing training, like `imageRy`, is crucial to enable students to approach ecological remote sensing in a simple and straightforward manner. In this book chapter, we described the GitHub version of the imageRy package; a CRAN version is also available at https://CRAN.R-project.org/package=imageRy.

References

Gillespie TW, Foody GM, Rocchini, D., Giorgi AP, Saatchi S (2008) Measuring and modelling biodiversity from space. Progress Phys Geography 32:203–221

Ince DC, Hatton L, Graham-Cumming J (2012) The case for open computer programs. Nature 482:485–488

McNutt M (2016) Taking up TOP. Science 352:1147–1147

Michener WK, Jones MB (2012) Ecoinformatics: supporting ecology as a data-intensive science. Trends Ecol Evol 27:85–93

Morin A, Urban J, Adams PD, Foster I, Sali A, Baker D, Sliz P (2012) Shining light into black boxes. Science 336:159–160

Naimi B, Araújo MB (2016) sdm: a reproducible and extensible R platform for species distribution modelling. Ecography 39:368–375

Oksanen J, Simpson GL, Blanchet FG, Kindt R, Legendre P, Minchin PR, O'Hara RB, Solymos P, Stevens MHH, Szoecs E, Wagner H, Barbour M, Bedward M, Bolker B, Borcard D, Carvalho G, Chirico M, De Caceres M, Durand S, Evangelista HBA, FitzJohn R, Friendly M, Furneaux B, Hannigan G, Hill MO, Lahti L, McGlinn D, Ouellette M-H., Ribeiro Cunha E, Smith T, Stier

A, Ter Braak CJF, Weedon J (2022) vegan: Community Ecology Package. R package version 2.6–2. https://CRAN.R-project.org/package=vegan
Ridout MS, Linkie M (2009) Estimating overlap of daily activity patterns from camera trap data. J Agricul Biol Environ Stat 14:322–337
Rocchini D, Neteler M (2012) Let the four freedoms paradigm apply to ecology. Trends Ecol Evol 27:310–311
Stokstad E (2011) Open-source ecology takes root across the world. Science 334:308–309
Stallman R (1985) The GNU manifesto
Turner W (2013) Satellites: make data freely accessible. Nature 498:37–37
Wulder MA, Masek JG, Cohen WB, Loveland TR, Woodcock CE (2012) Opening the archive: how free data has enabled the science and monitoring promise of Landsat. Remote Sens Environ 122:2–10

Part IV

Spatio-ecological Data Visualisation

The `cblindplot` R Package: Building New Bridges to Visually Inclusive Science for Colourblind People

Ludovico Chieffallo, Rossella D'Introno, Jakub Nowosad, Elisa Thouverai, Elisa Marchetto, and Duccio Rocchini

Abstract

In this chapter, we explore the realm of visual representations, encompassing images and maps, with a central focus on rendering them accessible to a broader audience, including individuals with colour vision deficiencies. Our overarching goal is to foster greater inclusivity within the realm of science communication. To achieve this objective, we showcase the R package `cblindplot` and explain its significance. We illustrate its genesis, based on the most common types of colour-blindness: deuteranopia, protanopia, and tritanopia. Furthermore, we undertake a series of practical, step-by-step examples to offer a comprehensive understanding of how `cblindplot` operates, ensuring a deeper grasp of its functionality. We

L. Chieffallo (✉) · E. Marchetto
Alma Mater University of Bologna, Bologna, Italy
e-mail: ludovico.chieffallo2@unibo.it

R. D'Introno
Sacco Hospital, Ophthalmology and Orthoptics, Milan, Italy

J. Nowosad
Institute of Geoecology and Geoinformation, Adam Mickiewicz University, Poznań, Poland

E. Thouverai
Department of Life Sciences, University of Trieste, Trieste, Italy

BIOME Lab, Department of Biological, Geological and Environmental Sciences, Alma Mater Studiorum University of Bologna, Bologna, Italy
e-mail: elisa.thouverai2@unibo.it

D. Rocchini
BIOME Lab, Department of Biological, Geological and Environmental Sciences, Alma Mater Studiorum University of Bologna, Bologna, Italy

Faculty of Environmental Sciences, Department of Applied Geoinformatics and Spatial Planning, Czech University of Life Sciences Prague, Praha, Suchdol, Czech Republic

D. Rocchini (ed.), *R Coding for Ecology*, Use R!,
https://doi.org/10.1007/978-3-031-99665-8_13

also describe some other R packages that may be useful for colourblind-friendly visualisations.

Keywords

Colourblindness · Computational ecology · Mapping · R · Scientific communication

1 Introduction

Ecology, as the science that studies the interactions between the environment and the organisms that inhabit it, employs a wide array of tools and techniques to depict and analyse the complexity of nature (Huang 2021). One of the primary resources at the disposal of ecologists is maps, which can showcase the outcomes of natural phenomena, from species distribution maps to ecosystem dynamics (Reza and Abdullah 2016). Over time, ecological cartography has played an essential role in understanding phenomena such as deforestation, biodiversity loss, and changes in precipitation patterns (Sturman 1996).

Thanks to these maps, areas of particular interest or concern have been identified, changes over time have been monitored, and future changes, such as climate shifts, have been estimated. These maps are crucial not just for scientific research but also for conservation planning, resource management, and environmental education (Miller and Rogan 2007).

Like any tool, maps can have certain shortcomings (Burrough 1986), some of which we will address in this chapter. Their inherently generalised nature means they are a simplification of reality (Palmer et al. 2008), which is vital to make data understandable but can also lead to potential ambiguities or misunderstandings. For instance, the visual perception of maps is highly subjective. Two or more individuals might view the same map and draw different conclusions based on their background, education, or even perception of colours (Popelka et al. 2012).

This is particularly pertinent in ecology, where slight differences in interpretation can lead to significant variations in management or conservation decisions. Cartography has made significant advancements in recent decades with the advent of technologies like Geographic Information Systems (GIS) and remote sensing. These new methods have allowed for an increasingly detailed representation of the natural world; however, the use of colour in maps remains a challenge, especially for the colourblind (Rocchini et al. 2023a, b).

This chapter focuses on colour perception, how it can mislead maps' interpretation, and how can we improve ecological data visualisations. We present the `cblindplot` package (Rocchini et al. 2023a, b), designed to smooth out these issues. Developed in R, this data visualisation tool is crafted to enhance map readability for the colourblind by reading an existing, non-colourblind-friendly, image and replacing its colours. This is an effective step towards creating more accessible ecological maps, marking an important move to make science more inclusive.

2 Colourblindness: An Overview

Colourblindness is a genetic condition affecting an individual's ability to perceive colours accurately. The main difficulty arises from the reduced or altered functionality of the cone photoreceptors in the retina, which are responsible for colour perception. There are three types of cones, each sensitive to different wavelengths of light, L (long), M (medium), and S (short), corresponding to the colours red, green, and blue, respectively. In colourblindness, there may be a deficiency or abnormality in one or more of these types of cones, leading to difficulty discriminating colours Albany-Ward (2015). This colour perception anomaly can range in severity from a mild difficulty in discerning shades to a complete inability to perceive colours (Vi'enot et al. 1995; Gordon 1998; Gegenfurtner 2003).

Colourblindness is often hereditary and is more prevalent in men than in women (Wan et al. 2017). This is because the responsible gene is linked to the X chromosome. Women have two X chromosomes and thus require two copies of the altered gene to exhibit colourblindness, while men, having only one X chromosome, need just one (Simunovic 2010; Birch 2012).

2.1 Types of Colourblindness

There are three main types of colourblindness (Fig. 1):

- Protanopia: This form of colourblindness results in the inability to perceive the colour red.
- Deuteranopia: Individuals with deuteranopia cannot recognise the colour green.
- Tritanopia: This is an inability to distinguish the colour blue.

There is also achromatopsia, which, though extremely rare, represents the inability to perceive any colour, seeing only in shades of grey.

2.2 Impact of Colourblindness

The inability to recognise specific colour scales, especially those with gradients from blue to green/yellow to red, can severely influence the perception of maps and other visual materials by colourblind individuals. For instance, the rainbow colour palettes often used in various contexts can pose challenges for those with colourblindness (Borland & Taylor Ii 2007; Crameri et al. 2020). The rainbow colour palette problem is compounded by the fact that they often feature smooth transitions between colours that are especially difficult for colourblind people to

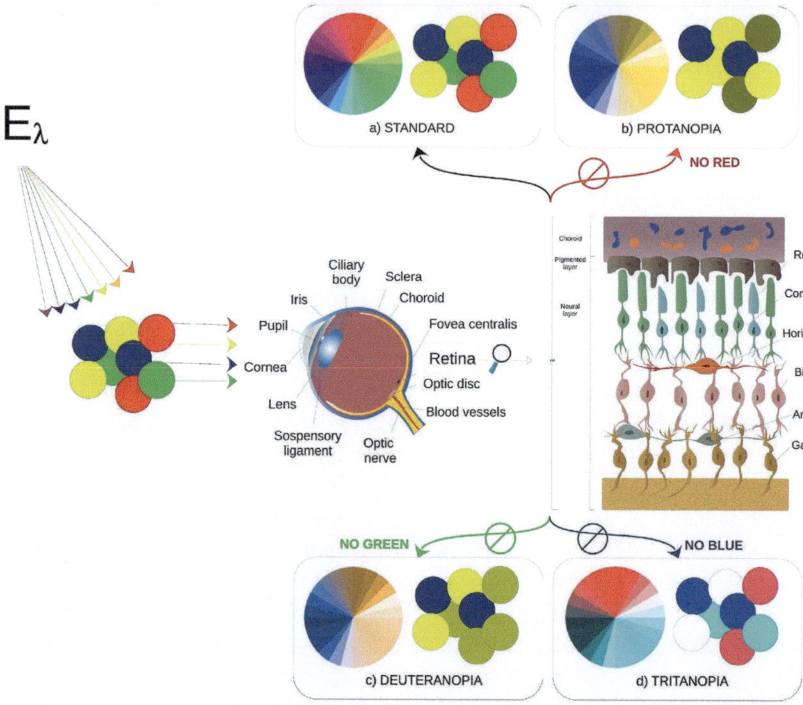

Fig. 1 Light rays coming from outside hit the cornea of the human eye; they pass through the lens and the vitreous humour and focus on the retina, which is a nerve structure. Here, the translation of the light into an electric signal takes place. This electrical impulse travels along the visual pathways to the occipital cortex, where it is converted into a visual image (Rocchini et al. 2023a, b)

distinguish, such as green and red. This can lead to misinterpretation of visual information, as shades that appear distinct to a person with normal colour perception may appear indistinguishable or very similar to a colourblind person (Silva and Santos 2011).

3 The R Package `cblindplot`

In this context, the `cblindplot` package was developed to facilitate the conversion of maps (or other images) with non-colourblind-friendly colour scales into colour schemes that are understandable for individuals with colour vision deficiencies. You can find this package at https://github.com/ducciorocchini/cblindplot. This tool is designed with the specific goal of streamlining and simplifying the process of converting these maps. It accomplishes this task through a user-friendly and intuitive function, ensuring a straightforward conversion process.

The `cblindplot` currently allows two approaches to create colourblind-friendly maps, described in detail below. The first approach is suitable when the input map uses a fairly simple colour scale (e.g. going from red through yellow to green). However, it is not recommended for maps with more complex colour scales (ones including many colours, e.g. blue, red, yellow, green, etc., such as rainbow colour palettes). The second approach works in such situations, but requires providing another image with the colour scale legend only. In this chapter, we will analyse the two approaches in depth with two examples that will clarify the basic ideas of the R cblindplot package.

3.1 Approach 1

The first approach works for simple cases, where the input colour palette is sequential, therefore, when composed of colours that vary gradually, usually from light to dark or the other way round.

The `cblind.plot()` function implements three fundamental steps when this approach is used (Fig. 2):

1. In the first step, it imports an image with multiple bands using `terra::rast()` (Hijmans 2023). Each band is represented by a matrix, and each cell represents a pixel value.
2. Next, the input data undergoes analysis using the `stats::prcomp()` function, which employs principal component analysis (PCA). PCA is a technique used to reduce the dimensionality of data while retaining most of the original information. In the context of spatial data and map colours, this technique is used to highlight dominant colour patterns and relationships in the data,

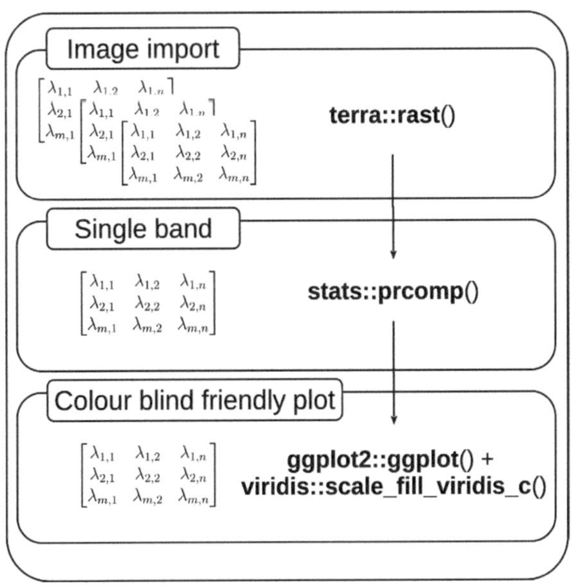

Fig. 2 The main steps of the first approach available in the `cblind.plot()` function (Rocchini et al. 2023a)

providing a clearer understanding of the distribution of colour across the map. This is particularly useful for identifying specific areas of the map where colour variations may be subtle but significant, helping to adjust the colour scheme to make these variations more perceivable to individuals with colour vision deficiencies. The first principal component is selected.

3. Finally, the values of the first principal component are displayed in a format that is easily interpretable by individuals with colour vision deficiencies. This is done by using the `ggplot2::ggplot()` function to create a plot and apply a colourblind-friendly colour scheme (e.g. using `viridis::scale fill viridis c()`). The viridis package (Garnier et al. 2023) provides colour scales that are uniformly perceived by human vision and that are printable on a black and white printer. This ensures that variations in the data are easily recognised even by those who are colourblind.

3.2 Approach 2

The second approach works for more complex cases such as a map with a rainbow colour palette, a palette not recommended precisely because it is one of the most difficult to interpret (Crameri et al. 2020). In this approach, an additional image is required to include a colour bar legend. This approach, unlike the first, does not use PCA, but samples the colours from the input legend sequentially, such as from bottom to top and then replaces these colours with colours from palettes visible and distinguishable by colourblind people.

4 Case Studies

To understand the functioning and potential of `cblindplot` well, it is useful to go step by step through practical examples that give us visible and tangible results.

We begin by installing the `cblindplot` package:

```
#  install.packages("remotes")
library(remotes)
remotes::install_github("ducciorocchini/cblindplot")
```

Once the package is installed, we will proceed to call it to make it usable.
within our workspace with the `library(cblindplot)` function:

```
library(cblindplot) #call cblindplot
```

Now, that we have our package loaded and ready to be used, we can load our image or map and check the functionality of the package.

4.1 Example 1

Our first example will be based on the dataset included in the `cblindplot` (Rocchini et al. 2023a) package, which represents the NDVI (Normalized Difference Vegetation Index) values of the central valley of California (United States) in 2023 extracted from *Copernicus* (https://land.copernicus.eu/global/products/ndvi). This region is one of the most productive agricultural areas in the world and produces a wide variety of fruits, vegetables, and nuts. Given that it is a very large area, with continuous challenges for the management of water resources also due to climate change, investigating the NDVI can be useful for identifying and monitoring areas at risk of drought or for monitoring the effects of climate change that impact on the biodiversity of the area.

To read this dataset, we need to get its filepath. This can be achieved with the `system.file` as our `NDVI_simple.png` file is included in the "`pic/`" folder of the `cblindplot` package:

```
image_path <- system.file("pic/NDVI_simple.png", package = "cblindplot")
```

Next, to read and view the image in RGB, we use the terra package:

```
image_RGB <- terra::rast(image_path) #we read the image
terra::plotRGB(image_RGB) #we get an RGB plot of our image
```

Can I Use My Own Images?

This approach works for any graphic image file. So you can use your files to test this R package:

```
user_image path <- "path/to/your/image.png" #here enter the
path to your file
user_image RGB <- terra::rast(user_image_path)
terra::plotRGB(user_image_RGB)
```

As we said, Fig. 3 represents the Copernicus Climate Change Service (2023) and presents a colour range with a predominance of red, orange, and green. These colours may pose the following challenges:

- For individuals with protanopia, red and orange hues might appear as a single colour, making it difficult to distinguish between the different intensities present in the image.
- For individuals with deuteranopia, green areas might not be easily distinguishable from orange or red areas, making it difficult to interpret the contrast between these zones.
- Even though the image has many yellow shades, individuals with tritanopia might have difficulty perceiving these shades, if present.

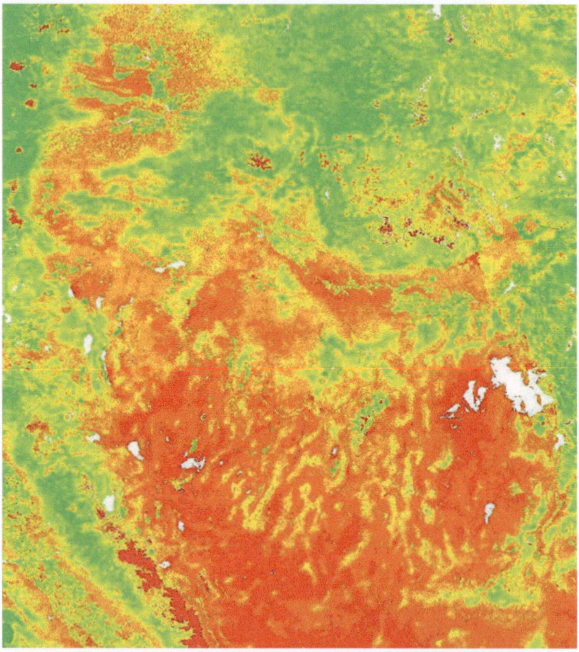

Fig. 3 The NDVI (Normalized Difference Vegetation Index) values of the central valley of California, United States (Rocchini et al. 2023a, b)

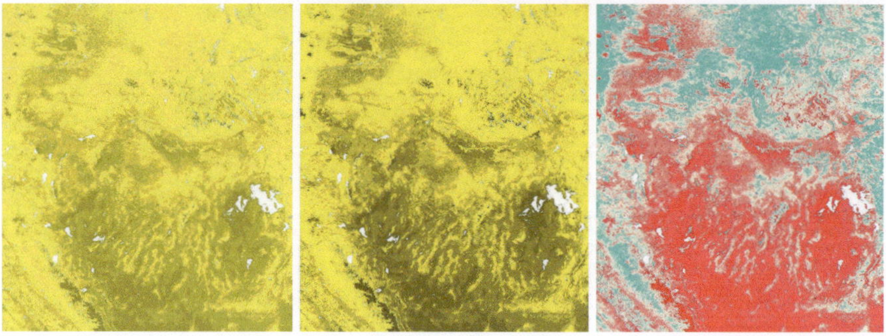

Fig. 4 Simulations of colourblindness types: deuteranopia, protanopia, and tritanopia

How Do Colourblind People See This Images?
We simulated, thanks to the `colourblindr` R package (https://github.com/clauswilke/colourblindr) (McWhite and Wilke 2023), how colourblind people see this image based on their visual defect.

Figure 4 shows the results.

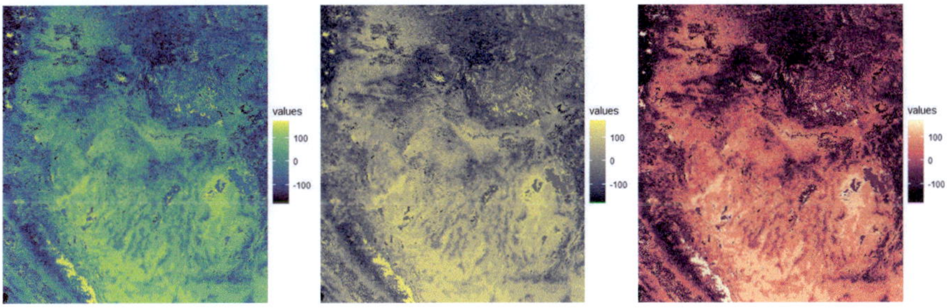

Fig. 5 The maps of NDVI (Normalized Difference Vegetation Index) values of the central valley of California, United States, prepared to improve interpretation by people with colour vision deficiencies (in order deuteranopia, protanopia, and tritanopia)

Now, as we have already obtained the RGB image and understood its limitations, we can finally apply the functionality of cblindplot to get images or maps that can be interpreted by colourblind people.

For that purpose, we use the cblindplot() function and provide it with the input image object and the colour vision deficiency (cvd) type:

```
cblind.plot(image_RGB, cvd = "deuteranopia")
cblind.plot(image_RGB, cvd = "protanopia")
cblind.plot(image_RGB, cvd = "tritanopia")
```

The outcome of the function is shown in Fig. 5.

4.2 Example 2

The data for our second example is available in the imageRy package, an R package designed to speed up teaching activities, streamlining image import/export processes, and providing a catalogue of different images coming from remote sensing (Rocchini et al. 2023b).

It also represents the same NDVI values as in the first example, but this image has an entirely different colour palette, that of rainbow colours.

Firstly, we need to install the imageRy package with the install github() function:

```
remotes::install_github("ducciorocchini/imageRy")
library(imageRy)
```

Now, we can locate the "NDVI rainbow.png" file with the system.file() function, and read and visualise it in its original rainbow colours with terra (Fig. 6). As we can see, the image presents a rainbow colour palette of ten colours, which is not recommended in scientific visualisations, as it can introduce ambiguity and distortions in the perception of the data:

```
map_path <- system.file("images/NDVI_rainbow.png", package="imageRy")
map <- terra::rast(map_path)
terra::plotRGB(map)
```

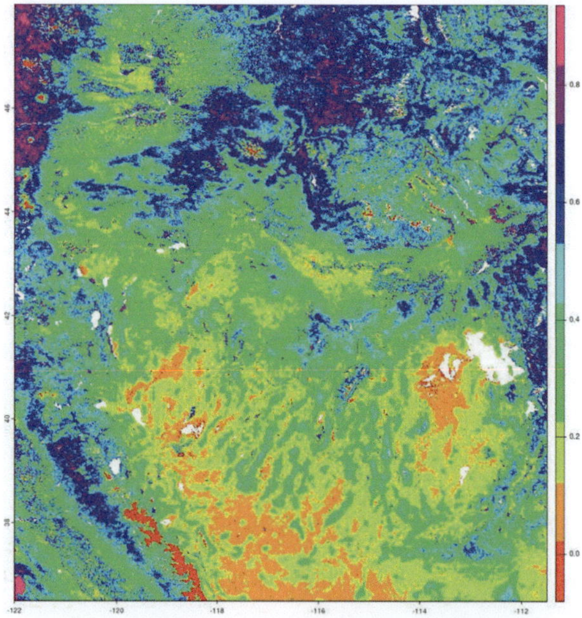

Fig. 6 The NDVI (Normalized Difference Vegetation Index) values of the central valley of California, United States, shown using a rainbow palette (Rocchini et al. 2023a, b)

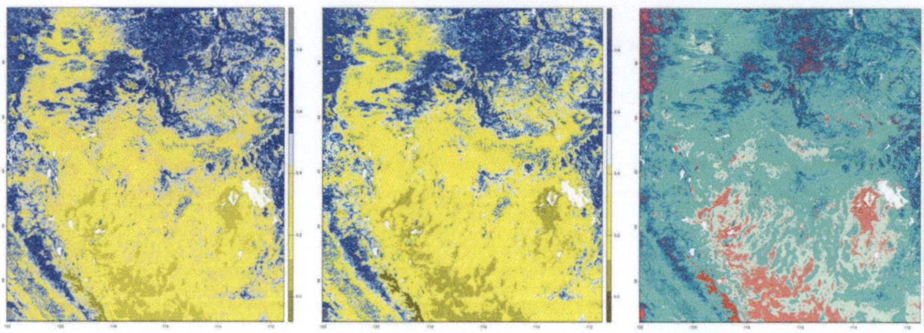

Fig. 7 In order deuteranopia, protanopia, and tritanopia

How Do Colourblind People See These Images?

We simulated, thanks to the `colorblindr` R package (https://github.com/clauswilke/colorblindr) (McWhite and Wilke 2023), how colourblind people see this image based on their visual defect. Here are the results (Fig. 7):

This, as we have seen, is a more complex case and therefore requires the second approach explained in this chapter.

Fig. 8 NDVI legend in the central valley of California, United States, with ten rainbow colour palettes

The first thing to do at this point will be to import our legend. It is cut out from our image with rainbow colours (you may perform this step with some external graphical tool, such as GIMP):

```
my_legend <- system.file("images/NDVI_rainbow_legend.png",
package="imageRy")
```

At this point, we can plot it to see its contents (Fig. 8):

```
my_terra_legend <- terra::rast(my_legend)
terra::plotRGB(my_terra_legend)
```

In this case, the order of the colours in the provided legend is used to update the colour palette of the input object when we use the `cblind.plot` function (Fig. 9):

```
cblind.plot(map, legend = my_legend)
```

You can also decide to insert a customised palette of your own such as "Subset" by providing a vector of colours. This may be achieved using the function `hcl.colors()` by specifying the number of colours and the palette you want to use (Fig. 10).

```
my_palette <- hcl.colors(7, palette = "Sunset")
cblind.plot(map, cvd = my_palette, legend = my_legend)
```

Please note that you can choose to use many other palettes instead of "Sunset", for example, "Viridis", "Inferno", "Magma", "Plasma", "Heat", "YlOrRd", "Blues", "YlGnBu", etc. Available colour palettes in the `hcl.colors()` function can be checked using `hcl.pals()`.

The complete code to reproduce the examples is available on GitHub at https://github.com/RCodingForEcology/II.7.cblindplot.

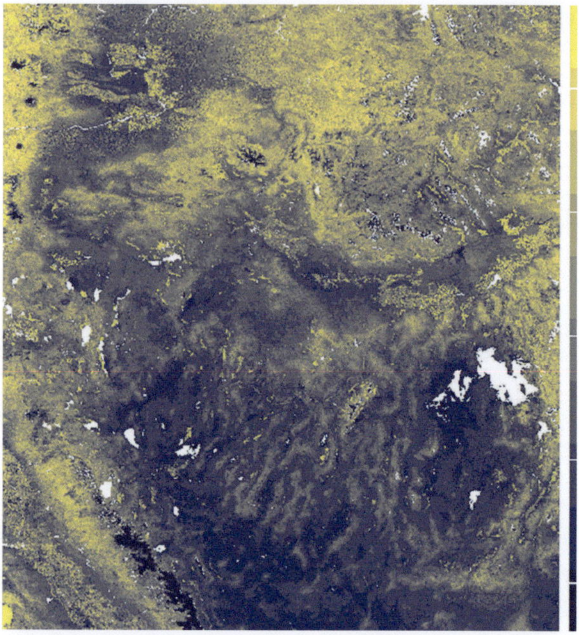

Fig. 9 The NDVI map reconstructed from the input image with a rainbow colour palette and a related legend. This version can be interpreted correctly by colourblind people

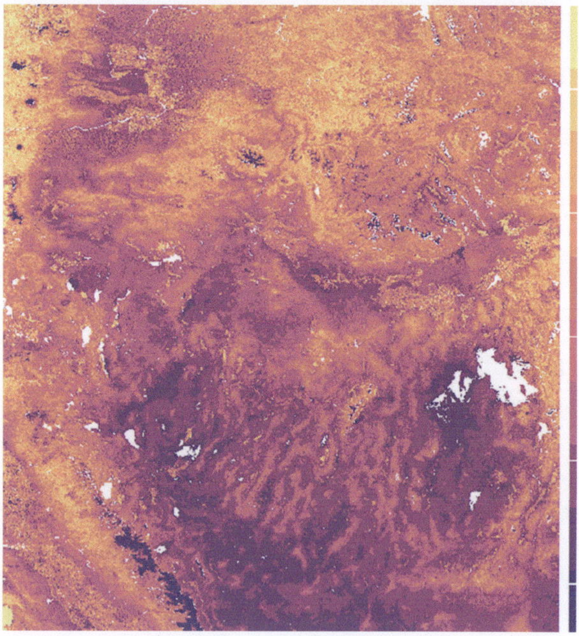

Fig. 10 The NDVI map reconstructed from the input image with a rainbow colour palette and a related legend. This version uses the "Sunset" color palette

5 Other Packages

This chapter focuses on the cblindplot package; however, several other R packages exist with a goal to support data visualisation for colourblind people. They include the following:

- colorspace: A set of tools for manipulating and assessing colour palettes, including simulation of how they would like for people with colour vision deficiencies. The package also has two accompanying websites with articles and online apps at https://colorspace.r-forge.r-project.org/ and https://hclwizard.org/ (Zeileis et al. 2020).
- colorblindr: The features provided by colorblindr allow users to edit and display their figures to reflect how they would appear to those with various types of colourblindness, thereby helping to identify and resolve potential readability issues before publishing or sharing visualisations (McWhite and Wilke 2023).
- colorBlindness: This package also provides a colour vision deficiency simulator and a collection of colourblind-friendly colour palettes. Additionally, it allows to replace problematic colours in R graphics with some alternatives, e.g. red with magenta or green with turquoise (Ou 2021). This approach can be helpful to distinguish objects on images, but it is not recommended when the input map has a related colour scale.
- colorblindcheck: The primary function in this package is palette check(), which creates a numerical and visual summary comparison between the provided colour palette and the colour vision deficit simulations. This feature can be used to identify problems in colour palettes related to colour vision and select the most accessible colour palette to apply in a given project (Nowosad 2019).

Moreover, R offers many packages with colourblind-accessible colour palettes, for example:

- viridis: Provides eight colour palettes to improve the readability of graphs for readers with common forms of colour vision deficiencies, including “viridis”, “cividis”, and “mako” (Garnier et al. 2023).
- rcartocolor: The purpose of the package is to provide colour schemes designed for maps by the CARTO company. The package includes four groups of colour palettes, aggregation, divergent, qualitative, and quantitative, most of which are colourblind-friendly (try to run display carto all (colorblind friendly = TRUE)) (Nowosad 2018).
- cols4all: This software contains a collection of many colour palette series. Moreover, it provides an interactive dashboard to check properties of each colour palette (e.g. if it is colourblind-friendly or can be used in low-contrast visualisations) (Tennekes 2023).

6 Conclusion

Colourblindness, a genetic condition affecting colour perception, poses a significant challenge in data visualisation and interpretation, particularly in the field of ecology. Accurately interpreting maps and graphs is crucial for understanding complex ecological phenomena, from species distribution to ecosystem dynamics. However, for colourblind individuals, traditional visual representations can often be misleading or incomprehensible. Within this chapter, we have underscored the significance of crafting tools and methodologies that consider the specific requirements of individuals with colour vision deficiencies when it comes to data visualisation. The R package `cblindplot` represents a pragmatic solution for enhancing the legibility of maps and graphs designed for those with colour vision impairments. Ultimately, as the science of ecology continues to evolve, the tools and methodologies used to communicate research findings must keep pace. Adopting solutions such as the `cblindplot` package and greater awareness of the needs of colourblind individuals represent important steps in this direction.

References

Albany-Ward K (2015) What do you really know about colour blindness? Br J School Nurs 10(4):197–199

Birch J (2012) Worldwide prevalence of red-green color deficiency. JOSA A 29(3):313–320

Borland D, Taylor Ii RM (2007) Rainbow color map (still) considered harmful. IEEE Comput Graph Appl 27(2):14–17

Burrough PA (1986) "Principles of geographical." Information systems for land resource assessment. Clarendon Press, Oxford

Copernicus Climate Change Service (2023) Global NDVI Dataset. Copernicus. https://land.copernicus.eu/global/products/ndvi

Crameri F, Shephard GE, Heron PJ (2020) The misuse of colour in science communication. Nat Commun 11(1):5444

Garnier S, Ross N, Rudis R, Camargo AP, Sciaini M, Scherer C (2023) viridis(Lite) -Colorblind-Friendly Color Maps for R. viridis package version 0.6.3

Gegenfurtner KR (2003) Cortical mechanisms of colour vision. Nat Rev Neurosci 4(7):563–572

Gordon N (1998) Colour blindness. Public Health 112(2):81–84

Hijmans R (2023) Terra: spatial data analysis. R Package Version 1:7–29. https://CRAN.R-project.org/package=terra

Huang G (2021) Ecology. In: Theory of mountain urbanology. Springer, Singapore. https://doi.org/10.1007/978-981-16-0819-39

McWhite C, Wilke C (2023) colorblindr: Simulate colorblindness in R figures. R package version 0.1.0. https://github.com/clauswilke/colorblindr

Miller JA, Rogan J (2007) Using GIS and remote sensing for ecological mapping and monitoring

Neitz M, Neitz J (2000) Molecular genetics of color vision color vision defects. Arch Ophthalmol 118(5):691–700. https://doi.org/10.1001/archopht.118.5.691

Nowosad J (2018) 'CARTOColors' Palettes. R package version 1.0.0, https://jakubnowosad.com/rcartocolor/

Nowosad J (2019) colorblindcheck: check color palettes for problems with color vision deficiency. R package version 0.1.0. https://jakubnowosad.com/colorblindcheck/

Ou J (2021). colorBlindness: safe color set for color blindness. R package version 0.1.9. https://cran.r-project.org/package=colorBlindness

Palmer MW, McGlinn DJ, Fridley JD (2008) Artifacts and Artifictions in biodiversity research. Folia Geobot 43:245–257. https://doi.org/10.1007/s12224-008-9012-y

Popelka S et al (2012) Advanced map optimalization based on eye-tracking. Cartography – a tool for spatial analysis. InTech, 17 Aug. 2012. Crossref, https://doi.org/10.5772/46190

Reza M, Abdullah S (2016) Developing ecosystem maps using eco-geological information for the sustainable management of natural resources. Open J Ecol 6:343–357. https://doi.org/10.4236/oje.2016.66033

Rocchini D et al (2023a) Scientific maps should reach everyone: the cblindplot R package to let colour blind people visualise spatial patterns. Eco Inform 76

Rocchini D, Nocera GA, Chieffallo L, Thouverai E (2023b) imageRy: modify and share images. R package version 0.1.0

Silva S, Santos BS (2011) Using color in visualization: a survey. Comput Graph 35(2):320–333

Simunovic MP (2010) Colour vision deficiency. Eye 24(5):747–755

Sturman VI (1996) Ecological mapping as a factor in the consolidation of the geographical sciences. Mapp Sci Remote Sens 33(4):283–290. https://doi.org/10.1080/07493878.1996.10642037

Tennekes M (2023) cols4all: colors for all. R package version 0.6, https://CRAN.R-project.org/package=cols4all

Vi'enot F et al (1995) What do colourblind people see? Nature 376(6536):127–128

Wan A, Sarah S et al. (2017) The reading of tajweed in surah Yaasin for red-green colour vision deficiencies. pp 357–363

Zeileis A, Fisher JC, Hornik K, Ihaka R, McWhite CD, Murrell P, Stauffer R, Wilke CO (2020) colorspace: a tool-box for manipulating and assessing colors and palettes. J Stat Softw 96(1):1–49. https://doi.org/10.18637/jss.v096.i01

The `cartogram` Package for Mapping Ecological Patterns

Elisa Marchetto, Sebastian Jeworutzki, and Jakub Nowosad

Abstract

This chapter explores the spatial representation of ecological data variables using cartograms. The cartograms are thematic maps that alter the actual geographical dimensions of areas or regions changing their size and shape according to a quantitative variable. Specifically, this chapter describes the use of the R package *cartogram* to create the main types of cartograms. In this chapter, three functions of the package are described through practical examples that aim to represent the spatial bias associated with the sampling effort of the species observations. We demonstrate the usefulness of cartograms as methods for illustrating the relationships between ecological data and geographic areas, as well as the impact of spatial bias on an ecological variable.

Keywords

Cartograms · Data quality · Sampling effort · Spatial bias · Species observations

E. Marchetto (✉)
Alma Mater University of Bologna, Bologna, Italy
e-mail: elisa.marchettophd@outlook.it

S. Jeworutzki
Faculty of Social Science, Ruhr-Universität Bochum, Bochum, Germany

J. Nowosad
Institute of Geoecology and Geoinformation, Adam Mickiewicz University, Poznań, Poland

D. Rocchini (ed.), *R Coding for Ecology*, Use R!,
https://doi.org/10.1007/978-3-031-99665-8_14

1 Introduction

Cartograms, i.e., thematic maps that modify the geographical dimensions of areas or regions, are usually used to precisely represent data by resizing areas according to a quantitative variable (Dorling 1996; Gastner and Newman 2004). Areas with higher values for the chosen variable appear proportionally larger on the map, while areas with lower values appear proportionally smaller. Cartograms are profoundly different from choropleth maps which are commonly used in cartography to show the data distribution across different geographic regions emphasizing spatial patterns and their variations (Skowronnek 2015). Indeed, choropleth maps maintain the original shape and size of the regions exhibiting the data values through changes in color intensity or shading. They also preserve the geographic aspect of regions saving the geographic positions and shapes of the boundaries. Cartograms, instead, focus on the magnitude of data values across regions stressing the data linkages over the geographic accuracy (Nusrat et al. 2016). This, on the other hand, can make the identification of the true geographic areas more difficult.

Starting from their conceptualization, cartograms have been applied in many different fields: sociopolitical studies (Bonifazi et al. 2021), transportation and urban planning (Ullah and Kraak 2015), public health (Yalcin 2022), economics and demographics (Li and Clarke 2012), and environmental science (Döll 2017), just to cite few. In the field of ecology, cartograms have primarily found use in depicting species diversity (Lomolino et al. 2015; Orr et al. 2021) and showing uncertainty of species distribution models (Rocchini et al. 2019), which are particularly linked in addressing issues related to sampling bias (Rocchini et al. 2023) and for visually conveying variations in the effort of biodiversity data collection (Hughes et al. 2021).

In recent times, the extensive examinations of biodiversity trends (Heberling et al. 2021) were driven by a significant surge in the accessibility of biodiversity data (Feng et al. 2022), accompanied by improved computational capabilities and modeling methods. Nevertheless, the data quality assessment remains a crucial factor to be undertaken (Maldonado et al. 2015; Wüest et al. 2020). Biodiversity databases can denote different sources of bias (e.g., taxonomic, spatial, temporal, and environmental bias), which impact data analysis and modeling (Meyer et al. 2016). Especially, the spatial bias, i.e., the uneven distribution of ecological data or observations across space, can determine a lack of knowledge regarding the geographic distribution of the species—the so-called Wallacean shortfall (Hortal et al. 2015)—preventing a comprehensive understanding of biodiversity patterns (Yang et al. 2013; García-Roselló et al. 2023). Different sources of errors, such as the opportunistic sampling method (Tessarolo et al. 2014), the geographic distortion introduced by observers or recorders (e.g., closeness to roads) (Oliveira et al. 2016), and limited funding and resources (Meyer et al. 2015), can originate the spatial bias. For instance, representing the sampling effort, which refers to the number of samples or observations collected, can lead to more robust and reliable results (Oliveira et al. 2017). Cartograms are one method for representing spatial bias associated with the sampling effort.

In this chapter, we focus on the representation of the spatial bias associated with the sampling effort of the species observations through the different types of cartograms using the R package *cartogram* (Jeworutzki 2025). More specifically, we depict the spatial bias through the continuous area cartogram, the noncontiguous area cartogram, and the nonoverlapping circles cartogram.

2 Data and Code

For the examples in this chapter, we rely on the observations of the presence of a virtual species, i.e., a simulated entity with known species-environment relationships (Meynard et al. 2019). Commonly, a virtual species is created from the response functions of a set of environmental variables representing the environmental suitability of the species (Meynard and Kaplan 2013). Once the presence of the species is generated virtually over space according to the environmental suitability (Fig. 1), the observations are collected setting the desired number of samples. The example virtual species has a global distribution, with observations being oversampled in certain regions (i.e., the USA, Mexico, Japan, China) by a factor of 20 compared to others.

To replicate the findings showcased in this chapter, you can retrieve the relevant data and code from the GitHub repository https://github.com/RCodingForEcology/cartogram . The main dataset is a dataframe

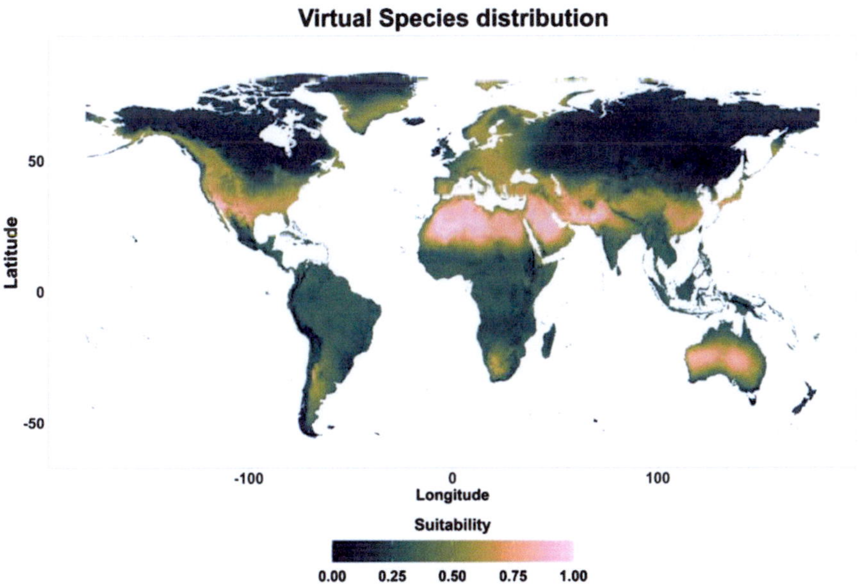

Fig. 1 Environmental suitability of the virtual species distribution. Areas showing the highest suitability are colored in pink while those with the lowest suitability are depicted in blue

with the sampled observations (i.e., the presence of the species) and the coordinates of the virtual species.

3 Continuous Area Cartogram

Continuous area cartograms distort the geographic area of regions or territories on a map to represent a specific data variable while maintaining a continuous and contiguous appearance (Dougenik et al. 1985). This is ensured by the cartogram algorithm which keeps the centroids of the regions approximately in their position while the regions are scaled and distorted proportionally to the data values. A founding principle of the creation process is a gradual distortion which guarantees a smooth transition between different regions. The rubber sheet algorithm proposed by Dougenik et al. (1985) uses a model of forces applied from the polygon centroid to each boundary point. In an iterative process, each point is forced toward its desired position, with forces decreasing with each iteration ("distance decrease").

To utilize the full range of features offered by the *cartogram* (v0.3.0) package and to prepare and manipulate data before and after processing, it is necessary to import multiple packages associated with spatial analysis and data handling. *sf* (v1.0.21) package provides methods for dealing with spatial vector data, *dplyr* (v1.1.4) provides tools for data cleaning and manipulation, while *tmap* (v4.1) allows for spatial data visualization.

The following lines of codes install the required packages and load them in R:

```
install.packages(c("cartogram", "sf", "dplyr", "tmap"))
library(cartogram)
library(sf)
library(dplyr)
library(tmap)
```

We will use a *World* map from the *tmap* package for the global examples:

```
data("World", package = "tmap")
```

Next, we need to download and load the dataframe with the previously derived virtual species data:

```
 download.file(
"https://github.com/RCodingForEcology/cartogram/raw/main/PresAbs.bias.rds"
, "PresAbs.bias.rds", mode = "wb")

PresAbs.bias <- readRDS("PresAbs.bias.rds")

head(PresAbs.bias)
           x        y Observed
1  140.58333 37.08333        1
2   36.41667 27.08333        1
3 -101.75000 26.41667        1
4  -49.08333 64.75000        1
5 -105.58333 28.91667        1
6   88.91667 34.25000        1
```

Now, our goal is to count the number of species in each country in the *World* object. For that purpose, we first need to convert the dataframe with the virtual species observations into an object of class *sf*. This can be done with the *st_as_sf* function, in which we need to specify *coords*—the columns in the dataframe that contain the coordinates of the observations and *crs*—the coordinate reference system of the observations.

```
PresAbs.bias <- st_as_sf(PresAbs.bias, coords =
c("x", "y"), crs = "EPSG:4326")
```

Next, we need to reproject both datasets, *World* and *PresAbs.bias*, to a common projected coordinate reference system. This is a vital step, as the *cartogram* does not handle working with coordinates in degrees (i.e., geographical coordinate reference systems). Importantly, the used projected coordinate reference system should be appropriate for the study area and preferably be equal-area. Here, we use the *st_transform* function to reproject the data to the Equal Earth projection (EPSG:8857):

```
World.ea <- st_transform(World, "EPSG:8857")
PresAbs.bias <- st_transform(PresAbs.bias, "EPSG:8857")
```

Now, we can count the number of our virtual species in each country. For that purpose, we use the *st_join* function to merge the information about each country ISO code from the *World* object with the *PresAbs.bias* object. Next, we summarize the number of observations for each ISO code. Finally, we join the summarized information with the *World* object. Countries that do not have any observations are assigned a value of 0 observations:

```
PresAbs.bias <- st_join(PresAbs.bias, World.ea["iso_a3"])

count.PresAbs.bias <- PresAbs.bias |>
    st_drop_geometry() |>
    group_by(iso_a3) |>
    summarize(countISO = sum(Observed))

World.bias <- World.ea |>
    left_join(count.PresAbs.bias, by = "iso_a3") |>
    mutate(countISO = ifelse(is.na(countISO), 0, countISO))
```

The function *cartogram_cont* creates the continuous area cartogram by a rubber sheet distortion algorithm (Dougenik et al. 1985). Here, we indicate *x* (a POLYGON or MULTIPOLYGON *sf* object), the name of the weighting variable in *x*, and the maximum iterations for the cartogram transformation.

The threshold parameter sets the minimum and maximum distortion ratio for an area. In the presence of a large number of zero values, the required ratio of area change becomes very large or very small or 0. As a result, the algorithm may not find a suitable distortion. To prevent this, the parameter *prepare = "adjust"* is used to adjust values below and above the percentile defined by the threshold to avoid such large ratios. In this example, the threshold parameter is set to 0.16 so that the

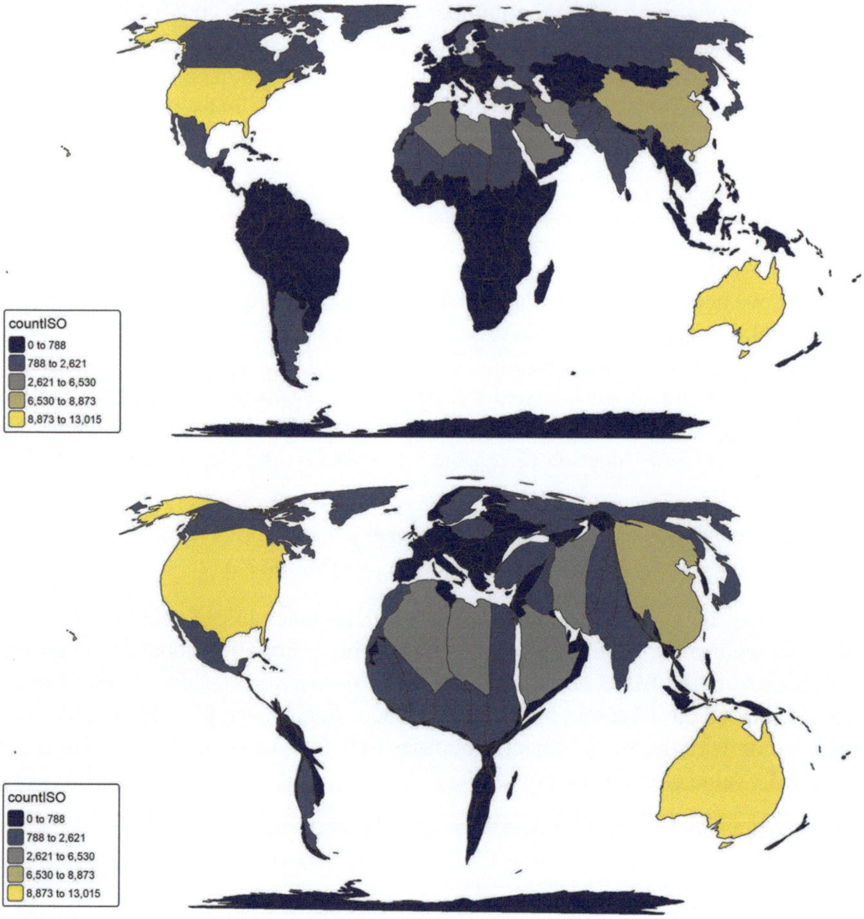

Fig. 2 Choropleth map and continuous area cartogram of sampling effort of virtual species observations. The maps show the areas where the sampling effort of the virtual species observations is higher (yellow color) and the areas where the sampling effort gradually decreases until it reaches lower values (blue color)

ratios of the 16 percent largest and 16 percent smallest values are adjusted. Another way to achieve better fitting results is to set the maximum number of iterations of the distortion algorithm to a higher value using the *itermax* parameter. To print the value of the "mean size error" for each iteration, a measure of the agreement between desired and achieved area for the regions, set the argument *verbose* = *TRUE*:

```
cart.cont <- cartogram_cont(World.bias, "countISO",
                            itermax = 30, threshold = 0.16,
                            verbose = FALSE)
```

cartogram_cont returns an spatial polygon object of the same class as *x*. Now, we can compare a standard choropleth map with the continuous area cartogram using the *tmap* package:

```
map.standard <- tm_shape(World.bias) +
    tm_polygons("countISO", fill.scale = tm_scale_intervals(style="jenks", values = "cividis")) +
    tm_layout(frame = FALSE, legend.position = c("left", "bottom"),
              legend.width = 9, legend.outside = TRUE)
map.carto1 <- tm_shape(cart.cont) +
    tm_polygons("countISO", fill.scale = tm_scale_intervals(style="jenks", values = "cividis")) +
    tm_layout(frame = FALSE, legend.position = c("left", "bottom"),
              legend.width = 9, legend.outside = TRUE)
tmap_arrange(map.standard, map.carto1)
```

The choropleth map (Fig. 2, top) helps to quickly identify the patterns of the distribution of the sampling effort around the world. On the other hand, from the continuous area cartogram (Fig. 2, bottom), we underline the relationship between the sampling effort value and the geographic area to which it refers. Thanks to the intensity of the distortion, we observe the countries and the geographic areas that are more sampled (e.g., the USA, Australia, China) compared to the less sampled ones (e.g., Brazil, Colombia). Hence, the continuous area cartogram, compared to the choropleth map, allows to visualize the effect of the biased sampling effort on space by means of the distorted countries. The representation of the biased sampling effort also provides a first partial view of the degree of error of the subsequent ecological modeling (e.g., in species distribution models) if the sampling bias of the species observations is not adequately corrected.

4 Noncontiguous Area Cartogram

Unlike continuous area cartograms, which aim to preserve the spatial relationships between neighboring regions, noncontiguous area cartograms distort the sizes of regions and maintain the shapes (Olson 1976). The regions are scaled independently so that the boundaries do not have to be adjacent and the topology is not preserved. The creation process is based on the density value of each region which is calculated by dividing the data value by the geographic area. The region with the highest density is designated as the anchor, meaning that its area stays constant while the areas of the other regions decrease proportionately to the specified values.

The noncontiguous area cartogram can be created using the function *cartogram_ncont*. The first both arguments define the map geometry and weighting variable. *k* is a scaling parameter to adjust the area of the new regions:

```
cart.ncont <- cartogram_ncont(World.bias,
                              weight = "countISO",
                              k = 1)
tm_shape(World.ea) +
    tm_borders() +
    tm_shape(cart.ncont) +
    tm_polygons("countISO", fill.scale = tm_scale_intervals(style="jenks", values = "cividis")) +
    tm_layout(frame = FALSE, legend.position = c("left", "bottom"),
              legend.width = 9, legend.outside = TRUE)
```

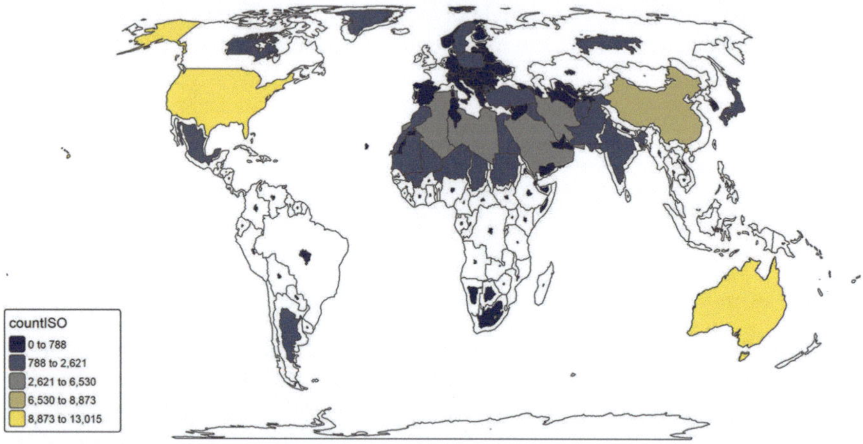

Fig. 3 Noncontiguous area cartogram of sampling effort of virtual species observations. The map shows the areas where the sampling effort of the virtual species observations is higher (yellow color) and the areas where the sampling effort gradually decreases until it reaches lower values (blue color)

cartogram_cont returns an object of the same class as *x*, but with resized polygon boundaries.

Compared to the continuous area cartogram, some countries maintain their size and shape in the noncontiguous area cartogram (Fig. 3). They correspond to geographic areas where the sampling effort was higher than in all the others. In particular, with this distortion technique, it is possible to quickly recognize the hotspots of sampling effort (North Africa, the Middle East, Australia, the USA, China) based on the size of the given country itself and the degree of separation from the neighboring countries. In contrast to the continuous cartogram, the results are closer to a choropleth map. The regional differences in the sampling effort are perceived more by the presence and absence of colored polygons than by the size of the regions.

5 Nonoverlapping Circles Cartogram

In contrast to conventional maps, nonoverlapping circles cartograms use circles to portray data related to distinct regions, guaranteeing that the circles do not intersect or cover one another (Dorling 1996). In nonoverlapping circles cartograms, the shape and topology of the original geographic areas are not preserved.

The circles are centered on the region's centroid and resized proportionally to the related data value. Higher data values are depicted by bigger circles; lower data values are represented by smaller circles. If neighboring circles overlap, the circles' locations and sizes are rearranged by an iterative process in order to avoid the intersection and maintain the topological relationships of the original regions.

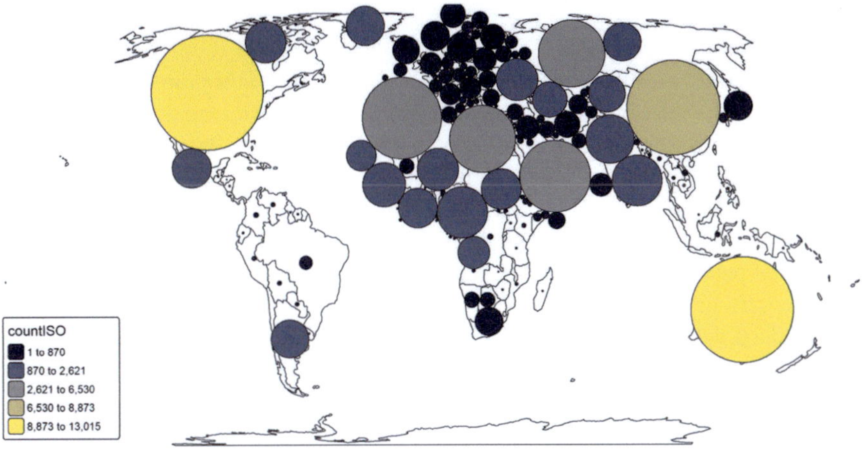

Fig. 4 Nonoverlapping circles cartogram of sampling effort of virtual species observations. The map shows the areas where the sampling effort of the virtual species observations is higher (yellow color) and the areas where the sampling effort gradually decreases until it reaches lower values (blue color)

Cartogram uses iterative pair-wise repulsion (see package *packcircles*) to find a nonoverlapping configuration.

The nonoverlapping circles cartogram is generated using the function *cartogram_dorling*:

```
cart.dorling <- cartogram_dorling(World.bias, "countISO")

tm_shape(World.ea) +
    tm_borders() +
    tm_shape(cart.dorling) +
    tm_polygons("countISO", fill.scale = tm_scale_intervals(style="jenks", values = "cividis")) +
    tm_layout(frame = FALSE, legend.position = c("left", "bottom"),
              legend.width = 9, legend.outside = TRUE)
```

cartogram_dorling also returns nonoverlapping circles of the same class as *x*.

The sampling effort of the species observations in the map in Figure (Fig. 4) is represented by the size and color of the circle in each centroid of the country. Visually, the identification of the sampling hotspots is based on the circle's size rather than the distorted size and shape of the countries. When using nonoverlapping circles cartograms, as compared to continuous and noncontiguous cartograms, the distortion of sampling effort on regions is visually more pronounced. Indeed, since the circles have a completely different shape from the underlying country boundaries, they induce the observer to focus more on the extent of bias rather than on the geographical position.

6 Conclusion

In this chapter, we showed the effectiveness of the cartograms as tools for representing the relationship between the ecological data and the geographic space and for depicting the spatial bias's effect on an ecological variable or measure. The geographic locations can sometimes be challenging to interpret when using cartograms; nevertheless, they are very helpful for showing how data properties vary between different regions emphasizing data variations. In this context, the *cartogram* package supplies three functions that allow the user to choose the more suitable cartogram for the spatial representation of the data variable. Here, we provided a practical example of three different types of cartograms, computed using the implemented functions in the package, which visually reproduce the sampling bias associated with the sampling effort of the species observations. Overall, all thematic maps agree that some countries and regions, such as the USA, Australia, and China, have been more sampled in virtual species observations than others, such as Brazil and Colombia.

In ecology, cartograms have been adopted in investigations of species diversity and species distribution models but also for representing the amount of scientific research in a specific ecological topic. However, given their effectiveness and versatility, they can be applied for depicting many other aspects related to ecology such as habitat fragmentation, population density, and dynamic, climate change effects or for the representation of the degree of error of a measurement.

References

Bonifazi C, Heins F, Licari F, Tucci E (2021) The regional dynamics of internal migration intensities in Italy. Popul Space Place 27(7):e2331. https://doi.org/10.1002/psp.2331

Döll P (2017) Cartograms facilitate communication of climate change risks and responsibilities. Earth's Future 5(12):1182–1195

Dorling D (1996) Area Cartograms: Their Use and Creation. Concepts and Techniques in Modern Geography. Environmental Publications, University of East Anglia

Dougenik JA, Chrisman NR, Niemeyer DR (1985) An algorithm to construct continuous area cartograms. Profess Geograph 37(1):75–81

Feng X, Enquist BJ, Park DS, Boyle B, Breshears DD, Gallagher RV, Lien A, Newman EA, Burger JR, Maitner BS, Merow C, Li Y, Huynh KM, Ernst K, Baldwin E, Foden W, Hannah L, Jørgensen PM, Kraft NJB et al (2022) A review 13 of the heterogeneous landscape of biodiversity databases: opportunities and challenges for a synthesized biodiversity knowledge base. Glob Ecol Biogeogr 31:1242–1260. https://doi.org/10.1111/geb.13497

García-Roselló E, González-Dacosta J, Lobo JM (2023) The biased distribution of existing information on biodiversity hinders its use in conservation, and we need an integrative approach to act urgently. Biol Conserv 283:110118. https://doi.org/10.1016/j.biocon.2023.110118

Gastner MT, Newman ME (2004) Diffusion-based method for producing density-equalizing maps. Proc Natl Acad Sci 101(20):7499–7504

Heberling JM, Miller JT, Noesgaard D, Weingart SB, Schigel D (2021) Data integration enables global biodiversity synthesis. Proc Natl Acad Sci 118(6):e2018093118. https://doi.org/10.1073/pnas.2018093118

Hortal J, de Bello F, Diniz-Filho JAF, Lewinsohn TM, Lobo JM, Ladle RJ (2015) Seven shortfalls that beset large-scale knowledge of biodiversity. Annu Rev Ecol Evol Syst 46:523–549

Hughes AC, Orr MC, Ma K, Costello MJ, Waller J, Provoost P et al (2021) Sampling biases shape our view of the natural world. Ecography 44(9):1259–1269. https://doi.org/10.1111/ecog.05926

Jeworutzki S (2025) cartogram: Create Cartograms with R. R package version 0.3.0. https://doi.org/10.32614/CRAN.package.cartogram

Li L, Clarke KC (2012) Cartograms showing China's population and wealth distribution. J Maps 8(3):320–323. https://doi.org/10.1080/17445647.2012.722792

Lomolino MV, Pijanowski BC, Gasc A (2015) The silence of biogeography. J Biogeogr 42(7):1187–1196. https://doi.org/10.1111/jbi.12525

Maldonado C, Molina CI, Zizka A, Persson C, Taylor CM, Albán J et al (2015) Estimating species diversity and distribution in the era of Big Data: to what extent can we trust public databases? Glob Ecol Biogeogr 24(8):973–984. https://doi.org/10.1111/geb.12326

Meyer C, Kreft H, Guralnick R, Jetz W (2015) Global priorities for an effective information basis of biodiversity distributions. Nat Commun 6(1):1–8

Meyer C, Weigelt P, Kreft H (2016) Multidimensional biases, gaps and uncertainties in global plant occurrence information. Ecol Lett 19(8):992–1006. https://doi.org/10.1111/ele.12624

Meynard CN, Kaplan DM (2013) Using virtual species to study species distributions and model performance. J Biogeogr 40(1):1–8. https://doi.org/10.1111/jbi.12006

Meynard CN, Leroy B, Kaplan DM (2019) Testing methods in species distribution modelling using virtual species: what have we learnt and what are we missing? Ecography 42(12):2021–2036. https://doi.org/10.1111/ecog.04385

Nusrat S, Alam MJ, Kobourov S (2016) Evaluating cartogram effectiveness. IEEE Trans Vis Comput Graph 24(2):1077–1090. https://doi.org/10.1109/TVCG.2016.2642109

Oliveira U, Paglia AP, Brescovit AD, de Carvalho CJ, Silva DP, Rezende DT et al (2016) The strong influence of collection bias on biodiversity knowledge shortfalls of Brazilian terrestrial biodiversity. Divers Distrib 22(12):1232–1244. https://doi.org/10.1111/ddi.12489

Oliveira U, Brescovit AD, Santos AJ (2017) Sampling effort and species richness assessment: a case study on Brazilian spiders. Biodivers Conserv 26:1481–1493. https://doi.org/10.1007/s10531-017-1312-1

Olson JM (1976) Noncontiguous area cartograms. Prof Geogr 28:371–380. https://doi.org/10.1111/j.0033-0124.1976.00371.x

Orr MC, Hughes AC, Chesters D, Pickering J, Zhu CD, Ascher JS (2021) Global patterns and drivers of bee distribution. Curr Biol 31(3):451–458

Rocchini D, Marcantonio M, Arhonditsis G, Cacciato AL, Hauffe HC, He KS (2019) Cartogramming uncertainty in species distribution models: a Bayesian approach. Ecol Complex 38:146–155. https://doi.org/10.1016/j.ecocom.2019.04.002

Rocchini D, Tordoni E, Marchetto E, Marcantonio M, Barbosa AM, Bazzichetto M et al (2023) A quixotic view of spatial bias in modelling the distribution of species and their diversity. NPJ Biodivers 2(1):10

Skowronnek A (2015) Beyond choropleth maps: a review of techniques to visualize quantitative areal geodata. Infovis Reading Group WS 16

Tessarolo G, Rangel TF, Araújo MB, Hortal J (2014) Uncertainty associated with survey design in species distribution models. Divers Distrib 20(11):1258–1269. https://doi.org/10.1111/ddi.12236

Ullah R, Kraak MJ (2015) An alternative method to constructing time cartograms for the visual representation of scheduled movement data. J Maps 11(4):674–687. https://doi.org/10.1080/17445647.2014.935502

Wüest RO, Zimmermann NE, Zurell D, Alexander JM, Fritz SA, Hof C et al (2020) Macroecology in the age of Big Data–Where to go from here? J Biogeogr 47(1):1–12. https://doi.org/10.1111/jbi.13633

Yalcin M (2022) Mapping the global spatio-temporal dynamics of COVID-19 outbreak using cartograms during the first 150 days of the pandemic. Geocarto Int 37(13):3791–3800

Yang W, Ma K, Kreft H (2013) Geographical sampling bias in a large distributional database and its effects on species richness–environment models. J Biogeogr 40(8):1415–1426. https://doi.org/10.1111/jbi.12108

MIX
Papier aus verantwortungsvollen Quellen
Paper from responsible sources
FSC® C105338

If you have any concerns about our products,
you can contact us on
ProductSafety@springernature.com

In case Publisher is established outside the EU,
the EU authorized representative is:
Springer Nature Customer Service Center GmbH
Europaplatz 3, 69115 Heidelberg, Germany

Printed by Libri Plureos GmbH
in Hamburg, Germany